Fluid Power

Hydraulics and Pneumatics

Second Edition

by

James R. Daines
Professor Emeritus
University of Wisconsin–Stout
Menomonie, WI

Publisher
The Goodheart-Willcox Company, Inc.
Tinley Park, Illinois
www.g-w.com

Library of Congress Catalog Card Number 2012007606

ISBN 978-1-60525-931-4

4 5 6 7 8 9 – 13 – 21 20 19 18

Front cover credit: vesilvio/Shutterstock

Library of Congress Cataloging-in-Publication Data

Daines, James R.
  Fluid power : hydraulics and pneumatics / by James R. Daines. --
  2nd ed.
    p. cm.
  Includes index.
  ISBN 978-1-60525-081-6
  1. Fluid power technology. 2. Hydraulic machinery. 3. Pneumatic
  machinery. I. Title.
TJ840.D34 2013
620.1'06--dc23   2012007606

Preface

This book is the outcome of many years of persistent interest in fluid power, beginning with the operation of farm equipment and involvement in a 4H tractor-maintenance project when I was a teenager, continuing throughout my college years and far into my professional life. Partly, the fascination came from the science foundations that are there and partly from the practical applications that brought those abstract concepts to life. There is orderliness to hydraulics and pneumatics that, beyond the work accomplished, has a sense of aesthetics. So the study of fluid power and the challenge of teaching others to use it has been a satisfying journey.

Our lives would be very different today if early civilizations had not recognized the potential of using air and water to do work. From the first rough waterwheels and windmills to the sophisticated applications we see today, fluid power has enabled us to do what was, in many cases, first thought to be impossible. Now, with computer interfaces, new materials, and imaginative technology to help us, we may be just at a new dawn of breakthroughs to come.

The purpose of this book is to introduce you to the field of fluid power. It lays the groundwork for building an understanding of the concepts involved and systematically guides the use of these concepts in designing functional circuits. Purposefully, the questions asked will require problem solving and creative thinking; that is what real life requires. Mastery of the subject matter is not easy, but careful study and thoughtful experimentation will make it happen.

Many individuals, companies, and organizations have contributed to this learning resource. I wish to acknowledge their assistance and express my sincere appreciation to them all. Likewise, I am deeply grateful to the publisher, Goodheart-Willcox, for continued encouragement and support while this was written. I hope the book will be a useful tool for the field.

—Jim Daines

Introduction

Fluid Power: Hydraulics and Pneumatics provides comprehensive exposure to the fluid power field, ranging from historical information to details on the design and operation of hydraulic and pneumatic components, circuits, and systems. Professional and standardizing organizations that actively support fluid power education, manufacture, distribution, and service activities also are introduced to provide an understanding of the background and magnitude of the field.

The content of the book is organized into three areas:

- Introduction to the fluid power field.
- Hydraulics systems.
- Pneumatics systems.

The introduction includes basic exposure to historical information, the development of hydraulic and pneumatic power principles, symbols, and fluid power support organizations. The hydraulic system and pneumatic system sections examine those specific areas in detail.

Content is organized to provide a logical sequence of topics for students who are being exposed to fluid power for the first time (Chapters 1 through 5). Specific hydraulic information (Chapters 6 through 13) and pneumatic information (Chapters 14 through 19) provide adequate detail for advanced study in those areas.

The approach used throughout the book is based on developing an understanding of the operation of fluid power component parts and circuits. Understanding component operation requires comprehension of fundamental principles of physics and basic mathematical formulas. These fundamentals are discussed with topics that require them for comprehension of component operation. It is important to note that mathematical applications required for the original design of complex fluid power components are beyond the scope of this book.

Textbook Features

This text includes many features to assist in learning. Each chapter begins with an introduction to the chapter content. This is followed by a list of learning objectives. The learning objectives identify the knowledge and skills you will obtain on successful completion of the chapter.

As you study the chapter, make note of all key terms. Key terms appear in the chapter in **_bold italic_** font. A list of selected key terms appears at the beginning of each chapter. Pay particular attention to these terms and be sure you understand each. It is recommended that you write out the definition for each key term as it is encountered in the chapter. You can then enter the definitions into a word processing document and use this for quick reference.

Also appearing at the beginning of each chapter is a list of Internet resources. These are related to the topics presented in the chapter. Each item provides the URL, the name of the site, and a brief description of the information found on the site. Refer to the next section for more information on this feature.

Throughout the book, you will encounter notes and cautions as boxed features. A note provides additional useful information about the topic being discussed. A caution, on the other hand, points out an important safety-related issue. There are many potentially dangerous situations that will be encountered when working with both hydraulics and pneumatics, not the least of which is the pressurized fluid. Be sure you fully understand the safety-related issues presented in a caution and follow any instructions provided. When in doubt, consult your instructor for further safety information.

Each chapter concludes with a chapter test. The test is designed to measure your knowledge of the chapter topics. Answer each question to the best of your ability. Then, review the chapter and verify or correct your answers.

At the back of the book is an extensive glossary of fluid power terms. Also included is an appendix. The appendix provides general information that you may need when working in the fluid power field, such as conversion charts, common drilled-orifice diameters, common pipe sizes, cylinder force, and addresses for fluid power organizations.

Internet Resources

Each chapter contains a list of Internet resources related to the chapter topic. These resources can be valuable as you explore fluid power systems. All Internet resources listed in this text have been verified at the time of publication. However, due to the dynamic nature of the Internet, the address of the provided resource or the content itself may change. If the page cannot be displayed, first make sure you have correctly typed the address. Also, some addresses are case sensitive, so match case to that shown in the text. If the page still cannot be displayed, it is likely the site holder has changed the address or removed the information. Try going to the home page of the site and searching for the correct page based on the description given in the Internet Resources section in this text.

The sites provided in the Internet resources are not affiliated with Goodheart-Willcox Publisher, nor does the publisher endorse any of the companies represented on the sites. Goodheart-Willcox cannot provide technical support for these sites, nor is the publisher responsible for the content on any of the provided Internet resources.

About the Author

James R. Daines is Professor Emeritus of Technology at the University of Wisconsin—Stout. He holds an Ed.D. degree from the University of Missouri and B.S. and M.S. degrees from Stout State College, Menomonie, Wisconsin.

Professional experience includes 25 years teaching power- and media-related courses to secondary, post-secondary, and adult students. In addition, 10 years were spent as president of a company specializing in developing courses and instructional materials for power and other related technologies. Those materials included extensive computer-based courses for hydraulics (60 hour) and pneumatics (40 hour). Other experiences related to fluid power involved working on the design and construction of training benches and instructional materials for a producer of fluid power training equipment.

Brief Contents

Expanded Table of Contents

Since the beginning of time, long before written history, humankind has searched for ways to conveniently transmit energy from its source to where it is needed and then convert the energy into a useful form to do work. This chapter introduces the fluid power field as an approach that provides an effective means of transferring, controlling, and converting energy.

Objectives

After completing this chapter, you will be able to:

- ■ Define the terms fluid power, hydraulic system, and pneumatic system.
- ■ Explain the extent of fluid power use in current society and provide several specific examples.
- ■ List the advantages and disadvantages of fluid power systems.
- ■ Discuss scientific discoveries and applications important to the historical development of the fluid power industry.

Chapter 1

Introduction to Fluid Power
The Fluid Power Field

Selected Key Terms

The following names and terms will be used in this chapter. As you read the text, record the meaning and importance of each. Additionally, you may use other sources, such as manufacturer literature, an encyclopedia, or the Internet, to obtain more information.

actuator
Archimedes
Bernoulli, Daniel
Boyle, Robert
Bramah, Joseph
Charles, Jacques
compact hydraulic unit
cup seal
da Vinci, Leonardo
fluid compressibility
fluid power
Hero
hydraulic
hydraulic accumulator
laminar
Pascal, Blaise
pneumatic
pump
Reynolds, Osborne
Torricelli, Evangelista
von Guericke, Otto
water screw
waterwheel
Watt, James
windmill

Internet Resources

www.ideafinder.com/history/inventors/watt.htm
The Great Idea Finder
Provides information on James Watt and other inventors who made major contributions to industrial development during the Industrial Revolution.

www.nfpa.com
National Fluid Power Association
A good overall review of the basic aspects of fluid power systems. Go to the "Our Industry" section of the site.

http://library.thinkquest.org/3044
Oracle ThinkQuest Education Foundation
This article provides insight into Leonardo da Vinci as an artist and inventor. Search the library for more information on Leonardo da Vinci.

www.en.wikipedia.org/wiki/hydraulic
www.en.wikipedia.org/wiki/pneumatic
Wikipedia: The Free Encyclopedia
These entries include information on the history and operation of hydraulic and pneumatic power-transmission systems.

Definition of Fluid Power

The basis of *fluid power* is pressurized fluids. These fluids may be either liquids or gases. The fluids are incorporated into physical hardware systems that generate, transmit, and control power in a wide variety of consumer and industrial applications. Today, it would be difficult to identify a product that has not been affected by fluid power at some point along the route from raw material to final installation.

Fluid power systems are versatile contributors to industry. Applications range from brute force needed in heavy industry to the sensitive positioning of parts in precision machining operations, **Figure 1-1.** The systems are generally grouped under the two broad classifications of *pneumatic* and *hydraulic*. Pneumatic systems use gas, usually air, while hydraulic systems use liquids, usually oil. Other fluids are often used in special applications.

Fluid power is one of the three types of power transfer systems commonly used today. The other systems are mechanical and electrical. Each of the systems transfers power from a *prime mover* (source) to an *actuator* that completes the task (work) required of the system.

Fluid power systems use the prime mover to drive a *pump* that pressurizes a fluid, which is then transferred through pipes and hoses to an actuator, **Figure 1-2.** Mechanical systems transfer power from the prime mover to the point of use by means of shafts, belts, gears, or other devices. Electrical systems transfer power using electrical current flowing through conductors. Typical applications in business, industrial, and consumer products and systems use combinations of fluid, mechanical, and electrical power transfer methods.

Fluid Power Industry

The fluid power industry is a complex entity. It includes education, design and manufacture of components, design and assembly of systems using those parts, and troubleshooting and maintenance needed to keep the systems performing efficiently, **Figure 1-3.** In addition, a complex sales and distribution system assures users access to replacement components and information concerning service, new and improved component designs, and new system applications.

Figure 1-1. Equipment used in construction and street maintenance is an example of a fluid power application commonly encountered in daily life. This backhoe is capable of producing the brute force needed to break and move concrete. (Deere & Company)

Figure 1-2. Many consumer items make use of fluid power in their operation. This garden tractor has a hydrostatic transmission, as well as several other fluid power accessories. (MDMA Equipment–Menomonie)

Figure 1-3. The service of fluid power systems in business and industry provides employment for many highly trained individuals. (Photo: Atlas Copco)

Growth of the fluid power industry has required a parallel growth in the number of people who understand and can work effectively with fluid power systems. These people range from engineers responsible for designing the components to mechanics responsible for maintenance and repair of fluid power equipment. The type of education and training available to prepare these people varies considerably. Formally organized programs exist in two-year technical and community colleges, four-year universities, and in programs offered by component manufacturers. Many individuals seek fluid power training after exposure to the field through their jobs.

The fluid power industry is a broad field and a key contributor to the success of many businesses and industries. Fluid power is extensively used in manufacturing, construction, transportation, agriculture, mining, military operations, health, and even recreation. The list is almost endless. Applications vary and components have different appearances in the various applications. System sizes range from miniature to massive, but fluid power principles provide the needed power, force, and control.

Fluid power has been a key contributing factor in the development of current agricultural equipment.

Modern farm equipment uses hydraulics extensively. These uses range from simple hydraulic cylinders that raise and lower implements to complex devices that maintain clearances, adjust torque, and provide easy control of speed and direction on tractors and a variety of specialized planting, harvesting, and processing equipment.

Fluid power is used in some form in all modern transportation systems designed to move people and products. These uses range from automobiles to complex, wide-body aircraft found on international flights. Specific examples of the application of fluid power principles include hydraulic and pneumatic braking systems, power-assisted steering found on most forms of wheeled vehicles, hydrostatic transmissions that provide almost unlimited speed and torque control, and suspension systems that use hydraulic and/or pneumatic dampening.

The construction industry is a very diverse industry. Construction activities include the building of residences and all types of commercial structures, roads and highways, irrigation systems, harbor facilities, and a wide variety of other construction-related activities. The industry makes use of many types of earth-moving equipment, material-handling equipment, and specialized fastening and finishing devices. Examples of typical applications that make use of fluid power include: backhoes for excavation; cranes for moving, lifting, and positioning materials; vibrators for consolidating concrete after it has been placed; and nail-driving apparatuses.

Manufacturing organizations extensively rely on fluid power. Applications range from huge presses in automobile body fabrication plants, which form body panels, to packaging equipment for miniature parts in electrical component manufacturing operations. These applications use hydraulics and pneumatics to make the equipment operate as needed. Required characteristics range from huge forces to draw metal into desired shapes, to a gentle nudge accurately positioning a part for machining, to the deliberate movement of sanders performing a final finish sanding. Fluid power can easily provide each of these characteristics. In many installations, the desired results can be obtained using off-the-shelf equipment. In many other situations, standard components may be used to assemble circuits and systems to produce the desired result, **Figure 1-4.**

Figure 1-4. Fluid power applications have been used during the manufacture or processing of most consumer products today. This carnival ride makes extensive use of fluid power. (Peter Hansen/Shutterstock)

Mining companies use fluid power both in open-pit and underground operations. Spectacular examples of an application in this industry are the huge shovels used in coal strip mining operations. These shovels remove the overburden from veins of coal that are near the surface. Some of these shovels are several stories high and they can remove multiple cubic yards of material during each pass of the scoop. The shovels use large numbers of fluid power systems and circuits for movement and control. Other open-pit mining operations use more standard front-end loader and truck designs for loading and moving ore. In addition, many fluid power applications, both pneumatic and hydraulic, can be found in mine drilling, crushing, and material-handling equipment. Fluid power applications are especially desirable in underground mining locations. Accumulations of gas may produce potentially explosive conditions, limiting or preventing the use of electrical devices.

Land, sea, and air defense forces use fluid power to assist in moving personnel, supplies, and equipment to support their operations. The military makes use of a full range of fluid power components and circuits. Many of these parallel

civilian commercial applications, but others are highly specialized and are not directly duplicated in commercial applications. Hundreds of applications exist, ranging from power-assisted steering of land vehicles to the precision positioning of rocket launchers for air defense, **Figure 1-5.**

Fluid Power Systems

Fluid power is a highly versatile power transmission system, as illustrated by the range of applications discussed earlier in this chapter. No system, however, is entirely suitable for all applications. All power-transmission systems have characteristics that are desirable in one application, but turn into disadvantages in other situations. A system cannot have every desired advantage without disadvantages. Understanding system characteristics as well as what is needed for a particular result will help in producing an effective and efficient application.

The range of applications that use fluid power makes the development of a simple list of advantages and disadvantages difficult, since examples that do not "fit" can easily be found. This problem

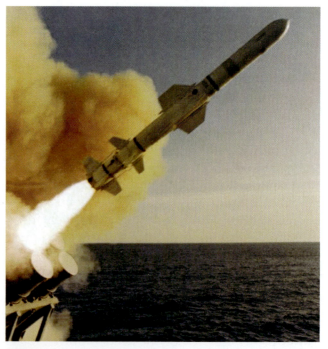

Figure 1-5. Complex defense systems make extensive use of fluid power. (©2007 Jupiterimages and its licensors. All rights reserved.)

is further complicated by the inherent differences of the two major divisions of the fluid power field: hydraulics and pneumatics.

System Characteristics

Although hydraulic and pneumatic systems share the characteristics of energy transfer by means of fluid pressure and flow, differences affect how and where they are applied. These differences include:

- Accuracy of actuator movement
- Operating pressure
- Actuator speed
- Component weight
- Cost

Accuracy of movement

Fluid compressibility is the inherent characteristic that produces the difference between hydraulic and pneumatic systems. A gas is compressible, while a liquid can be compressed only slightly. Hydraulic systems, therefore, can produce more accurate, easily controlled movement of cylinders and motors than pneumatic systems. Compressibility produces a more "spongy" operation in pneumatic systems that is not suitable where highly accurate movement is required.

Operating pressure

Hydraulic systems can operate at much higher pressures than pneumatic systems. Hydraulic system operating pressure ranges from a few hundred pounds per square inch (psi) to several thousand psi. Pressures of more than 10,000 psi are used in special situations. Pneumatic systems, in contrast, normally operate between 80 to 120 psi. Extremely high–pressure pneumatic systems normally are not used.

Actuator speed

Pneumatic systems are commonly used when high-speed movement is required in an application. Rotation speeds of over 20,000 revolutions per minute (rpm) are possible. Rapid-response cylinder operation is also possible with pneumatic systems. These designs are generally found in situations involving lighter loads and lower accuracy requirements.

Component weight

System operating pressure affects the structure of components. Hydraulic systems operate at higher pressures, requiring the use of stronger materials and more-massive designs to withstand the pressure. Pneumatic systems operate at much lower pressures and, therefore, can be manufactured using lightweight materials and designs that minimize the amount of material.

Hydraulic applications tend to involve equipment that handles heavier weights, requiring both higher system operating pressure and physical strength of machine parts. **Figure 1-6** shows a front-end loader. The cylinders used to operate the bucket must be of a construction that can withstand high system pressure and the heavy load of the bucket and its contents.

Pneumatic systems tend to involve applications where ease of handling and lightweight are critical for effective operation of the tool or system. **Figure 1-7** shows a pneumatic grinder being used on a large metal casting. The grinder is lightweight and very portable. The tool is easily manipulated by an individual and constructed to provide a long service life.

Figure 1-6. Hydraulic systems are commonly used in applications where high system pressures are needed to complete the required work. (Deere & Company)

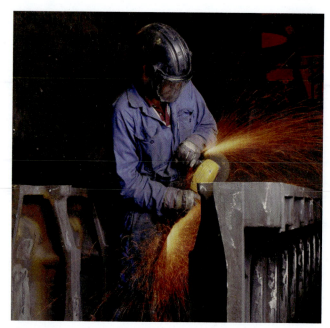

Figure 1-7. Pneumatics are commonly used where high speeds and lightweight tools are needed. (Photo: Atlas Copco)

Cost

The cost of fluid power systems ranges widely. A variety of situations exist and a number of solutions are available for each one. The solution selected to solve the problem directly affects the cost. Understanding system advancements, basic characteristics of hydraulic and pneumatic systems, and knowing which standard components are available are necessary to produce a system that does the best job at the lowest cost.

The cost of system operation is a factor that must be considered. Generally, pneumatic systems are more expensive to operate than hydraulic systems. This cost can be directly associated with the compression, conditioning, and distribution of air. Careful maintenance to eliminate leakage can greatly reduce operating cost.

Advantages and Disadvantages of Fluid Power Systems

Fluid power systems have several advantages and disadvantages when compared with mechanical and electrical power transfer systems. Several of the important advantages and disadvantages of fluid power systems are presented in the next two sections.

Advantages

The following list of advantages applies to both hydraulic and pneumatic systems, except as noted.

- An easy means of multiplying and controlling force and torque.
- Infinitely variable speed control for both linear and rotary motion.
- Overloading the system simply stalls the actuator without damage to the components.
- Provides an easy means of accurately controlling the speed of machines and/or machine parts.
- Provides the ability to instantly stop and reverse linear and rotary actuators with minimal shock to the system.
- Systems easily adapt to accommodate a range of machine sizes and designs.
- Systems readily adapt to external control methods, including mechanical, pneumatic, electrical, and electronic systems, **Figure 1-8.**
- Systems can easily provide component lubrication.

Figure 1-8. Both hydraulic and pneumatic systems can be readily adapted to external control systems, including state-of-the-art electronic designs. (Photo: Atlas Copco)

- Large volumes of compressed air may be easily stored in pneumatic systems to provide energy for intermittent, heavy system demand, **Figure 1-9.**
- Pneumatic systems provide clean operation with minimal fire hazard.

Disadvantages

The following list of disadvantages applies to both hydraulic and pneumatic systems, except as noted.

- Higher safety factors associated with high-pressure oil and compressed air.
- Susceptibility to dirty environments, which can cause extreme component wear without careful filtration.
- Fluid leakage and spills cause a slippery, messy work environment around hydraulic equipment.
- Fire hazard with hydraulic systems using combustible oils.

- Special handling and disposal procedures for hydraulic oil required by environmental regulations.
- High cost of compressing and conditioning air for use in pneumatic systems.
- Reduced accuracy in actuator speed control in pneumatic systems caused by compressibility of air.
- Noise level of pneumatic systems when air is directly exhausted to the atmosphere from components.

Historical Perspective of Fluid Power

Awareness of the historical background of our civilization, our particular culture, and our nation are generally considered an important part of the preparation to be a responsible citizen. This awareness provides an appreciation of our current situation by providing information about the origin and development of cultural, religious, and political ideas, principles, and systems. Likewise, an awareness of the historical development of a technical field of work should provide an appreciation of what we have, some idea of how it was achieved, and an appreciation of the cultural changes associated with the technical field.

This section provides a brief discussion of some aspects of technical development and innovation as viewed by Western culture. Volumes of information are available on developments throughout history involving Islamic and Eastern contributions as well as those reflecting Western thought. Fluid power, as the term is currently used in our society, describes a relatively new field. However, the roots of this field extend far back into history.

Historical Awareness

The use of fluid power has developed along with civilization. The natural movements of air and water were probably the first sources of power used by early humans. It is speculated that the use of crude sails to reduce the effort to move boats was the first attempt to harness this natural power, **Figure 1-10.** These early applications were followed by the development of more sophisticated systems, which eventually lead to the development of the windmill.

Figure 1-9. Pneumatic systems can easily store large volumes of compressed air to meet the intermittent high demand of some systems. The horizontal tank under the compressor serves as the storage area for compressed air.

Figure 1-10. People have used the natural movement of both air and water throughout history to reduce work and aid in transportation. Using sails on ships is an example of using fluid power (wind). (©2007 Jupiterimages and its licensors. All rights reserved.)

Flowing water in rivers and streams was also used to assist in transporting boats and materials. The Egyptians, Persians, and Chinese built dams, ditches, and gates to form elaborate water-control and irrigation systems. These various applications eventually led to the development of variations of the waterwheel to lift water for irrigation purposes and turn simple mills. These early uses of fluid power were dependent on vast quantities of low-pressure air and water supplied by nature. They also were subject to variations in the weather, which made them somewhat unpredictable and only partially under the control of the operators.

The windmill and waterwheel were extensively used as power generating devices before and through the early years of the Industrial Revolution. Their power-generating capabilities were very limited, however, usually in the 5–10 horsepower range with a maximum of 30 horsepower. The development of large industrial complexes needed a larger and more-controllable power generating system, such as the steam engine developed by James Watt. This steam engine was not constructed until 1775.

It is doubtful whether the physical principles associated with modern hydraulics and pneumatics were fully understood in early history or even at the beginning of the Industrial Revolution. The physical principles and design features that we consider basic to the operation of fluid power systems were developed over several centuries. Many of these developments were the products of tinkerers more than due to an understanding of advanced scientific principles. Application of the scientific method played an ever-increasing role in the development of both fluid power theory and machinery design. The *scientific method* dictates accurate measurement, controlled testing, reproducibility of results, and systematic demonstration and reporting of results.

Individuals in History

The identity of individuals involved in the original development of the significant fluid power scientific principles has been lost to time. One of the factors that makes the identification of individuals difficult is that many of the practical solutions were considered below the dignity of "pure" science and, therefore, never recorded. Several individuals, however, need to be recognized as having made important discoveries. Other individuals contributed by bringing together a group of known elements to form a principle that has become basic to the fluid power field.

Archimedes is credited with the discovery of the principle of buoyancy in the third century BC. Folklore relates that discovery to the need to establish the amount of gold in the king's crown. Archimedes did extensive work with mathematics and other scientific principles. Two volumes of his known manuscripts deal with hydrostatics and floatation. He is also credited with the invention of the water screw, **Figure 1-11.** This device has been used to lift and move water for centuries. It has also greatly influenced the development of pumping devices in the fluid power field.

Hero is another individual who is recognized as contributing basic knowledge to the area of fluid mechanics. His work appears to have been done in the second century BC, although no specific dates have been established. Historians often considered him to be a recorder of information, rather than an originator. Nevertheless, Hero is popularly identified with a rotating, steam-jet device. At the time,

Figure 1-11. Archimedes is credited with the design of the water screw, which has been used to move and lift water since the third century BC. The basic principles found in this device are still used in modern fluid power components. (©2007 Jupiterimages and its licensors. All rights reserved.)

Figure 1-12. Blaise Pascal developed during the 1600s what is now called Pascal's law. He is one of a number of individuals who worked during the fifteenth through the eighteenth century identifying and proving basic principles by using the scientific method. (©2007 Jupiterimages and its licensors. All rights reserved.)

the device was considered to be an interesting toy, rather than a practical machine. His more practical and enduring contributions relate to basic concepts of fluid flow and the study of the principles of the siphon.

Leonardo da Vinci is a name commonly associated with the fine arts, but he also must be considered an equal genius in many other fields. During the latter part of the fifteenth and early part of the sixteenth centuries, he worked extensively in the engineering and architectural fields in Italy and France. He made many contributions in the fields of mechanical design and fluid mechanics that involved the flow of water, hydraulic machinery, and principles that closely match the later work of Pascal. As with many other early scientists, his work was recorded as private notes, many of which have been discovered only in recent years.

Blaise Pascal is usually credited with the basic principle that is the foundation of the fluid power industry, **Figure 1-12.** During his relatively short life (1623–1662), he experimented extensively with liquids and mechanical devices, including siphons, syringes, and tubes. This work produced the proof that many previous mathematicians and scientists had sought:

> In a fluid at rest, pressure is equally transmitted in all directions.

He is also credited with demonstrating and clearly defining the principles involved in the hydraulic press: multiplication of force, and piston movement relationships.

The English scientist *Robert Boyle* (1627–1691) was one of the first to work with the characteristics of gases. He referred to this as the "springiness of air." By direct measurement, he was able to establish:

> If the temperature of a dry gas is held constant, its volume inversely varies with its pressure.

Today, this principle is referred to as *Boyle's law.*

As in most other situations where new laws or principles are under development, Boyle was not the only person to do fundamental work in the area of gases. *Otto von Guericke* (1602–1686) from Germany was also involved in experimentation with gases. He developed a vacuum pump in 1650. He used this pump to demonstrate the pressure of the atmosphere by pumping the air from a sphere constructed of two equal parts. Once the air had been removed, the pieces could not be pulled apart by horses.

An Italian, *Evangelista Torricelli* (1608–1647), is recognized as having identified the principles that affect fluid flow. He also did extensive work in the field of mechanics and had an influence on the development of many scientific principles. Today, he is generally associated with the development of the barometer. However, close analysis of historical data indicates the basic principles were tested by others with Torricelli properly interpreting the results.

The French scientist *Jacques Charles* (1746–1823) provided another key element to understanding fluid power principles. He developed a law relating to the effect of temperature on the volume of a gas:

> All gases expand or contract in
> direct proportion to the change in
> absolute temperature.

Charles' law combined with Boyle's law form the *general gas laws,* which are fundamental to any calculations done for gases today.

Daniel Bernoulli (1700–1782) is credited with laying the foundation of hydrodynamics. Although his name is applied to the Bernoulli theorem, some historians contend that his 1738 publication *Hydrodynamics* does not include specific formulas. He extensively studied both static and dynamic fluid phenomena. His contemporaries and successors considered his work to be the first specific principles on fluid movement.

James Watt (1736–1819) was a productive inventor who is usually thought of with respect to the steam engine, **Figure 1-13.** His inventive genius went well beyond this device, however. He impressed his contemporaries as the individual who provided mankind with devices that had the potential of producing unlimited power. As new designs were developed to meet the power demands of the Industrial Revolution, Watt developed, or caused to be developed, new manufacturing techniques that greatly influenced the Industrial Revolution.

English engineer *Joseph Bramah* (1748–1814), with his assistant *Henry Maudslay* (1771–1831), invented the cup seal. This seal greatly contributed to the practical application of fluid power devices. The cup seal enabled the development of devices requiring pressurized liquids and gases to operate with little leakage. Bramah went on to build the first functional hydraulic press, which had extensive application in the fluid power industry.

Figure 1-13. James Watt perfected the steam engine and greatly influenced the development of the Industrial Revolution. Many of his ideas were used in the development of pumping devices. (©2007 Jupiterimages and its licensors. All rights reserved.)

Maudslay went on to develop numerous devices and is generally considered to be the father of the machine-tool industry.

Lord William Armstrong (1810–1900) developed a hydraulic accumulator that made a major contribution to the development of the early fluid power industry. An *accumulator* stores excessive pressurized fluid from the pump until needed during peak system operation. The Armstrong accumulator was basically a large, vertically positioned cylinder with a weighted ram. These devices were extensively used during the late nineteenth century in large, centralized, pressurized-fluid systems found in major cities throughout Great Britain.

Osborne Reynolds (1842–1912) established by direct observation that two types of fluid flow exist: *laminar* (smooth and steady) and *turbulent* (chaotic and rough). The observations he made in 1883 eventually led to the development of a formula that produces a dimensionless number today called the Reynolds number. The *Reynolds number* is considered one of the basic parameters of fluid mechanics.

Applications of Fluid Power through History

Fluid power as it is know today has been under development for thousands of years. It has experienced rapid growth since the onset of the Industrial Revolution. The previous section introduced some of the many individuals credited with identifying the basic concepts that resulted in continued industrial growth. This section presents some of the concepts and applications of fluid power from a historical perspective.

Antiquity

The origins of fluid power are difficult to identify. Recorded evidence does not exist and the relationships that eventually produced important developments are not evident in most cases. Archeology gives us many hints, but speculation still plays a major role in our current thinking about how existing scientific principles and mechanical devices were developed.

Three items that are very essential to the existence and comfort of humankind are associated with fluid power developments:

- Transportation.
- Movement of water.
- Generation and transmission of power.

These elements have had far-reaching effects on the development of our industrial society.

Artwork from Egyptian tombs that date to 2500 BC shows reed boats with bipod masts and sails designed for the special problems associated with river sailing. *Sails* for the propulsion of ships continued to be refined through the ages. Many sailing principles were applied to the development of windmills and, eventually, pneumatic components.

Movement and control of water for irrigation, flood control, and municipal water systems also produced ideas that eventually were used in fluid power systems. Controlling the annual flood of the Nile River and moving water for irrigation purposes during the remainder of the year produced a better understanding of water flow in channels. Archimedes is credited with inventing pumping devices that were used in the Nile Delta for irrigation purposes. Applications for moving and pumping water were also developed in the Islamic world, especially in Persia. The most common design, the *water screw,* is still used in parts of Egypt and other areas of the world for moving or lifting small volumes of water. The Roman Empire constructed great aqueducts for the movement of water. Ruins of these systems can still be found throughout the territories the Romans controlled, **Figure 1-14.**

It is generally believed that the energy of running water was first effectively harnessed during the first century BC. Early *waterwheels* used a horizontal wheel rotated by a running stream. These designs typically developed one-half horsepower or less. Roman engineers developed vertical wheels that produced up to three horsepower. These wheels used both undershot and overshot designs. Undershot wheels were placed directly in a stream. Overshot wheels used water directed on the top of the wheel through sluice channels from dams or natural waterfalls. Waterwheels revolutionized the grinding of grain and were gradually adapted to other purposes.

The use of *windmills* for doing work did not appear until after the decline of the Roman Empire. The earliest record of a windmill-like device is from central Asia. These wind-driven devices were used to turn prayer wheels. Historians believe that the first "real" windmills were developed and used in central Asia about 400 AD.

Figure 1-14. An understanding of basic scientific principles has developed slowly throughout civilization. Ruins of the aqueduct system of the Roman Empire illustrate an early need for a practical understanding of fluid flow.

Middle Ages

During the Middle Ages following the fall of the Roman Empire, the development of new technology was extremely slow in Western civilization. In the Western countries of Europe, existing Roman structures were allowed to gradually deteriorate as major emphasis was placed on defense, conquest, development of monastic (religious) orders, and support of the Crusades. The Islamic world, however, continued to make progress on public water supply systems, public baths, and power-driven mills. Many of these ideas were introduced in Europe during the twelfth and thirteenth centuries by returning Crusaders.

At first, waterwheels were just a means of turning millstones for grinding grain for local consumption. Waterwheels began to be recognized as an important source of power for other uses during the fourth and fifth centuries. They were slowly adapted for use in sawmills, paper mills, iron mills, and mining operations. An example of the slow, but pervasive, growth of water power as a prime mover is that the earliest recorded use in England was in the eighth century with over 5000 in use by the eleventh century, **Figure 1-15.** Water mills were profitable operations and found in every community that had a suitable stream.

Windmills continued to be developed in several forms in the Near East, where easy access to water was limited. Records indicate that windmills were extensively used in Afghanistan and Persia, with a number of designs developed to control speed. Speed control was required because of strong winds in the area. Use of wind power spread throughout the Islamic world during the period.

The first recorded use of windmills in Western Europe was during the twelfth century, **Figure 1-16.** These mills were initially used for grinding grain, but were gradually modified for use in other applications, such as sawing lumber, pumping water, and manufacturing. Holland and the other "low countries" of Europe were prime users of these mills because of their flat terrain and consistent winds.

Windmills gradually developed with improvements in design and increases in size. Power output ranged from 4 to 5 horsepower for the average mill to over 15 horsepower for larger mills. At the peak of windmill use, Holland had over 8000 mills, with many of them having sail spans as large as 100 feet.

Figure 1-15. Thousands of watermills have been used throughout history as prime movers for grinding grain and powering early machines. Modern water turbines for electric generation and fluid power motors have evolved from these early designs. (©2007 Jupiterimages and its licensors. All rights reserved.)

Figure 1-16. Windmills have played an important part in the development of civilization. Their use is limited to coastal and plains areas with consistent winds. This design was used on a farm from the early to mid-1900s.

Many mechanical improvements were made as water and windmills became prime sources of power. These improvements included methods of power transmission through shafts, cogged wheels, crude gears, and cams. These devices were needed to transmit the power and transform it into the type of motion needed to do the desired work. Many ingenious devices were developed that have contributed to the body of knowledge as the world moved toward the Industrial Revolution.

Industrial Revolution

The eighteenth and nineteenth centuries are often designated as the period in which an "industrial revolution" occurred. That period of history produced tremendous changes in world society, particularly in Great Britain. Great Britain had established extensive world trade, a strong financial base, large local reserves of iron ore and coal, and key individuals interested in the practical development of scientific principles. These factors combined to lead to the *Industrial Revolution.*

Changes during the Industrial Revolution occurred much more quickly than those during the Middle Ages. However, identification of specific dates or events that started the "revolution" is not possible. It must also be emphasized that the industrial change was not just tied to mechanical devices, but involved changes as diverse as world trade and application of scientific methods. Nevertheless, development of new prime movers must be considered to be the key to the Industrial Revolution, **Figure 1-17.** The *steam engine,* internal combustion engine, and gas turbine all were developed during the period. Unlike water and windmills, these devices were mobile and not dependent on local weather or terrain conditions. Even with those new power development devices, water and windmills continued to play important roles in industrial power generation throughout the middle 1800s and much later in smaller, local and rural applications.

The use of machines instead of hand tools was another key to the changes of the period. These machines exceeded the capabilities of skilled craftsmen to develop products. They allowed production rates that met demands resulting from expanded trade areas.

The development of manufacturing increased the need for methods and devices to move water and transmit power. The procedures and devices

Figure 1-17. The steam engine provided a reliable, portable source of power. Its development is considered one of the key factors in the development of the Industrial Revolution. (©2007 Jupiterimages and its licensors. All rights reserved.)

developed to satisfy these needs are where fluid power, as we think of it today, began to emerge. Many of the early fluid power elements involved pumping water from mines or for manufacturing uses. James Watt designed steam-driven, reciprocating pumps that were used in the late 1700s and early 1800s. By the mid-1800s, these pumps were widely used. The East London Water Company operated a unit with a 100″ diameter cylinder and a stroke of 11 feet. By the 1870s, patents were issued for variable-stroke, piston pumps. The centrifugal pump was invented in the late 1600s, but was not generally used until the mid-1800s. The jet pump was also produced during this same period.

A key factor that allowed the development of many practical fluid power components was the *cup seal.* This type of seal uses the pressure within the system to force a sealing collar against a shaft or ram to prevent fluid loss, **Figure 1-18.** Invented in 1795, the cup seal allowed hydraulic presses to be built that both did not leak and could be continuously used.

Two other devices that were especially important to the development of fluid power during the 1800s were the *hydraulic accumulator* and the hydraulic intensifier. Large, weighted accumulators were used to store pressurized fluid from

Figure 1-18. Modern sealing devices are often taken for granted. However, their development in the late 1700s by Bramah and Maudslay was instrumental to the practical application of fluid power principles. (Photo courtesy of Apple Rubber Products, Inc.)

a pump during the idle time of a press or other machine. During peak operating time the charged accumulator assisted the pump by supplying high volumes of fluid needed for operation.

The *hydraulic intensifier* was used to more easily obtain high system pressures for use in bailing, metal forming, forging, or other applications. The intensifier, which usually consisted of some combination of large and small diameter rams, used the area differences of the rams to boost system pressure above the capability of the pump. This principle allowed the use of higher pressure even when the technology of the time did not allow pumps to operate at extremely high pressure.

Pressurized water was extensively used to distribute power to businesses and manufacturers in several cities in Great Britain. By 1900, it was considered economical to transmit power up to 15 miles from a centralized pumping station. These systems provided water under several hundred pounds per square inch of pressure that could be directly used for the operation of presses, hoists, and water motors. These systems continued in use until electrical generating and distribution systems were perfected.

Fluid power development and use during the nineteenth century was very extensive. This use involved the generation of power through the design of effective water turbines; the transmission of power using central power stations and elaborate distribution lines; and the use of fluid power in construction, manufacturing, and material distribution systems.

Recent history

The emphasis on the development of fluid power applications decreased as the use of electricity grew in the late 1800s and early 1900s. Development concentrated more in the heavy industrial and mobile areas where fluid power applications appeared to be most practical. Generally, these factors have promoted progress in fluid power applications in recent years:

- Development of new materials.
- Miniaturization of components.
- Effective electrical/electronic control.

Improvements were made in sealing devices and machining techniques that reduced both internal and external leaks. This, in turn, improved system efficiency. Reducing external leaks also allowed use in applications where cleanliness was an important factor. Water was replaced in hydraulic systems by petroleum-based fluids, which improved lubrication and eliminated the danger of freezing in cold climates.

Military applications contributed to the development of fluid power. A milestone occurred in 1906 when hydraulic systems first appeared on a warship, the battleship USS Virginia. These hydraulic systems replaced many mechanical and electrical systems. Since that time, all branches of the military have incorporated fluid power on numerous devices to solve problems in gunnery, materials handling, navigation, and support services.

The development and refinement of the concept of **compact hydraulic units** in the 1920s had a far-reaching effect, extending to today's fluid power applications, **Figure 1-19.** These self-contained systems include the power source, pump, and reservoir. The units have been applied to a full range of industrial and consumer applications. Central hydraulic and pneumatic power systems have remained in large industrial applications, but the more-compact direct system adds flexibility.

Figure 1-19. Compact power sources such as these hydraulic units have added flexibility to the application of fluid power. (Continental Hydraulics)

It has been the key to the development of today's large mobile hydraulics field.

The development of new materials and manufacturing techniques has promoted the design of new fluid power concepts and allowed practical application of old ideas, **Figure 1-20.** The miniaturization of components has produced new applications, while the combining of electrical/electronic control and fluid power systems has produced more effective machines. Two components that illustrate these factors are hydrostatic transmissions and servo systems. These units are heavily used in both industrial and consumer applications, **Figure 1-21.** Neither could exist in their present form without the sophistication of the present manufacturing community.

Today's automobile can be used as an example of the use of fluid power. The body of the vehicle is formed by huge hydraulic-powered presses. Hydraulically-controlled resistance welding equipment assembles those parts, while untold numbers of other hydraulic and pneumatic tools are used in the production of the additional parts and during the final assembly process. Intricate fluid power systems are also used in the steering, braking, and ride control systems of the vehicle, which promote safety and comfort of the driver and passengers.

Figure 1-20. Central power systems are often used in large manufacturing facilities to produce compressed air for pneumatic applications. (Photo: Atlas Copco)

Figure 1-21. An automobile assembly line illustrates the diverse use of fluid power in industry today. A full range of both pneumatic and hydraulic applications can be found in such operations. (Photo: Atlas Copco)

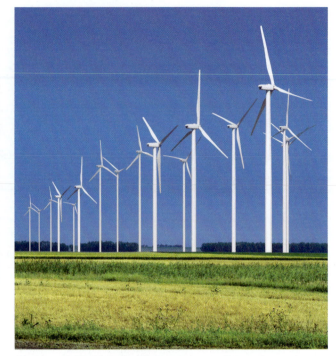

Figure 1-22. Windmills have experienced a resurgence as prime movers in the windmill farms currently used for electrical generation. As energy costs rise, so does interest in harvesting this "free" energy. (ssuaphotos/Shutterstock)

To say the least, fluid power has grown tremendously during the 1900s and into the 2000s. Current space technology, manufacturing industry demand, and consumer interest indicate that this growth will continue into the twenty-first century, **Figure 1-22.**

Chapter Test

Answer the following questions. Write your answers on a separate sheet of paper.

1. Fluid power systems use _____ fluids to transmit power.

2. The physical components in a fluid power system are used to _____, _____, and _____ power to produce the desired results in an application.

3. The three power transfer systems commonly used today are _____, _____, and _____.

4. Name six industries in which fluid power applications contribute to daily operations and long term business success.

5. The _____ fluid power system can usually provide the best solution where lightweight and easily handled tools are a requirement.

6. The pneumatic fluid power system is generally considered to be the most expensive to operate because of the cost of _____, _____, and _____ the air.

7. Clean operation with minimum fire hazards are characteristics of which fluid power system(s)?

8. The earliest records indicating the use of natural air and water movement to reduce effort comes from archaeological discoveries in _____.

   A. Egypt

   B. Germany

   C. Holland

   D. Great Britain

9. It is generally believed by historians that much of the early development of fluid power was based on _____, rather than on an understanding of scientific principles.

10. James Watt, who perfected the steam engine, is credited with greatly influencing the Industrial Revolution through his development of new and innovative _____ techniques.

11. _____ and _____ are credited with the invention of the cup seal, which led to the development of the first functional hydraulics press.

12. Artwork from Egyptian tombs indicates that sails were used to assist in the propulsion of boats as early as _____ BC.

13. In early times, the movement and control of water for _____, _____, and _____ also produced ideas that eventually were used in fluid power systems.

14. The invention of new _____ is usually considered to be the key that allowed the development of the Industrial Revolution.

15. List three factors that have promoted progress in fluid power applications in recent years.

Imaginative designs involving fluid power with mechanical and electrical components may produce robots that assume a human appearance. (Atlas Copco)

The general definition of a system is a set or group of principles, beliefs, things or parts that form a whole, such as a system of government, an accounting system, a mountain system, or the digestive system. The parts of a system work together to produce a desired end result. This chapter introduces the basic design, structure, and operation of fluid power systems.

Objectives

After completing this chapter, you will be able to:

- Explain the functions of fluid power systems.
- Identify the basic structure of fluid power systems.
- List the basic component groups involved in the structure of fluid power systems.
- Describe the function of the components involved in basic fluid power systems.
- Describe the similarities and differences of hydraulic and pneumatic systems.
- Explain the operation of basic hydraulic and pneumatic systems.

Chapter 2

Fluid Power Systems
The Basic System

Selected Key Terms

The following terms will be used in this chapter. As you read the text, record the meaning and importance of each. Additionally, you may use other sources, such as manufacturer literature, an encyclopedia, or the Internet, to obtain more information.

actuators
adapters
component groups
compressor
conductors
control valves
cylinders
directional control valves
filters
fittings
flow control valves
fluid conductor
fluid maintenance devices
heat exchanger
hoses
lubricator
mechanical coupler
motors
pipes
power unit
pressure control valves
pressure regulator
prime mover
pump
receiver
reservoir
separator
system functions
tubes

Internet Resources

www.eaton.com/Eaton/ProductsServices/index.htm
Eaton Corporation
Choose the "Hydraulics" product category to view resources available from this company in the area of hydraulics.

www.science.howstuffworks.com
How Stuff Works, Inc.
Search the site for "hydraulics" to return a variety of articles on how hydraulic machines work. Includes basic concepts, as well as short videos on machine operation.

http://hydraulicspneumatics.com
Hydraulics and Pneumatics Magazine
Contains articles covering many aspects of both hydraulic and pneumatic components and systems.

http://www.ingersollrandproducts.com/am-en/solutions
Ingersoll-Rand Company
Choose "Compressed Air Solutions" to review the impact of compressed air equipment in various industries.

www.ifps.org
International Fluid Power Society
To access downloadable papers devoted to various fluid power topics, select the "Education and Training" tab, and then choose "Technical Papers". The materials available have been developed by a number of individuals and companies.

www.nfpa.com/Education/Edu_SelfPaced.asp
National Fluid Power Association
Presents online resources for self-paced learning. The "Fundamentals of Pneumatics" program introduces basic concepts, terminology, applications, and automation processes used throughout the fluid power industry.

Functions of Fluid Power Systems

Fluid power systems are made up of *component groups* containing parts designed to perform specific tasks. These component groups act together to perform the work desired by the system designer. The work may involve simple or complex tasks, but the component groups perform specific *system functions* that are basic to all fluid power systems. There are five functions that are basic to system operation of any fluid power system, **Figure 2-1:**

■ Energy conversion.

■ Fluid distribution.

■ Fluid control.

■ Work performance.

■ Fluid conditioning.

Each of these functions must be performed by a fluid power system if the system is to operate efficiently and provide a reasonable service life. The operating environment, power output, and complexity of the system establish the number of components required to perform a particular function.

Energy Conversion

Fluid power systems do not *generate* energy, but *transform* it into a form that can be used to complete a task. The process begins with a prime mover pressurizing a fluid. It ends with an actuator using the energy stored in the pressurized fluid to perform work.

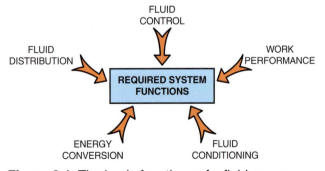

Figure 2-1. The basic functions of a fluid power system.

Fluid Distribution

The operation of fluid power systems requires the distribution of fluid to the components in the system. Various types of lines are involved in this function. Valves and other components also serve to assist in fluid distribution.

Fluid Control

Fluid power systems require the control and regulation of the fluid in the system to perform the tasks desired by the system designer. A number of different components are used to control fluid flow rate, direction, and pressure in a system. Control of these three elements allow the system to provide the desired operating characteristics.

Work Performance

Using the energy stored in the pressurized fluid of the system is the primary function of a fluid power system. This process involves actuators that convert the energy stored in the fluid to linear or rotary motion to perform the desired work.

Fluid Conditioning

Fluid power system performance and service life require a fluid that is clean and provides lubrication to system components. This function involves storing fluid, removing dirt and other contaminants, and maintaining proper system operating temperature.

Structure of Fluid Power Systems

The physical appearance of fluid power systems vary considerably, depending on the type of fluid used, application, and power output. However, each system is structured using component groups that perform the various system functions. The structure of typical fluid power systems involves five component groups, **Figure 2-2:**

■ Power unit.

■ Actuators.

■ Conductors.

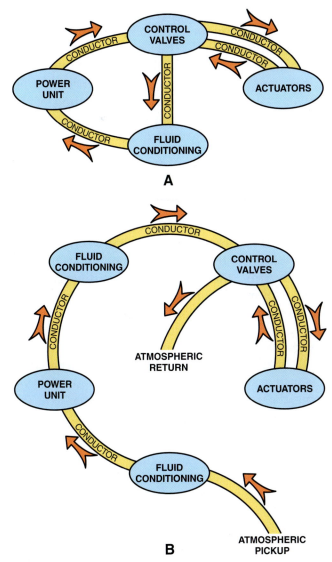

Figure 2-2. The structure of fluid power systems. A—Hydraulic system. B—Pneumatic system.

■ Control valves.

■ Fluid conditioning.

Each component group has primary tasks to perform that relate to one of the five system functions. However, each group also performs secondary tasks relating to one or more of the other functions.

Power Unit

The power unit group of components deals primarily with the energy-conversion function of the system. The unit consists of a prime mover, pump, and reservoir. The prime mover is the source of

energy for the system. The energy produced by the prime mover turns the pump, which produces fluid flow that transmits energy through the system. The reservoir serves as a storage unit for system fluid. It also performs fluid maintenance functions.

Actuators

The actuators group of components performs the work done by the system. These components convert the energy in the system fluid to linear or rotary motion. The basic actuators are cylinders for linear motion and motors for rotary motion. A variety of cylinder and motor designs are used to produce the specific motion needed to complete the work required of the system.

Conductors

Fluid distribution is the primary function of the conductors group of components. Pipes, hoses, and tubes serve as the conductors that confine and carry system fluid between the pump and other system components. Conductors perform such tasks as the intake of fluid for the pump, distribution of fluid to and from control valves and actuators, transmission of sensing and control pressures, and the draining of liquids that have accumulated in components.

Control Valves

Three different types of valves are required to perform the fluid control function in a fluid power system. The valves in the control valves group are:

■ Directional control valves.

■ Pressure control valves.

■ Flow control valves.

Directional control valves provide control over fluid flow direction in sections of a system to start, stop, and change the direction of actuator movement. Pressure control valves are used to limit the maximum pressure of the system or in a section of the system. Flow control valves provide control over fluid flow rate in a section of a system to control the rate of movement of an actuator.

Fluid Conditioning

The fluid conditioning group involves maintaining and conditioning system fluid. This requires removal of dirt and moisture from the fluid and assuring proper operating temperature. Specially designed components are available to perform these tasks. However, basic systems often maintain the fluid using other system components that perform the task as a secondary function. *Filters* are used to remove dirt and moisture from systems, although a properly designed reservoir can perform this task under certain conditions. Maintaining the proper system operating temperature can require the use of a *heat exchanger.* However, this task is usually performed by dispensing heat through the reservoir, system lines, and other components.

Basic System Components

Fluid power systems are constructed of various components that perform the specific system functions. The number and appearance of components required to perform these functions varies considerably depending on the complexity and accuracy of the work performed, the environment in which the system operates, and the manufacturer of the component. Understanding the operation of fluid power systems requires an understanding of the construction and operation of the components that form the structural groups of the system. Refer to **Figures 2-3** and **2-4.**

Power Unit

The appearance and structure of the power unit in a fluid power system can vary considerably, depending on the application of the system. In basic systems, the *power unit* consists of a prime mover, pump, reservoir or receiver, mechanical coupler, and the fluid conductors required to make the unit operational. Terminology and design vary somewhat between pneumatic and hydraulic systems, but direct comparisons can be made between the two systems. For example, in pneumatic systems, the power unit is generally referred to as the *compressor,* **Figure 2-5.** On the other hand, in hydraulic systems, the power unit is called a *pump,* **Figure 2-6.**

Figure 2-3. A basic hydraulic fluid power system. Note the components and their relationships to each other.

Figure 2-4. A basic pneumatic fluid power system. Note the components and their relationships to each other.

Figure 2-5. A typical pneumatic system compressor unit (power unit). The prime mover in this application is a gasoline engine.

The *prime mover* of a system is usually an electric motor or an internal combustion engine. However, gas or steam turbines may be used to power large industrial operations. System size, operating environment, and mobility are factors that determine the device selected as the prime mover. Prime movers range from fractional horsepower electric motors and small single cylinder gasoline engines to gas turbines producing thousands of horsepower.

The pump or compressor is the heart of a fluid power system. These components produce the fluid

Conductors

Conductors confine the system fluid as it is distributed throughout the system. Pipes, tubes, and hoses are the three general types of conductors used in fluid power systems. Special manifolds consisting of multiple passageways are used in systems where space and weight are important factors. The type of conductor used depends on the type of fluid, system pressure, required component movement, and the environment in which the system operates. This group of components also includes a variety of fittings and adapters to allow easy assembly while assuring a system free from leaks.

Pipes are rigid conductors commonly used in stationary fluid power installations. Piping is lower in cost than most other conductors with comparable specifications. The pipe designed for use in fluid power systems is made from mild steel. It is manufactured as seamless.

Tubes are similar to pipes, but are considered to be semirigid. Tubing for use as a conductor in fluid power installations is made from thin steel. Tubing is lightweight, easy to install, has a good appearance, and develops few leaks during system operation, **Figure 2-11.**

Hoses are flexible conductors made from layers of materials. They contain system fluid while withstanding system pressure and allowing easy movement of system components. Hose construction includes an inner tube to contain the fluid, a middle section of braided fabric or wire to withstand pressure, and an outer layer of material for protection from dirt and abrasion, **Figure 2-12.**

Fittings and adapters are needed to assemble conductors and other system components. *Fittings* are considered the parts needed to assemble similar conductors, while *adapters* are required when connecting different types of conductors or when attaching conductors to system components.

Control Valves

Control valves are the components that make it possible to establish the direction of movement and the maximum force and speed of a system actuator. These characteristics are achieved by controlling fluid flow direction, pressure, and flow rate in the system. Many different designs of valves are available to achieve these characteristics. These designs range from simple, basic valves

Figure 2-11. Tubing is used in both hydraulic and pneumatic systems for transfer of pressurized fluids. Here, both tubing and hoses are used to allow required machine member movement.

Outer layer   Braided reinforcement material   Inner tube

Figure 2-12. Hydraulic hose is made using multiple layers of synthetic rubber, braided fabric, and wire to provide a flexible high pressure conductor.

to extremely complex devices. However, each of these designs can be placed under a basic classification of directional control, pressure control, or flow control.

Directional control valves regulate the direction of actuator movement by creating flow paths to and from the actuator. A basic control valve consists of a valve body, internal elements that open and close fluid flow paths through the valve, an external device for shifting the internal elements, ports to connect the valve to the system, and sealing devices to prevent fluid leakage, **Figure 2-13.**

When a directional control valve is used in a fluid power system, it directs fluid to one side of the internal elements of an actuator, causing the actuator to move in one direction. Fluid from the other side of the actuator's internal elements flows through the valve and is returned to the system (or atmosphere). Shifting the valve creates a flow path to the opposite side of the internal elements of the actuator, causing the actuator to move in the opposite direction.

Pressure control valves regulate pressure in the system or a part of the system. These components are used to control maximum system pressure, limit pressure in a part of the system, or delay the movement of an actuator until a desired pressure is reached. A basic pressure control valve consists of a valve body, internal elements that control fluid flow through the valve, an external device to allow adjustment of the valve, ports to connect the valve to the system, and appropriate sealing devices.

In a hydraulic system, the basic pressure control device is a closed valve that does not open to allow fluid to flow through it until a desired pressure is reached, **Figure 2-14.** When the system reaches the desired pressure, the internal elements of the valve move opening a passageway through the valve. Fluid that is not needed to maintain the desired system pressure then passes through the valve to a lower-pressure section of the system. The amount of fluid that passes through the valve varies according to the amount of fluid needed to maintain the desired system pressure. The smaller the amount of fluid needed to maintain the desired system pressure, the higher the flow through the valve to the low-pressure section of the system.

In a pneumatic system, the basic pressure control device is called a *pressure regulator*, **Figure 2-15.** Regulators are located at each workstation in a pneumatic system. A regulator is an open valve that allows air to move through without resistance until a desired system pressure is reached. Once the desired pressure is reached, the internal elements of the regulator begin to close, allowing only enough air to pass through to maintain the desired pressure.

Flow control valves regulate fluid flow in the system or a part of the system. These valves control the volume of flow by varying the size of an orifice in the passageway through which the system fluid flows. A basic flow control valve consists of a valve body, internal elements to control flow through the valve, an external mechanism to allow adjustment of the valve, ports to connect the valve into the system, and appropriate seals to prevent fluid leakage, **Figure 2-16.**

The passageway in a basic flow control valve provides an open route for fluid flow through the valve. The volume of fluid flow through the valve

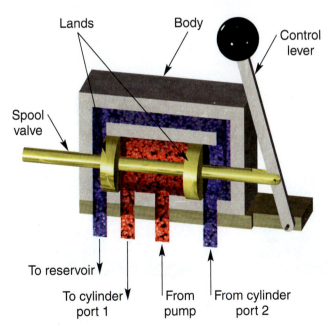

Figure 2-13. Cutaway of a two position directional control valve showing the flow passageways that allow cylinder extension and retraction.

Figure 2-14. The internal construction of a hydraulic pressure control valve used to set the maximum operating pressure of a system.

is controlled by the size of the passageway opening. Adjusting the size of the passageway varies the flow volume, which controls the speed of system actuator(s). The larger the opening, the higher the volume of flow through the valve and the higher the speed of the actuator(s).

Fluid Conditioning Devices

Various devices are used to assure the system fluid is free of contaminants and operating within an acceptable temperature range. Proper care of the fluid has a major effect on the performance and service life of components and the overall system. The type of conditioning devices required in a system is directly related to the complexity of the components and the environment in which the system operates.

Filters are the most common conditioning devices incorporated in fluid power systems, **Figure 2-17.** They remove foreign particles and other contaminates. Filters are located in various places of a system to safeguard specific components, as well as the overall system. Three typical locations are:

- In the intake line for the system.
- In high-pressure working lines.
- In the lines that return fluid from system actuators.

A basic filter consists of a filter housing, filter element, bypass valve, and ports to connect the filter into the system.

Many different filter designs are used in fluid power systems. In a typical basic filter, fluid enters the filter housing through the inlet port, is routed

Figure 2-15. A pressure regulator is used to set the operating pressure of a pneumatic system workstation. (Fairchild Industrial Products)

Figure 2-16 Fluid flow through a flow control valve is adjusted by varying the size of an orifice in the valve. A—In the closed position, fluid does not flow. B—With the orifice partially open, restricted flow is allowed through the valve. C—When the orifice is fully open, flow through the valve is not restricted.

Figure 2-17. Filters, such as this compressor inlet filter, are used in several locations in both hydraulic and pneumatic systems to remove contaminates, which increases both the performance and service life of system components. (Photo: Atlas Copco)

Figure 2-18. A lubricator in a pneumatic system adds a fine mist of oil to the compressed air before it is used at a system workstation. (Photo: Atlas Copco)

through the filter element, which removes contaminates, and is then directed back into the system through the outlet port. If the filter element becomes clogged with contaminants, the bypass valve opens and routes fluid around the filter element.

A pneumatic system includes two additional fluid conditioning devices: a separator and a lubricator. The *separator* removes water droplets from the compressed air. The filter and separator are often combined in one unit called a separator-filter. The *lubricator* adds a fine mist of oil to the air for lubrication of system components, **Figure 2-18.**

The hydraulic system reservoir also serves as an important component for fluid maintenance. A well-designed reservoir helps control temperature by dissipating heat, trapping contaminants, and allowing air trapped in the system oil to separate and escape.

The receiver also serves as a fluid maintenance device in the pneumatic system. The receiver allows water droplets and other contaminates to settle out of the compressed air. The device also helps control the temperature of the system by allowing heat to dissipate.

Basic System Operation

Individual fluid power components must be assembled into a system to convert, distribute, and control the energy provided by the prime mover and produce the work desired from a machine. The basic systems described in this section include components from each of the five component groups and provide directional, pressure, and speed control of an actuator. Basic system examples are provided for both hydraulic and pneumatic systems to illustrate the structural characteristics of each system and the inherent differences between the systems.

Basic Hydraulic System

A basic hydraulic system is illustrated in **Figure 2-19.** The system includes an electric motor that serves as the prime mover. The prime mover produces the energy needed to operate the system and perform the work required of the actuator. The system contains basic components from each of the component groups to allow operation and control of a cylinder.

As the electric motor turns the pump, pressure differences are created within the system.

Figure 2-19. The components of a basic hydraulic fluid power system that can provide directional, flow, and pressure control for a cylinder.

As a result, oil moves from the reservoir into the intake line of the pump and through a system filter where impurities are removed. Continued rotation of the pump forces oil into the system lines (conductors) where it is distributed to other system components.

The first system component the oil encounters as it is forced through the system lines is the pressure control valve, which is used to set system pressure. Next is the directional control valve. This valve directs the oil to the actuator (a cylinder in this example). Forcing oil into the actuator causes the actuator piston and rod to move (either extend or retract). Oil already in the actuator on the other side of the piston is moved out into system lines and returned to the reservoir through the directional control valve, **Figure 2-20.** Shifting the directional control valve directs the oil flow to the other side of the actuator piston, which forces the piston and rod to move in the opposite direction, **Figure 2-21.** Oil from the other side of the piston is returned to the reservoir. The extension and retraction of the actuator does the work that the system was designed to perform.

When the actuator is fully extended or retracted, or when it encounters a load heavier than it can move, there is no place for the oil that the pump continues to force into the system. This condition could cause pressure in the system to rise to a level high enough to damage system components. A pressure control valve located between the pump and the directional control valve protects the system from this potentially damaging high pressure. The pressure control valve is set at a pressure that is safe and appropriate for the system. The internal elements of the pressure control valve begin to open when system pressure approaches the valve setting. Oil that cannot be moved into the actuator passes through the pressure control valve and is returned to the reservoir. See **Figure 2-22.**

The speed at which an actuator moves depends on the rate of oil flow into the actuator. In a basic system, actuator extension speed is controlled by placing a flow control valve in the line between the directional control valve and the actuator. Closing the valve reduces the size of the orifice, which restricts oil flow though the valve. This restricted flow reduces the rate of actuator movement. Opening the valve enlarges the size of the orifice, which increases oil flow and the rate of actuator movement.

Restricting flow using a flow control valve effects the system in ways other than controlling actuator speed. The pump moves more oil into the system in front of the flow control valve than the lines and valves in that part of the system can accommodate. Attempting to force the oil into an inadequate space causes system pressure to rise.

High-pressure oil flow

Low-pressure oil flow

Figure 2-20. Oil flow through a basic hydraulic system during cylinder extension.

High-pressure oil flow

Low-pressure oil flow

Figure 2-21. Oil flow through a basic hydraulic system during cylinder retraction.

The pressure increase causes the pressure control valve to open, returning excess oil to the reservoir. The flow control valve continues to supply a set volume of oil to the actuator until the actuator is fully extended or the load on the actuator slows or stops movement.

During retraction of the actuator the oil returning to the reservoir must pass through the flow control valve. Oil movement through the valve orifice is reversed, but continues to be at a reduced flow rate, which slows the retraction speed of the actuator. A one-way check valve is often included in a flow control valve to allow free flow of oil around the restrictive orifice. This design allows the actuator to be retracted at maximum speed.

This system illustrates the structure and operation of a typical basic hydraulic system. The size, arrangement, and complexity of components varies from system to system, but the basic system functions remain the same.

Basic Pneumatic System

A basic pneumatic system is illustrated in **Figure 2-23.** The system includes an electric motor as the prime mover, which provides the energy to compress the air used for system operation. The system also contains basic components from each of the other component groups to allow operation and control of a cylinder.

Figure 2-22. Oil flow through a basic hydraulic system when the cylinder is fully extended or retracted or when it encounters a load heavier than it can move. A—Fully extended or stalled on extension. B—Fully retracted or stalled on retraction.

As the electric motor turns the compressor, pressure differences are created within the system. As a result, atmospheric air enters the compressor through an intake-line filter that removes airborne dirt. The compressor compresses the air and forces it into an air receiver. The air is held in the receiver in compressed form until it is distributed through lines to other system components. See **Figure 2-24.** A pressure switch controls the prime mover. The compressor functions only when additional air is needed to maintain the desired pressure in the receiver and the distribution lines leading to system workstations.

The first system component the compressed air encounters after it moves out of the receiver through system lines to a workstation is a separator-filter. See **Figure 2-25.** This component removes droplets of condensed water and any remaining dirt particles. The air then moves to a regulator, which controls the operating pressure at the workstation. The operating pressure maintained by the regulator is always less than the pressure at the receiver.

Figure 2-23. The components of a basic pneumatic system that can provide direction, flow, and pressure control for a cylinder.

Figure 2-24. Airflow during compressor operation in a basic pneumatic system.

The compressed air next moves to the lubricator. This device adds a fine mist of oil to the air. The oil lubricates any components through which the air flows.

A directional control valve is the next component in the basic pneumatic circuit. This valve directs the air to the actuator, such as a piston in a cylinder, causing the actuator to move. Existing air from the other side of the piston is returned through a system line to the same valve where it is exhausted to the atmosphere. Shifting the directional control valve directs airflow to the opposite side of the piston, which reverses the actuator. Air from the other side of the piston is exhausted to the atmosphere through the valve. See **Figure 2-26.**

When the actuator is fully extended or retracted, it simply stops moving. The actuator also stops moving when it encounters a load heavier than it can move. This is possible because of the compressibility characteristic of air and the overall design of the pneumatic system.

Figure 2-25. Airflow through a basic pneumatic system during cylinder extension.

Figure 2-26. Airflow through a basic pneumatic system during cylinder retraction.

The receiver stores compressed air at a preset maximum system pressure established by a pressure-operated electrical switch, which controls the operation of the compressor. The regulator located at the workstation sets a working pressure at a level lower than the maximum system pressure. Air moves from the higher-pressure area of the receiver to the lower pressure area of the workstation. When the actuator is overloaded or has reached the end of its travel, the pressure in the workstation area reaches the setting of the regulator. The regulator then blocks airflow from the receiver, protecting workstation components from excessive pressure.

Accurate control of actuator speed in a pneumatic system is difficult because of the compressibility of air. To provide the best control in a basic system, the flow control valve is placed in the return line between the actuator and the directional control valve. Adjusting the flow control valve changes the size of an orifice in the valve. The larger the orifice, the higher the airflow and the higher the actuator speed. Reducing the orifice size reduces actuator speed.

During actuator operation, air is routed through the flow control valve before it is returned to the directional control valve, where it is then exhausted to the atmosphere. This placement of the flow control creates pressure on both sides of the actuator piston, which produces a more-uniform actuator speed. During actuator retraction, system airflow must pass through the flow control valve before entering the actuator. Flow through the valve orifice is reversed, but the size of the orifice still controls the airflow rate, slowing the retraction speed of the actuator. A one-way check valve allowing free flow of air around the restrictive orifice is often used to allow the actuator to be retracted at maximum speed.

This system illustrates the structure and operation of a typical pneumatic system. The size, complexity and arrangement of the components varies from system to system, but the basic system functions remain the same.

Chapter Test

Answer the following questions. Write your answers on a separate sheet of paper.

1. Fluid power systems are made up of _____ containing parts designed to perform specific tasks.

2. The tasks of fluid power system component groups can be classified under one of five functions. List these five functions.

3. The fluid control function of a fluid power system controls system pressure and fluid flow _____ and _____.

4. The number and appearance of components in a fluid power system is influenced by the type of _____, application, and _____.

5. The basic actuators used in fluid power systems are _____ and _____.

6. Fluids are distributed throughout a fluid power system by components referred to as _____.

7. Shifting a(n) _____ control valve creates a new fluid flow path in the system, which controls actuator movement.

8. The basic pressure control valve in a hydraulic system is a(n) _____ valve that does not _____ until a desired system operating pressure is reached.

9. The maximum pressure in a basic pneumatic system is set by a(n) _____ that remains _____ until the desired pressure is reached.

10. Flow control valves control the volume of flow by varying the size of a(n) _____ through which system fluid flows.

11. Fluid power system filters may be located in one or more of three typical locations. List these three locations.

12. Removal of water droplets from pneumatic system is the responsibility of the _____.

13. During retraction of a cylinder, free flow around a flow control valve is provided by a one-way _____.

14. In the pneumatic fluid power system, after the pressurized system air has completed its work, it is exhausted to the _____.

15. In a pneumatic system, a preset maximum system pressure is established by a(n) _____ electrical switch, which controls the operation of the compressor.

Fluid power is used in a wide variety of devices to aid in the performance of work. These devices range from very simple tools to complex machines. However, the operation of each device is based on fundamental scientific concepts. This chapter discusses these concepts in order for you to develop a general understanding of their application to fluid power systems and the tools and machines that are operated.

Objectives

After completing this chapter, you will be able to:

- ■ Identify and explain the design and operation of the six basic machines.
- ■ Describe the factors that affect energy in fluid power systems.
- ■ Explain how the potential power of a fluid power system is established.
- ■ Describe the nature of heat and how it is measured in fluid power systems.
- ■ Explain the nature of electricity and the basic methods used for electrical measurement.
- ■ Identify and explain the operation of basic electrical circuits.
- ■ Name and describe the characteristics associated with the fluids used in fluid power systems.

Chapter 3

Basic Physical Principles
Applications to Fluid Power Systems

Selected Key Terms

The following terms will be used in this chapter. As you read the text, record the meaning and importance of each. Additionally, you may use other sources, such as manufacturer literature, an encyclopedia, or the Internet, to obtain more information.

absolute pressure
alternating current
atmosphere
buoyancy
conduction
electrical potential
electrons
force
general gas law
head
horsepower
ideal gas laws
latent heat
lever
magnetism
mass
mechanical advantage
parallel circuit
Pascal's law
pressure
sensible heat
specific gravity
thermodynamics
torque
vacuum
wheel and axle

Internet Resources

www.efunda.com/formulae/fluids
eFunda, Inc. (Engineering Fundamentals)
Provides information on fluid mechanics, including theory, measurement, calculators, and diagrams.

www.electronics-lab.com/articles/
basic_electronics.html
Electronics Lab
Select "Electronics for Beginners" for an explanation of basic electronic circuits.

www.engineersedge.com/thermodynamics/
table_contents.htm
Engineers Edge
Provides a broad base information on the basics of energy, work, and heat. Introduces properties and laws of thermodynamics.

http://enginemechanics.tpub.com/14037/
Integrated Publishing—Engine Mechanics
Provides extensive information on simple machines, including basic concepts, illustrations, and summary statements.

www.lightandmatter.com/lm
Light and Matter
Online version of Benjamin Crowell's book *Electricity and Magnetism*. Provides extensive details on electricity and magnetism.

http://wikis.lawrence.edu/display/CHEM/
Ideal+Gas+Laws--Adrienne
Enterprise Wiki—Ideal Gas Laws
Provides information on various gas laws.

www.wikipedia.org/wiki/Bernoulli's_principle
Wikipedia: The Free Encyclopedia—Bernoulli's Principle
An article that provides information on Bernoulli's principle, including equations.

Mechanics

Fluid power systems are used in a wide variety of equipment to do work. Many different designs are used in the construction of this equipment. These designs make use of the basic elements of mechanics, including the principles of:

- Simple machines.
- Energy transfer.
- Force.
- Power.

An understanding of these principles is required to fully comprehend the operation of hydraulic and pneumatic systems used in industry today.

Simple Machines

The number of different designs of complex machines used throughout the world today is unknown. It would be safe to estimate the number is in the millions. While there are many different designs of machines used, only six basic *types* exist, **Figure 3-1.** The *basic machine types* include the lever, inclined plane, wedge, screw, pulley, and wheel and axle. Each of these simple machines serves a specific purpose that cannot be duplicated by any of the others. The simple machines provide mechanical advantages that allow difficult tasks to be completed with a minimum of effort. They also provide for a gain in movement in exchange for a reduction in force.

Lever

The *lever* is constructed of a simple, rigid bar that pivots on a fixed axis called a *fulcrum.* It is one of the most commonly used machines. The parts of a lever include an effort arm, load arm, and fulcrum. The *lever arm* extends from the fulcrum to the point at which force is applied. The *load arm* extends from the fulcrum to the location of the load. The material from which a lever is

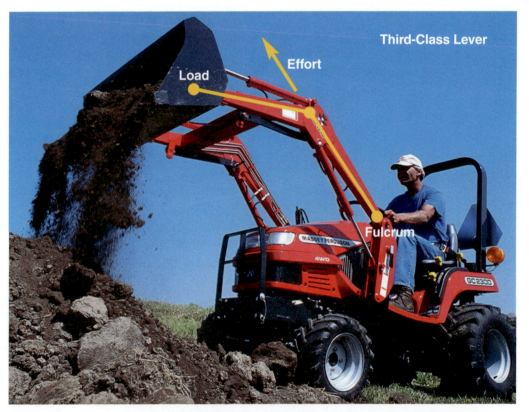

Figure 3-1. Although fluid power equipment may be complex in its design, several simple machines such as the lever, screw, or wheel and axle are usually easily identifiable in its structure. (AGCO Corporation)

constructed may vary considerably as it must only be solid and rigid enough to withstand the forces applied. See **Figure 3-2.**

Levers are divided into three classes determined by the placement of the fulcrum in relation to the effort and load arms. See **Figure 3-3.**

■ First-class levers.

■ Second-class levers.

■ Third-class levers.

A *first-class lever* is one in which the fulcrum is located between the effort and load arms. A common pair of pliers is a good example of this class of lever. A *second-class lever* has the load arm between the effort arm and the fulcrum. A common example of this lever is a nutcracker. A *third-class lever* is one in which the effort arm is placed between the load arm and the fulcrum.

Figure 3-3. Levers are grouped in three classes, depending on the location of the load and effort arms in relation to the fulcrum. A—First-class lever. B—Second-class lever. C—Third-class lever.

Salad tongs are a good illustration of a third-class lever. **Figure 3-4** shows two examples of levers in fluid power systems.

Inclined Plane

The *inclined plane* is a slope used to increase the distance an object is moved while it is being raised or lowered by rolling or sliding. Loading ramps used to roll barrels up gradual inclines are excellent examples of incline planes. A ramp for handicapped individuals allowing ready access to different levels that normally require an elevator is another example of an inclined plane, **Figure 3-5A**. In each of these examples, the total effort to make the change in height is

Figure 3-2. Levers are often used in conjunction with fluid power cylinders to obtain the movement required by the machine designer.

A

B

Figure 3-4. The application of simple machines in fluid power equipment may be as simple as the control handle on a manually operated directional control valve or as massive as the steering linkages on large earthmoving equipment. A—Directional control valve with manual actuator: first-class lever. B—Mobile equipment articulated steering: third-class lever.

not changed by the incline plane. However, a mechanical advantage is obtained as the maximum effort needed at any given time is reduced due to the distance traveled on the incline.

Wedge

The *wedge* is made-up of two inclined planes that share a common base. See **Figure 3-5B.** Wedges are generally considered small tools used against great resistance to separate objects. Splitting blocks

A

B

Figure 3-5. A wedge consists of two inclined planes that share a common base. A—An incline plane. B—A wedge. (Photo courtesy of handiramp.com)

of firewood using a wedge and sledgehammer is an example of an application of this simple machine. Other examples include axes, chisels, and other cutting devices, as well as nails that use a wedge shape to penetrate boards and structural timbers.

Screw

The screw is a form of the inclined plane. In a *screw,* the inclined plane is wrapped around a rod in a continuous spiral to form threads. The gradual slope of the threads provides a very high mechanical advantage. The principle of the screw is, therefore, often used when high force is required in the operation of a device. Bolts, wood screws, jar lids, and worm gears are examples of the screw.

Pulley

The *pulley* allows the direction of a force to be changed. When a single, fixed pulley is used, neither speed nor intensity of the force is changed. The distance over which the effort is applied is the same distance the load is moved. A single, moveable pulley provides a mechanical advantage as the distance over which the effort is applied is twice the distance that the load moves. Multiple pulley sets are often used to provide an increased mechanical advantage when a high force is required to move or lift an object. Construction cranes make use of pulleys.

Wheel and Axle

The *wheel and axle* can be used to increase force or movement. The device is constructed of a wheel attached to an axle. This device is actually a form of a second- or third-class lever, depending on where the effort is applied. The fulcrum is the common center of the wheel and axle. The second-class lever principle applies when force to the wheel increases the force at the axle. The third-class lever principle applies when force is applied to the axle. The force at the surface of the wheel decreases, but the surface movement increases. The steering wheel of an automobile and the pulley drives on engine accessories are examples of using the wheel and axle to increase force. The drive wheels of a vehicle may be used to illustrate how movement is increased. Force applied to the axle turns the wheels, producing increased

movement, but with a reduced force. Machines also use gears to apply the wheel and axle principle when increased force or speed is desired.

Basic Principles of Mechanics

Principles of mechanics are used extensively in machines that apply hydraulics and pneumatics in their operation. An understanding of these principles is essential to the successful conversion, transmission, and control of energy in fluid power systems. This section discusses the principles of energy, inertia, friction, and force.

Energy

Energy is defined as the capacity to do work. Energy is classified as potential energy or kinetic energy. The operation of fluid power systems involves both types.

Energy stored in a form that allows its release and use is *potential energy.* It is found in natural fuels such as crude oil and coal, **Figure 3-6.** Man-made devices such as springs and hydraulic accumulators also store potential energy. In these devices, the potential energy is created by work that has been performed on the equipment, **Figure 3-7.**

Energy in motion is *kinetic energy.* Its source is potential energy that has been released. Potential energy becomes kinetic energy as a tightly wound spring unwinds, a hydraulic accumulator discharges, or gasoline is burned in an engine.

Potential and kinetic energy exist in many forms. Fluid power systems are affected by a number of these forms, including electrical, heat, and mechanical. Other forms of potential and kinetic energy also exist, such as those related to chemicals, light, and sound. These forms may be included in specialized, accessory fluid power systems.

Energy cannot be created or destroyed, but it may be converted from one form to another. The wise use of energy is essential. Wasted energy cannot be salvaged and is lost as heat to the atmosphere. Careful planning of energy use is essential for effective fluid power system operation. Ineffective energy use increases the cost of system operation and is detrimental to the environment.

A

B

Figure 3-6. The coal burned in electrical generating plants (A) or fuel consumed by internal combustion engines (B) provides the energy used in the operation of most fluid power systems. The potential energy in these fuels is converted to several forms before it generates fluid flow and pressure in a hydraulic or pneumatic system. (Deere & Company)

Figure 3-7. A hydraulic accumulator is used to store potential energy. Energy accumulates in the device as it is pressurized, is stored under system operating pressure, and is then released in the form of kinetic energy during discharge of the unit. (MTS Systems Corporation)

Inertia

Inertia is the tendency of a body to remain at rest or, if in motion, to maintain its speed and direction. An outside force must be applied to initiate the movement of an object or to change its movement. Designers must be aware of inertia so they can design equipment that provides appropriate power for startup, braking, and directional control for effective and safe machine operation.

Friction

Friction is the resistance to the sliding movement of one or both of two surfaces in contact with each other. Friction is found, to some degree, in all machines. It converts kinetic energy to heat, which reduces the efficiency of the machine. Dry surfaces usually produce greater friction than wet surfaces. The careful lubrication of bearing surfaces reduces friction and increases the efficiency of equipment.

High levels of friction may be desired in specialized machine systems. A brake system is an example of a system that uses friction to perform work. Care must be taken in these systems to remove the generated heat to prevent damage to component parts.

Force

Force is defined as the effort that produces, changes, or stops the motion of a body. Force may be considered the effort needed to complete a task, but it may be applied without causing movement of an object. The force exerted to complete a task must overcome an opposing force that is produced by gravity, friction, or inertia, **Figure 3-8.** For example, lifting a weight involves an effort equal to the force of gravity on that weight. The startup of a machine requires a force equal to the resistance of friction and the inertia of all the parts in the device.

Mechanical Measurements

Fluid power systems involve a number of basic principles of mechanics that include measurements and the units used to express those measurements. These principles include factors relating to:

- Force.
- Movement.
- Efficiency.

An understanding of these concepts and the measurements involved is essential to anyone who works with either hydraulics or pneumatics.

Pressure

Pressure is force exerted over a unit area. All materials apply pressure to the objects on which they rest. The weight, velocity, and temperature of materials effect the pressure they exert. Pressure is determined by dividing the force exerted by the area over which it is applied:

pressure = force ÷ area

A common unit of measure for pressure is *pounds per square inch.* For example, a box weighing 1000 pounds that rests on a 10″ × 10″ surface exerts a pressure expressed as 10 pounds per square inch, or 10 psi. See **Figure 3-9.**

Figure 3-8. Cylinders are the devices commonly used to develop force in fluid power systems. The pressurized fluid in hydraulic and pneumatic systems acts on the face of the cylinder piston to generate the force needed to overcome a resistance and move a load.

1000 lbs

10″

10″

Force = 1000 lbs

Area = 10″ × 10″ = 100 in²

$$\text{Pressure} = \frac{1000 \text{ lbs}}{100 \text{ in}^2} = 10 \text{ psi}$$

Figure 3-9. Pressure is defined as force divided by the area over which it is applied. It is expressed in units such as pounds per square inch (psi).

Pressure may also be stated using several other systems of measurement. *Atmospheres* and *head* are commonly used in the fluid power industry to indicate pressure. Both of these units of measurement indicate the pressure over an area created by the weight of a fluid. These systems of measurement are discussed in the Fluid Power Transmission section of this chapter.

Torque

Torque is defined as a turning or twisting force applied to a shaft, **Figure 3-10.** It is basically a rotary force that does not require motion. Torque is measured as the force applied to the radius of a circle centered on the shaft. The distance from the center of the shaft to the point where the turning force is applied is called the *torque arm.* Torque is calculated by multiplying the applied force by the length of the torque arm:

torque = force × length of torque arm

The unit of measure for torque is the foot-pound or inch-pound. For example, a force of 10 pounds applied to a torque arm 1 foot in length produces a torque of 10 pound-feet or 120 pound-inches. Hydraulic motors apply torque as they perform work, **Figure 3-11.**

> **Note:** Torque is commonly stated in foot-pounds or inch-pounds. However, you may also find it stated in pound-feet or pound-inches. There is no difference between the designations. You may also find torque expressed in the metric system as Newton-meters.

Work

Work is the application of force through a distance. Work is produced only when the force that is applied moves an object or device or causes a physical change in an object. It is measured by multiplying the force applied by the distance through which it is applied:

work = force × distance

The unit of measure for work is the foot-pound. For example, a force of 100 pounds applied over a distance of 10 feet produces 1000 foot-pounds of work.

Figure 3-10. A torque wrench allows bolts to be uniformly tightened by measuring the turning force applied. The length of the wrench serves as the torque arm and the scale on the wrench indicates the torque in pound-feet or pound-inches.

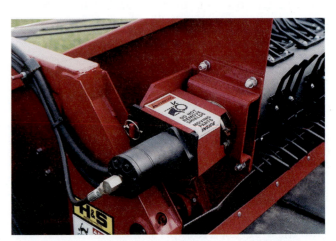

Figure 3-11. Motors are the devices commonly used in fluid power systems to develop torque. Pressurized fluid acting on internal parts applies a turning force to the output shaft of the motor to rotate machine parts.

Power

Power is the rate at which work is performed or energy is expended. Power output is calculated by dividing the amount of work completed by the time needed for its completion:

power = (force × distance) ÷ time

The unit of measure for power is foot-pounds per minute. For example, moving an object requiring a force of 100 pounds a distance of 10 feet in 1 minute produces 1000 foot-pounds per minute of power.

Horsepower

Horsepower is a commonly used unit of work. The term is historically credited to James Watt. It was originally used to compare the work output of Watt's steam engine to horses, which were a primary source of power at the time. One horsepower is equal to 33,000 foot-pounds of work per minute. The horsepower output of a device is calculated by dividing its power output in foot-pounds per minute by 33,000:

horsepower = power ÷ 33,000

The calculations for determining the horsepower requirements of prime movers and the output horsepower of fluid power systems are discussed later in this text.

Velocity

Velocity is defined as the distance traveled in a basic unit of time. It is calculated by dividing the distance traveled by the total units of time required for the travel:

velocity = distance ÷ time

The unit of measure for velocity commonly used in the fluid power field is inches per second, although inches per minute, feet per second, or feet per minute may also be used. For example, a cylinder that requires 10 seconds to extend 10 inches is operating at a velocity of one inch per second, or 1 in/sec.

Mechanical Advantage

Mechanical advantage is an expression of the relationship between the effort expended and the resistance that is overcome. Humans use machines to develop mechanical advantages in order to complete tasks that require forces greater than the direct application of their strength will produce. See **Figure 3-12.** Mechanical advantage may be calculated using either the forces or the distances involved:

mechanical advantage = output force ÷ input force

mechanical advantage = effort distance ÷ resistance distance

Both formulas provide identical results. Mechanical advantage is expressed as a numeric ratio without units of measure.

Figure 3-12. A hydraulic jack develops a high mechanical advantage through the use of both the area differences of the input and output pistons and the advantage gained in the first-class lever (the handle) operating the input piston.

For example, a device is used to move a 500 pound weight a distance of one foot. An effort of 100 pounds applied over a five foot distance is required to produce that movement. This device functions with a mechanical advantage of 5 to 1. Either the forces (weight and effort) or the distances involved can be used for calculation:

500 lbs ÷ 100 lbs = 5:1

5 ft ÷ 1 ft = 5:1

Mechanical Efficiency

Mechanical efficiency is a comparison of the work input to the work output of a machine. Appropriate design and proper lubrication can increase the efficiency of a machine. Mechanical efficiency is calculated by dividing the work output of a machine by the work input:

mechanical efficiency = (work output ÷ work input) × 100

It is expressed as a percentage and is always less than 100% because of energy losses caused by friction.

Principles of Heat Transfer

The operation of a fluid power system involves the transfer and conversion of energy. Any movement of a solid, liquid, or gas produces heat, which is a form of kinetic energy. Excessive heat results in a waste of energy and a reduction in the efficiency of the system. Understanding and controlling heat in fluid power systems is desirable as heat is a form of energy not totally usable in the systems.

Thermodynamics

Thermodynamics is a science concerned with relationships between the properties of matter, especially those affected by temperature, and the conversion of energy from one form to another. Many systems for generating or transmitting power are influenced by the thermodynamic behavior of the fluids involved in their operation, **Figure 3-13.** Both hydraulic and pneumatic fluid power systems are affected by these principles.

The principle of the conservation of energy states:

> Energy can be transformed from one type to another, but can be neither created nor destroyed.

According to this principle it is impossible to continuously extract energy from a source without replenishment. A good example of this principle is a battery, which must be recharged after it has been operating an electrical device. This basic principle is also known as the *first law of thermodynamics.*

Another principle critical to thermodynamics is less precise. It is known as the *second law of thermodynamics.* This law states:

> Heat only flows between two bodies when one body has a higher temperature than the other, with heat moving from the warmer to the colder body.

This law is applied to the movement of heat throughout fluid power systems, but especially to heat exchangers used to maintain proper operating temperatures in the systems.

Heat

Heat is energy in transit due to a difference in temperature between two areas. It is kinetic energy. Heat is indicated by rapid molecular movement in a substance and is present in all materials that contain energy. Molecular movement may be slow

A   B

Figure 3-13. All devices that involve energy transfer follow the laws of thermodynamics in their operation. Heating water for making coffee (A), as well as a heat exchanger in a complex fluid power system (B), must follow these basic laws.

and orderly, as in a substance at room temperature, or it may be forceful, as in a substance heated to a high temperature.

Fluid power systems generate considerable heat during operation. Heat is developed in these systems whenever a pressure drop that does not produce work occurs. Heat is generated by pressure drops caused by resistance to flow in pipes, hoses, and fittings; flow over relief valves and through flow control valves; and leakage in pumps and motors. See **Figure 3-14.**

Heat energy is measured in British thermal units (Btu) or calories. One ***Btu*** is the heat energy required to raise one pound of water one degree Fahrenheit at atmospheric pressure, **Figure 3-15.** One *calorie* is the heat energy required to raise one gram of water one degree Celsius at atmosphere pressure. These values are used in calculating energy losses in systems and the sizing of heat exchangers to maintain proper system operating temperature.

High-pressure fluid from pump →

Fluid returned to reservoir at low pressure, but at a higher temperature

Figure 3-14. The relief valve in a hydraulic system generates considerable heat in the system. The energy in the fluid returned to the reservoir through the valve is not doing useful work and is converted to heat as it passes through the valve.

1 lb water at 70°F   1 lb water at 71°F
+
1 Btu   =

Figure 3-15. One Btu of heat is required to raise the temperature of one pound of water one degree Fahrenheit.

The ability of a substance to absorb heat is known as *specific heat.* It is expressed as a number that compares the quantity of heat required to raise the temperature of one unit weight of the substance one degree in temperature to the heat needed to raise an equal weight of water one degree.

Temperature

Temperature is the measure of hotness or coldness of a substance. Temperature is determined by the rate of molecular movement within the substance. The more rapid the molecule movement, the higher the temperature and the greater the kinetic energy.

Temperature is indicated by units called *degrees* and is measured using a thermometer or electronic device. Several degree scales are used to indicate temperature. The Fahrenheit scale is commonly used in the United States. The Celsius scale is used in the remainder of the world. The freezing point of pure water at atmospheric pressure is 32 degrees on the Fahrenheit scale and the boiling point is 212 degrees. The Celsius scale indicates the freezing point at 0 degrees and the boiling point at 100 degrees. See **Figure 3-16.**

Fahrenheit temperatures can be converted to the Celsius scale using the formula:

Celsius temperature = 5/9 (Fahrenheit temperature – 32 degrees)

Celsius temperatures may be converted to the Fahrenheit scale using the formula:

Fahrenheit temperature = 32 degrees + 9/5 Celsius temperature

For example, converting the 212 degree boiling point of water on the Fahrenheit scale to Celsius temperature is done as such:

5/9 (212 degrees – 32 degrees)

Completing the calculations produces 100°C, which is the boiling point of water on the Celsius scale.

Absolute temperature systems are also used in calculations in the scientific and engineering fields. These scales use absolute zero as their base. *Absolute zero* is the point where it is thought no molecular movement occurs in a substance, which indicates complete absence of heat. Two absolute temperature systems are used. See **Figure 3-17.** The Rankine system uses Fahrenheit degrees and the Kelvin system uses Celsius degrees. The freezing point of water on these scales is 491.67 degrees Rankine (459.67 + 32) and 273.16 degrees Kelvin.

Forms of Heat

Two forms of heat are encountered in the operation of fluid power systems. The most common is sensible heat. It is also the form we are most aware of in our daily life. *Sensible heat* can be measured with a thermometer and is

Figure 3-16. Comparison of the Celsius and Fahrenheit temperature scales.

Figure 3-17. Comparison of the Kelvin and Rankine absolute temperature scales.

encountered when a room is too hot or too cold or when a substance feels cold, warm, or hot when it is touched, **Figure 3-18.** Changes in the application of heat energy in these situations results in changes in temperature, but does not usually change other factors.

A second form of heat is *latent heat.* This is heat that changes the state of a substance without changing its temperature. A common example of the application of latent heat is when water freezes or turns to steam. Both of these conditions are changes of state and involve latent heat without a change in temperature. See **Figure 3-19.** When the temperature of a substance reaches the point at which a change of state occurs, the temperature remains constant even though additional quantities of heat are being added or subtracted from the substance.

For example, 970 Btus of heat must be *added* to one pound of water at 212°F to produce one pound of steam at 212°F. See **Figure 3-20.** The quantity of heat required to produce this type of change

Figure 3-19. Latent heat is the heat that changes the state of a substance without changing its temperature.

Figure 3-18. Sensible heat can be measured with a thermometer and sensed by people as hot or cold.

Figure 3-20. The amount of heat that must be added to or subtracted from water to change its state from ice (solid) to liquid to steam (gas).

(liquid to vapor) is known as the *latent heat of vaporization.* The same amount of heat must be *removed* to change steam to water.

To produce one pound of ice at 32°F, 144 Btus of heat must be *removed* from 1 lb of water at 32°F. The hidden heat that produces this change (liquid to solid) is known as the *latent heat of fusion.* The same amount of heat must be *added* to convert ice to water.

Modes of Heat Transfer

There are three modes by which heat is transferred: conduction, convection, and radiation. Each mode provides a distinct means by which heat can flow from a warmer to a cooler substance. More than one mode may be used at a given time to produce a heat gain or loss. Fluid power systems employ all three transfer methods to maintain desired operating temperatures.

Conduction

The transfer of heat using molecular activity in solids is called *conduction.* A metal bar heated at one end will gradually increase in temperature at the other end as the bar transfers heat by conduction. See **Figure 3-21.** This transfer occurs as molecule movement on the heated end stimulates increased molecular activity along the bar. Molecules do not travel along the bar, but heat is gradually transferred from one portion of the substance to another as molecular activity increases in the metal and the bar warms up.

Substances conduct heat at different rates. Metals such as copper are good conductors. Liquids and gases conduct heat to some degree, but are not considered conductors. Substances that do not readily transfer heat are considered *heat insulators.* Substances that contain microscopic chambers filled with a gas, such as foam plastics, are generally good heat insulators. Designers must be aware of the heat transfer characteristics of materials in order to produce component parts and systems that are efficient and provide an appropriate service life.

Convection

The transfer of heat using the physical movement of a gas or liquid to dispense the heat is called *convection.* The fluid is usually warmed by the heat source through direct contact using conduction. The warmed portion rises because it is less dense than the cooler portion. The cooler portion of the substance then moves to occupy the space vacated by the warmed material. This produces fluid movement or flow as the heating and replacement sequence continuously repeats. See **Figure 3-22.**

The principle of convection is used to aid in the cooling of fluid power systems. As the reservoir and other components in the system warm, they in turn warm the air they contact. The warm air rises and is replaced by cooler air, which in turn is warmed. This cycle is continued during the operation of the system and is a primary source of cooling in many basic systems.

Figure 3-21. Heat transfer through solid objects is called conduction and involves increased molecular activity along the route of the heat movement.

Warm gas rises

Heat source

Cool gas drops and replaces the warm gas

A

Warm liquid rises

Cool liquid descends

Heat source

B

Figure 3-22. Heat transfer by convection involves the physical movement of a liquid or gas. Movement occurs when the lighter, heated molecules rise and are replaced by cooler, heavier molecules. A—Convection in a gas. B—Convection in a liquid.

Figure 3-23. Heat transfer by radiation involves electromagnetic waves that can transfer heat without the aid of a solid or liquid medium.

Radiation

The transfer of heat using electromagnetic waves is called *radiation.* See **Figure 3-23.** The electromagnetic waves travel at the speed of light and can be transmitted even in a perfect vacuum, unlike conduction and convection that require a transfer medium. Radiant heat warms only the substance that adsorbs the energy, not the material through which it travels.

All bodies emit radiant heat to some degree, but the emissions do not become significant unless temperatures are very high, such as in combustion chambers. The sun provides the most familiar example of radiant energy transfer, although electric radiant heating systems are available for home and commercial applications. Radiation plays only a small role in the operation of fluid power systems and is usually ignored when calculating heat transfer.

Electricity and Magnetism

Electricity is a form of energy that is used extensively in the operation of fluid power systems. It is used to power the motors that serve as the prime movers for many of the systems. It is also used to operate sensing devices and valves that control the systems. Finally, it is the power source for safety devices and temperature control equipment that assure safe and efficient system operation.

Nature of Electricity

An understanding of the nature of electricity requires an understanding of the basic structure of a substance. All substances are made up of *molecules,* which are compounds of the basic chemical elements. The structure and chemical elements that make-up each molecule of a substance are identical. Changing the combination of the elements in a molecule produces an entirely different substance.

Molecules are composed of *atoms,* which are the smallest identifiable parts of an element. Atoms are composed of parts called electrons, neutrons, and protons. These atomic parts are the same in all elements. All protons are alike, all neutrons are alike, and all electrons are alike. Even though each element is structured from identical basic atomic parts, the characteristics vary because the number and arrangement of the parts are different for each.

Protons and neutrons form the *nucleus,* or center, of the atom. See **Figure 3-24.** The electrons follow orbital paths around the nucleus, much like the planets orbit the Sun. Electrons, protons, and neutrons are very small even when compared to the size of an atom. Microscopic as these parts are, they play a major role in electrical theory.

Electrons, protons, and neutrons possess different electrical charges. *Protons* are positively charged particles. *Electrons* are particles with a negative charge. *Neutrons* are particles that have no charge. *Charge* refers to a characteristic of these particles that attracts or repels other particles. Like charges repel each other, while unlike charges attract each other. When positively and negatively charged particles are attracted to each other, they are "in balance" and neither attract nor repel other particles.

The atoms of substances are normally in electrical balance as each atom has an equal number of electrons and protons. The electrons continuously move, but the atoms are in electrical balance and stable because of the attraction of equal numbers of negatively charged electrons and positively charged protons. However, it is possible for electrons in the outer shell of the structure to modify their orbit and, in some cases, to actually become free. These free electrons are attracted to the outer shell of adjoining atoms. An atom is considered *charged* whenever it gains or loses electrons. An atom is considered negatively charged if it gains electrons and positively charged if it loses electrons.

Electrons move from a negatively charged substance to one that is positively charged. *Electrical potential* is the difference between the electron pressure levels of these substances. The level of electrical potential is established by the number of electrons in the substance, how easily the electrons

in the outer shell can move, and the number of free electrons available. See **Figure 3-25.**

Electrical flow occurs when electrons begin to leave the outer shell of the atoms in a substance and move other free electrons with them through the substance. A substance that permits the flow

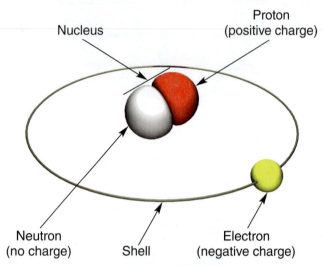

Figure 3-24. The structure of an atom involves positively charged protons, negatively charged electrons, and neutrons that have no charge.

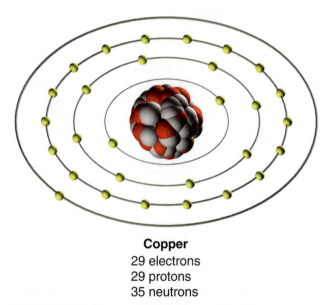

Copper
29 electrons
29 protons
35 neutrons

Figure 3-25. The outer electron orbit of the copper atom contains only one electron, which is some distance from the nucleus. Copper is a good conductor because this electron can be easily forced out of its orbit as a free electron, which increase the conductivity of the substance.

of electrons is called an *electrical conductor,* while one that resists the movement of electrons is known as an *electrical insulator.*

Electrical devices use electron flow to do work. Electric motors that operate fluid power pumps, solenoids that shift directional control valves, and electronic sensing devices that automatically complete complex machine operations all function using this basic theory.

Nature of Magnetism

Applications in the fluid power field that use electricity also heavily depend on the principles of magnetism. The two phenomena interact in many situations. Magnetism is used to generate electricity. On the other hand, electricity is used to form the magnetic fields that operate electric motors, solenoids, and many other devices.

Magnetism is the characteristic of a substance that causes it to attract iron, **Figure 3-26.** Lodestone (magnetite), a magnetic iron oxide, is the only naturally occurring material that possesses polarity and has the ability to attract iron objects. It is possible, however, to magnetize certain other materials on either a temporary or permanent basis. Iron and iron alloys are most often used for this purpose as their *permeability,* or lack of resistance to magnetization, is high.

The ability of a magnet to attract another substance is concentrated in two areas called the *magnetic poles.* The poles are identified as the north, or north-seeking, pole and the south, or south-seeking, pole. Similar poles repel each other, while dissimilar poles attract each other. This is much like the negative and positive polarity factor associated with the electrons and protons of atoms.

Each magnetic pole is surrounded by a magnetic field. This is the area influenced by the attracting and repelling forces of the magnet. The field is made up of lines of force, or *flux.* In theory, the lines of force extend from the north pole to the south pole with movement occurring from north to south. The lines of force create a magnetic circuit similar to the movement of electrons when electrical potential exists between two points in a substance. See **Figure 3-27.**

Natural magnets are rare and usually quite weak. They are not considered satisfactory for commercial use. The materials are not suitable and the physical size required to produce adequate

Figure 3-26. A permanent magnet has the ability to attract and hold iron substances. This attraction is concentrated in two areas of the magnet called poles.

Figure 3-27. A magnet is surrounded by a magnetic field that forms paths between the magnetic poles.

magnetic field strengths is not practical. However, it is possible to readily create suitable magnets in almost unlimited shapes and sizes. Artificial magnets made from either iron or metal alloys can be produced that have either temporary or permanent magnetic states.

Creating a magnet

A magnet can be created by exposing a hardened steel bar or other special metal alloy to a strong magnetic field. This may be done by simply rubbing the bar against an existing magnet or by exposure to a magnetic field produced by passing electrical current through a wire that has been coiled around the bar. When electricity is passed through a wire, a magnetic field is generated, **Figure 3-28.** In either

Figure 3-28. A wire carrying electricity is surrounded by magnetic lines of force.

A

B

Figure 3-29. A—An electromagnet can be produced by coiling wire around a soft iron core. When electricity is flowing through the coiled wire, the metal concentrates the lines of magnetic force, producing a temporary magnet called a solenoid. B—Solenoids are used to operate many control devices in fluid power systems. (emel82/Shutterstock)

case, the lines of force (flux) rearrange the atoms in the metal and magnetize the bar to produce a permanent magnet.

When wire is coiled around a metal bar, a solenoid is created, **Figure 3-29.** Devices using solenoids are commonly found in industry and the home. These solenoids use a soft iron bar or other material rather than hardened steel for a core. This produces an electromagnet that quickly becomes magnetized when electrical current flows through the coil, but loses magnetism almost instantly when the electricity stops flowing. The strength of magnetism in these devices can be easily varied by controlling the number of turns of wire in the coil and the amount of electricity that flows through the coil. The larger the number of turns in a coil and higher the electrical flow, the stronger the magnetism.

Producing electricity

Magnetism is also used to produce electricity. The principle involved is called *electromagnetic induction* and involves the cutting of magnetic lines of force by an electrical conductor. Electron flow is created in a conductor when it physically cuts (moves) through magnetic lines of force or when the lines cut (move) across the conductor. The amount of electricity produced by induction depends on the rate of movement of either the conductor or the magnetic field and the number of lines of force cut. Electrical output increases as both the speed at which the lines of force are cut and the number of lines increases.

Electrical Measurement

Working with electrical equipment and accessories in the fluid power area requires familiarity with the measurement of electricity, **Figure 3-30.** Although a wide variety of designs are used by the equipment manufacturers, operating theory and terminology are often similar. The basic electrical terms and the units of measurement are identical. These terms and measurements are related to the basic concepts of electrical current, potential, and resistance.

Electrical *current* can be defined as electron flow in a conductor. The unit of measure for current flow is the ampere (A). The amperage of the electricity flowing in an electrical circuit is measured by an ammeter. Electrical *potential* can be defined as the pressure that causes electrons

Figure 3-30. Electrical amperage, voltage, and resistance may be compared to fluid flow, pressure, and resistance to flow in a hydraulic fluid power system.

to move through a conductor. The unit of measure for electrical potential is the volt (V). The voltage of electricity flowing through a circuit is measured with a voltmeter. Electrical *resistance* can be defined as the opposition to electrical flow through a conductor. The resistance of a substance is determined by its basic structure, cross-sectional area, and physical temperature. The unit of measure for electrical resistance is the ohm (Ω). Electrical resistance is measured using an ohmmeter.

Electrical current (amperage), potential (voltage), and resistance (ohms) are mathematically related. This relationship is stated in an equation referred to as *Ohm's law.* Ohm's law can be stated in any of the following forms, where E = potential, I = current, and R = resistance.

$$I = E \div R \qquad \text{amperes} = \text{volts} \div \text{ohms}$$

$$E = I \times R \qquad \text{volts} = \text{amperes} \times \text{ohms}$$

$$R = E \div I \qquad \text{ohms} = \text{volts} \div \text{amperes}$$

These equations show that the current flow in a circuit is directly proportional to the voltage and inversely proportional to the resistance. This means that the current flowing in a circuit will increase when the voltage increases and decrease as the voltage decreases. In contrast, when the resistance in a circuit increases, the current flow decreases. Current flow increases when the resistance decreases.

The usefulness of electricity in fluid power systems relates to its ability to provide power and do work. The unit of measure for electrical power is the watt (W). One watt of work is produced by a current flow of one ampere at a pressure of one volt. The wattage used in an electrical circuit is measured by a wattmeter, but can be easily calculated by multiplying the amperage of the circuit by the voltage:

$$W = I \times E \qquad \text{wattage} = \text{amperes} \times \text{volts}$$

Types of Electrical Current

Two types of electric current flow are found in electrical systems. These are referred to as direct current (dc) and alternating current (ac). The difference between the two types is the manner in which the electrons flow in a conductor carrying the electricity. In *direct current* systems, electrons flow in only one direction. In *alternating current* systems, electrons flow first in one direction, then stop and reverse their direction.

Direct current maintains a fairly constant voltage. However, the voltage of alternating current fluctuates following the peaks and valleys of the electron flow. See **Figure 3-31**. These peaks and valleys are inherent in the start, stop, and reversing cycle. Alternating current electrical systems used in the United States commonly operate using 60 cycles per second, or 60 hertz (60 Hz).

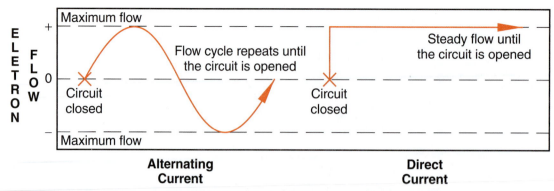

Figure 3-31. Direct current begins at zero and flows in one direction at a steady rate until it is stopped. Alternating current begins at zero, reaches a peak flow in one direction, decreases to zero, reverses its direction of flow, and reaches a peak in that direction before again returning to zero and repeating the cycle.

Both direct and alternating currents are used in fluid power systems. Stationary fluid power equipment usually uses alternating current supplied by the local power company. The alternating current is converted to direct current by rectifying units to supply the dc needs of accessory systems. Internal combustion engines provide the power needed to drive mobile fluid power systems. These engines also drive alternators (providing ac) that, through a rectifier, maintain the charge of batteries, which in turn provide direct current for accessory system operation.

Basic Electrical Circuits

Circuits provide channels or routes for electrical current to follow as it performs work or completes a control function in a system. See **Figure 3-32.** A circuit must form a complete route that provides for the return of electrons to their source. The return route is called the *ground.* The flow of electrons is stopped if a break occurs at any point in the circuit.

Circuit components

A circuit is made up of four basic elements:

■ Conductors.

■ Control devices.

■ Loading devices.

■ Source of electrical energy.

Conductors can be any substance that permits the flow of electrons. In most circuits, the conductors are made from metals such as copper or aluminum. In some designs, they may be a special metal alloy that readily conducts electricity.

Control devices vary with the design of a circuit. These devices are used to start, increase, decrease, or stop the flow of electrons. They include such devices as switches, fuses and circuit breakers, and rheostats. Switches complete and break a circuit. Fuses and circuit breakers protect the circuit from overloading. Rheostats allow the resistance in a circuit to be varied without breaking the circuit.

Loading devices are the elements that use electrical energy to perform work or convert the energy into another desired form. An electric motor or a valve-operating solenoid are loading devices common in the electrical circuits of fluid power systems. See **Figure 3-33.** Indicator lights or heating elements are other typical examples of electrical loads.

The source of electrical energy in fluid power systems is typically the local power company. This is the source of power for most stationary systems. Batteries and engine-driven alternators are typically the source of power for mobile applications. Both of these energy supplies are adequate to provide the electricity required to operate the electrical systems used with a fluid power system.

Circuit designs

The components of a circuit may be connected using either a series or a parallel configuration. A *series circuit* provides only one path for the current to follow as it moves through the circuit,

Figure 3-32. An electrical circuit is made up of a source of electrical energy, conductors to distribute the energy, a loading device to perform the desired work, and control devices to obtain the desired operating characteristics.

Figure 3-33. A number of electrical components can act as loading devices in electrical circuits found in fluid power systems. These include motors (A), solenoids (B), and indicator lights (C).

Figure 3-34. If the circuit is made up of several loads, the current must flow through each, one after the other, in order to complete the circuit. The total resistance or load of this circuit design is high. The total load is equal to the sum of all of the loads. A problem with this design is the fact that failure of any one of the loading devices causes electron flow to stop throughout the circuit. Thus, the remaining loads no longer function. Think of a string of holiday lights where if one bulb burns out, the remaining bulbs cease to shine.

A *parallel circuit* provides a structure where the current can follow individual routes to each load simultaneously, **Figure 3-35.** This design allows the current to be divided among the loads instead of being forced to pass through each load, as in the series circuit. A common primary conductor delivers current to the loads and returns it to the source after it has passed through the loads. The resistance in a parallel circuit is far less than in a series circuit with identical loads. The current flow through each of the multiple routes is affected only by the resistance in that route plus the resistance of the primary conductor.

In a series circuit, the voltage drops across each load, while the amperage remains the same throughout the circuit. Connecting the same loads in parallel causes the current to divide equally among the loads, while full circuit voltage is available to each load.

Series and parallel circuits are often combined in electrical equipment. For example, electrical switches are placed in series with the loads they control. Multiple loads in a single circuit are usually placed in parallel to obtain the most efficient performance. A number of series, parallel, and series-parallel circuits are typically found in the electrical equipment used in conjunction with fluid power systems.

Figure 3-34. A series circuit design provides a single path for the electrical current to follow through a circuit.

Figure 3-35. A parallel circuit provides individual routes through an electrical circuit for multiple electrical loads. Each load has a complete and separate route from the power source, through the control devices, and back to the power source.

Fluid Power Transmission

The primary topic of this text is the use of fluid power in society, especially in industrial operations. However, the design and operation of many consumer products also make use of the properties of liquids and gases. These fluids enable industrial designers to provide effective methods of power transmission to meet a variety of needs. The result is a range of applications from a simple hydraulic jack to complex applications in automated manufacturing lines.

General Characteristics of Fluids

A fluid is a substance that can flow and adjust its shape to fit the form of the container in which it is stored, **Figure 3-36.** A fluid may be either a liquid or a gas. Both of these states have a number of characteristics that are very similar, if not identical. Both liquids and gases possess weight, occupy space, and can be poured from a container. However, neither can possess a fixed shape. Both also have the ability, when pressurized, to exert that pressure on all the surfaces they contact.

Important differences also exist between liquids and gases. Liquids are visible and often colored. Gases are frequently colorless and odorless, making them invisible and difficult to detect. The molecules of liquids are tightly held together providing them a stable volume. This makes it possible to store liquids in open containers. The liquid varies little in volume and stays in the container, except for the amount lost through evaporation. Gases, however, cannot be held in open containers. Their molecules will migrate to fill the container, spill over into the surrounding area, and be lost to the atmosphere. See **Figure 3-37.**

Another important difference between liquids and gases is compressibility. A liquid reacts much like a solid. It does not undergo a significant volume change when pressurized. Gases, on the other hand, compress as the pressure on them increases. See **Figure 3-38.** Both the noncompressibility of liquids and the compressibility of gases are important characteristics in the operation of hydraulic and pneumatic systems.

Figure 3-36. A liquid assumes the shape of the container in which it is stored. Liquid has a stable volume and can be stored in open containers.

Figure 3-37. A gas assumes the shape of the container in which it is stored. However, it does not have a stable volume and will migrate to the atmosphere if the container is not covered.

Liquid

Gas

Figure 3-38. A gas is compressible, while a liquid resists compression and is generally considered incompressible.

Basic Properties of Materials

All materials possess a number of properties that serve to describe them and aid in their identification. These properties include a number of factors that primarily deal with appearance, rather than the structure of the material. However, another set of factors deals with basic properties that permit comparisons and allow the identification of materials suitable for specific situations.

Mass

Mass is a fundamental concept in physics, mechanics, and other sciences. In the simplest of definitions, it is how much matter is in an object. The mass of an object is the same wherever the object is located. This applies to Earth-bound objects subjected to gravity or objects floating in space effected very little by gravity. Common units of mass that may be encountered are pounds, kilograms, and slugs.

Weight

Weight is one of these basic properties that allow comparisons. The *weight* of a device, structure, or body is the result of its mass being acted on by gravity. Weight is stated as a single figure, which represents the total gravitational pull. Common units of measure for weight in the fluid power field are pounds and kilograms.

Specific weight

The weight of a specific volume of a substance at a specific temperature and pressure is known as its *specific weight.* The need for the specific temperature and pressure reference is critical in calculations for gases as they are greatly affected by these factors. However, liquids and solids are only slightly changed by pressure and temperature. The term *incompressible fluid* has evolved as temperature and pressure are often ignored when working with fluids because of the slight affect.

specific weight = total weight of substance ÷ total volume of substance

Specific weight is usually expressed as pounds per cubic foot. The maximum specific weight of water is 62.4 pounds per cubic foot. Similarly, dry air has a specific weight of .08 pounds per cubic foot at a temperature of 32°F and a standard atmospheric pressure of 14.7 psi.

Specific gravity

Specific gravity of solids and liquids is a comparison of the weight of a substance to the weight of an equal volume of water:

specific gravity = weight of test substance ÷ weight of equal volume of water

Water has a specific gravity of one and is used as the standard to measure other substances.

The specific gravity of substances varies depending on their atomic and molecular structure. Solids tend to possess higher specific gravities, while liquids tend to be lower. The specific gravity of liquids can be easily tested using an instrument called a hydrometer, **Figure 3-39.**

The density of a substance changes with temperature. This change can influence the results obtained when using a hydrometer or other device to test specific gravity. Be aware of the temperature of the sample and the procedure for correcting the test instrument readout.

Specific gravity in operating compressed air and gas systems is often defined as the specific weight of the air or gas in the system compared to that of dry air at the same pressure and temperature. Be aware of this variation as you collect data from scientific data tables for system calculations.

Figure 3-39. The specific gravity of liquids can be quickly determined by using a hydrometer.

Buoyancy

Buoyancy is another basic factor that affects the design and operation of many fluid power systems. *Buoyancy* is the force vertically exerted on a body by the fluid in which it is partially or wholly submerged. The force is equal to the weight of the fluid displaced by the body. A floating body displaces only enough fluid to equal its weight. In fluid power systems, float-operated valves that control fluid levels in tanks and reservoirs depend on the principle of buoyancy for their operation.

Pressure Measurement in Fluids

Several different scales are used to measure and designate the pressure of fluids, **Figure 3-40.** Remember, pressure is defined as force exerted over a unit area. All materials, including fluids, apply pressure to the objects on which they rest. The amount of pressure a fluid exerts depends on the specific weight of the fluid, when in an open container, or the force applied to it, when in a closed container.

Gauge pressure scale

The one fluid that affects all substances on earth is the atmosphere. The weight of the various gases that make up the atmosphere produces a pressure of approximately 14.7 pounds per square inch at

Figure 3-40. A variety of pressure gauges are available to provide accurate readings of pressure under many different operating conditions. (Dresser Instruments; Dresser, Inc.)

sea level. This is considered *standard atmospheric pressure.* Pressure measurements that are made using standard atmospheric pressure as the base are referred to as *gauge pressure.* Standard atmospheric pressure registers 0 psi on the gauge pressure scale. Gauge pressures are identified as pounds per square inch gauge (psig). Most fluid power gauges are calibrated for gauge pressure.

Absolute pressure scale

Absolute pressure scale readings are required in many fluid power calculations. The *absolute pressure* scale uses a theoretical "zero pressure" as the base. Standard atmospheric pressure registers as 14.7 psi on this scale. For example, 100 pounds per square inch gauge equals 100 psi plus 14.7, or 114.7 pounds per square inch absolute (psia). Refer to **Figure 3-41.** The units used in the gauge and absolute systems are equal in size.

Vacuum scale

Pressure below standard atmospheric pressure is called a *vacuum.* Refer to **Figure 3-42.** Zero pressure, or the theoretical "perfect vacuum," exists in a space in which there is no matter in either solid, liquid, or gaseous form. This condition has never been achieved, although it is not difficult to

Test Condition	Reading	
	psig	psia
100 psi gauge	100	114.7
100 psi absolute	85.3	100
Atmospheric pressure	0	14.7
Full vacuum	–14.7	0

Figure 3-41. Comparison of the absolute and gauge pressure scales.

Test Condition	Reading		
	Inches Hg	psig	psia
Atmospheric pressure	0	0	14.7
	10.18	–5.0	9.7
	20.36	–10.0	4.7
Full vacuum	29.92	–14.7	0

Figure 3-42. Comparison of the vacuum scale to the absolute and gauge pressure scales.

produce a space that contains only minute quantities of matter. The concept of a vacuum is important to understand as lower-than-atmospheric pressure plays a large role in both hydraulic and pneumatic systems.

The intensity of a vacuum is usually measured by the height of a column of mercury it will support. Standard atmospheric pressure supports a mercury column 29.92″ tall in a tube that is closed on the top end. The intensity of the vacuum is stated in inches of mercury, with 2.035 inches of mercury (Hg) equal to 1 psi.

Other scales

Pressure may also be stated using several other systems. *Atmospheres* and *head* are commonly used in the fluid power industry to indicate pressure. Both of these systems indicate the pressure created by the weight of a fluid. One **atmosphere** is the pressure created by the weight of the atmosphere at sea level. The average atmospheric pressure is considered to be 14.7 pounds per square inch. The pressures generated during the operation of fluid power equipment may be stated in atmospheres or parts of an atmosphere. For example, a pressure of one atmosphere is equal to 0 psig.

Head is stated as a unit of distance representing the height of a column of fluid. For example, a pressure may be stated as 25 feet head of water. In certain situations, head may be stated in inches of mercury. The concept of head can be better understood if pressure is examined in a container forming a one-foot cube. The container holds one cubic foot of liquid. If the liquid is water, the liquid weighs 62.4 pounds. The pressure generated by the water at the bottom of the cubic container can be stated as a one foot head of water. Dividing the weight of the water by the area of the bottom surface of the cube (62.4 pounds ÷ 144 square inches), you find that the water produces a pressure of .433 pounds per square inch. This is equivalent to one foot head of water.

Fluid Reactions to Pressure and Temperature

Fluids are able to effectively move energy from one location to another in hydraulic and pneumatic systems. The practical applications of this concept are based on several basic scientific principles, theorems, and laws. The following sections discuss several of these.

Pascal's law

Pascal's law reveals the general principle of how fluids move energy. This law states:

> Pressure applied to a confined, nonflowing fluid is transmitted, undiminished, to all points in the fluid with the pressure acting at right angles to all of the surfaces of the components in which it is confined.

Refer to **Figure 3-43.** The type of fluid—liquid or gas—and the shape of the components do not alter this law. The law is critical to the whole concept of energy transmission in fluid power systems.

Pascal's law is also important in terms of the amplification of force in systems. The law allows a small force applied to a small area in one part of a system to produce a proportionally larger force in another part of the system, **Figure 3-44.** For example, a force of 10 pounds applied to a piston with a surface area of one square inch produces a pressure of 10 pounds per square inch throughout the system. This allows a working cylinder with a piston area of 10 square inches to produce a force of 100 pounds.

force of 10 pounds ÷ 1 in² piston area = 10 psi system pressure

10 in² working cylinder piston area × 10 psi system pressure = 100 pounds force

Figure 3-43. Pascal's law is basic to the operation of fluid power systems. The law accounts for the ability of a confined, nonflowing fluid to transmit energy.

Figure 3-45. Pascal's law allows the small internal working surfaces of hydraulic pumps and pneumatic compressors to develop maximum pressures with realistic power input.

Figure 3-44. A direct application of Pascal's law is the force amplification provided by the equal application of pressure to all surface areas contacted by pressurized fluid in a common chamber.

However, Pascal's law does not create energy. What is gained in force is lost in distance moved. In the example above, the piston of the working cylinder moves only 1/10 of the distance of the piston receiving the input force. This ratio is established by the volume of the fluid displaced by the input piston and the volume needed to move the working cylinder piston.

Pascal's law is applied throughout fluid power systems. For example, based on this law, the small internal working area of a pump can be used to generate higher pressure and realistic power output. See **Figure 3-45.**

Bernoulli's theorem

Bernoulli's theorem establishes a relationship between fluid velocity, pressure, and elevation in a system. Details of the theorem have many implications for component and system designers. Equipment service personnel and operators, as well, need to be aware of the energy conversion concepts involved and how they can affect operating systems.

Bernoulli's theorem states:

> In a stream of steadily flowing, frictionless fluid, the total energy contained in a given volume of the fluid is the same at every point in its path of flow.

This assumes that energy is neither created, destroyed, nor transferred out of the system. The energy exists in both kinetic and potential forms. The kinetic energy is related to fluid weight and velocity. The potential energy consists of fluid pressure caused by the pressurization of the fluid by a prime mover in the system or fluid elevation.

In a fluid power system, the principles of the theorem mean that each time the diameter of a line changes, the fluid flow velocity and pressure change to maintain an energy balance in the system. See **Figure 3-46.** A decrease in line or orifice size causes an increase in velocity (kinetic energy) and a decrease in pressure (potential energy). These situations occur in an actual operating system, but many times the changes are small and go unnoticed by a machine operator. If this information is needed, formulas and example calculations may be found in the Internet Resources section at the beginning of this chapter.

A problem may occur in operating equipment because system fluid is not frictionless and resists being pumped through the system. This resistance to flow causes problems when an excessive amount of the potential energy is converted to heat, resulting in overheated fluid and component parts. In other situations, increased fluid velocities through valves may cause them to malfunction.

Ideal gas laws

An ideal gas, also referred to as a perfect gas, is a theoretic gas with infinitely small molecules

Reduced Pressure

Increased Velocity

Figure 3-46. Bernoulli's theorem concerns the flow of fluid through a system. The velocity of the fluid is higher in restrictions to maintain flow rate, while the pressure is lower in those areas in order to maintain energy equilibrium.

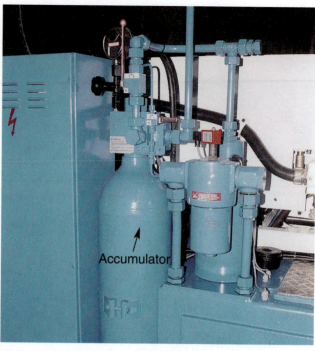

Accumulator

Figure 3-47. The principles established by the gas laws are used in sizing and determining the gas charge for hydraulic system accumulators.

that exert no force on each other. The *ideal gas laws* are mathematical equations based on this theoretic gas that provide a good approximation of how a real gas reacts to pressure, temperature, and volume changes. All gas law calculations are completed using *absolute* pressure and temperature figures.

Several laws and equations are related to the gas laws. These laws bear the names of the individuals historically responsible for their development. The following sections describe the laws and equations commonly involved in fluid power system calculations. The laws apply to many situations in pneumatic systems and to hydraulic systems where gases are used in component operation, such as gas filled accumulators. See **Figure 3-47.** The symbols in the equations represent these terms:

P1 = Initial pressure

P2 = Final pressure

V1 = Initial volume

V2 = Final volume

T1 = Initial temperature

T2 = Final pressure

Boyle's Law. *Boyle's law* deals with the changing of a gas from one pressure and volume condition to another while holding the temperature constant. The law states:

> When the temperature of a gas is held constant, the volume varies inversely with the absolute pressure of the gas.

This relationship is shown by the equation:

$$P1 \div P2 = V2 \div V1$$

The equation can be rearranged in these forms:

$$V2 = (P1 \times V1) \div P2$$

or

$$P2 = (P1 \times V1) \div V2$$

Charles' Law. *Charles' law* deals with the changing of a gas from one temperature and volume condition to another while holding the pressure constant. The law states:

> When the pressure of a gas is held constant, the volume of the gas is directly proportional to the absolute temperature.

This relationship is shown in the equation:

V1 ÷ V2 = T1 ÷ T2

The equation can be rearranged in these forms:

V2 = (V1 × T2) ÷ T1

or

T2 = (V2 × T1) ÷ V1

Gay-Lussac's Law. *Gay-Lussac's law* deals with the changing of a gas from one pressure and temperature condition to another while holding the volume constant. The law states:

> When the volume of a gas is held constant, the pressure exerted by the gas is directly proportional to the absolute temperature.

This relationship is shown in the equation:

P1 ÷ P2 = T1 ÷ T2

The equation can be rearranged in these forms:

P2 = (P1 × T2) ÷ T1

or

T2 = (P2 × T1) ÷ P1

General Gas Law. A *general gas law* can be developed by combining the elements of Boyle's, Charles', and Gay-Lussac's laws. Combining these equations yields the equation:

(P1 × V1) ÷ T1 = (P2 × V2) ÷ T2

Any of the terms can be solved for through the use of algebraic substitution. For example, the final volume of a gas can be found through manipulating the equation to produce this formula:

V2 = (T2 × P1 × V1) ÷ (T1 × P2)

Chapter Test

Answer the following questions. Write your answers on a separate sheet of paper.

1. Although millions of complex machines exist in the world today, they are constructed using the principles found in the _____ basic types of simple machines.

2. Examine the descriptions below and then identify the basic machine:

    A. Two inclined planes fastened together.

    B. A spiral inclined plane.

    C. A circular form of a lever.

3. A wheelbarrow is an example of a _____-class lever.

4. *Energy* is defined as the capacity to _____.

5. Energy in motion is known as _____ energy.

6. The tendency of a body to remain in motion once it has achieved operating speed is known as _____.

7. Force exerted over a unit area is known as _____.

8. An example of torque is the _____ applied to the shaft of a machine by a prime mover.

9. The distance traveled in a standard basic unit of time, such as 3″ per second, is known as _____.

10. The principle that heat always moves from a warmer to a colder body is known as the _____.

11. Name four ways in which heat is generated in a fluid power system.

12. A ratio of the heat required to raise the temperature of equal weights of a substance and water the same amount is known as _____.

13. The temperature scale that is used in scientific work such as gas law calculations is the _____ scale.

14. The three modes of heat transfer are _____, _____, and _____.

15. The three parts of the atom are the _____, _____, and _____.

16. A substance that resists the movement of electrons is known as a(n) _____.

17. The two factors that control the amount of electricity produced by electromagnetic induction are _____ and _____.

18. The unit of electrical power is the _____, which is equal to the work produced by a current flow of one ampere at a pressure of one volt.

19. Name the four basic elements of a basic electrical circuit.

20. The electrical _____ in a parallel circuit is always less than in a series circuit with equal loads.

21. The weight of a standard volume of a substance at a specific temperature and pressure is known as its _____.

22. The pressure produced by 10 feet head of water is equal to a gauge pressure reading of _____ psi.

23. The basic concept of _____ law is that pressure applied to a confined, nonflowing fluid is transmitted equally to all points in the fluid.

24. According to Bernoulli's theorem, each time the diameter of a line in a fluid power system decreases, the velocity of the fluid in the line _____ and the pressure _____.

25. Boyle's law assumes that the _____ of the gas is held constant between the two situations expressed in the problem.

The principles of levers are used in this machine. How many levers can you identify in the arm? Can you identify the fulcrum and load for each lever? Which classification is each lever? (Atlas Copco)

The use of fluid power systems is common in many industries in this country and throughout the world. This widespread use results in the involvement of many people in the design, operation, and maintenance of both components and systems. The need for guidelines so these individuals have common, basic rules to follow in their work has resulted in specialized standards for the fluid power industry. These standards have become a part of the documents available from national and international standardizing organizations. This chapter discusses the standardizing groups involved in the fluid power industry and the structure of standards, with a special emphasis on component symbols and circuit diagrams.

Objectives

After completing this chapter, you will be able to:

- Describe the meaning of a standard and the importance of standardization in an industry.
- Identify the primary groups that provide standards for the fluid power industry and describe the standards information maintained by each group.
- Provide specific examples of standards from standardizing groups involved in the fluid power industry.
- Explain the symbols used to designate components in fluid power circuit diagrams.
- Identify and explain the parts of circuit diagrams typically used with fluid power equipment.

Chapter 4

Fluid Power Standards and Symbols
Language of the Industry

Selected Key Terms

The following terms will be used in this chapter. As you read the text, record the meaning and importance of each. Additionally, you may use other sources, such as manufacturer literature, an encyclopedia, or the Internet, to obtain more information.

accumulators
air receiver
American National Standards Institute (ANSI)
American Petroleum Institute (API)
American Society of Mechanical Engineers (ASME)
ASTM International
circuit diagrams
control mechanisms
cutaway symbols
de facto standard
due process
electrical controls
energy conversion devices
feedback controls
Fluid Power Distributors Association (FPDA)
Fluid Power Educational Foundation (FPEF)
general consent standard
graphic symbols
International Electrotechnical Commission (IEC)
International Fluid Power Society (IFPS)
International Organization for Standardization (ISO)
legal regulation standard
mechanical controls
muscular controls
National Fluid Power Association (NFPA)
pictorial symbols
pressure controls
standard
symbols
Underwriters Laboratories, Inc. (UL)

Internet Resources

www.ansi.org
American National Standards Institute (ANSI)
Provides detailed information about ANSI, including an overview of the organization, structure and management, membership, and standards activities.

www.cagi.org
Compressed Air and Gas Institute (CAGI)
Provides complete information about the organization, which is dedicated to promoting improved performance of air and gas compressors and related equipment.

www.efunda.com/formulae/fluids
eFunda, Inc. (Engineering Fundamentals)
Provides information on fluid mechanics, including theory, measurement, calculators, and diagrams.

www.engineersedge.com/hydraulic/symbols/hydraulic_iso_schematic_symbols.htm
Engineers Edge
Provides ISO hydraulic schematic symbols organized in several categories.

www.iso.org
International Organization for Standardization (ISO)
Provides links to the general structure of the organization, standards development, current news and media, frequently asked questions, and products.

www.nfpa.com
National Fluid Power Association
Provides an overview of the association, including structure of the organization, details of the industries served, membership, sponsored events, and involvement in fluid power education.

www.oberginnovation.se/uploads/pdf/downloads/tabeller/Symboler.pdf
Norgren Graphic Symbol Library
This document provides an extensive library of ISO pneumatic symbols.

Standardization

Standardization is a phenomena that touches most aspects of our lives. Manufactured products that we buy use standardized, interchangeable parts; our clothing comes in standard sizes; and the food we buy must meet health standards while being grown on the farm and processed and transported by a distribution company. See **Figure 4-1.** A variety of processes are used to set these standards. Governmental bodies often become involved when health or environmental issues are a factor. Standards and the standardizing processes vary among industries and governmental agencies. However, the processes and factors discussed in this chapter are typical of those used to establish standards.

Figure 4-1. Standardization is the basis for modern industrial applications worldwide. This phenomena extends from intricate machined parts in aircraft engines to our retail stores that provide standard sizes as an aid in purchasing ready-to-wear clothing.

Definition of a Standard

Standards can be defined in many different ways. A generalized dictionary definition of a *standard* is anything used as a basis for comparison. These comparisons may be easily made on items that involve size, shape, weight, and strength or any combination of these or other factors. In an area that involves a manufactured product, standards are usually a set of specifications that define a product or product part. However, standards also relate to procedures such as the accounting methods that are commonly used in a wide variety of businesses. Also subjected to standardization are cultural concerns that are often not as easily defined. These include such factors as social behavior, treatment of the environment, and health and safety issues for workers, **Figure 4-2.**

Standards are established using a number of different processes. The methods used to establish standards are based on:

- Custom or de facto basis.
- General consent of an affected group.
- Legal regulation based on existing law.

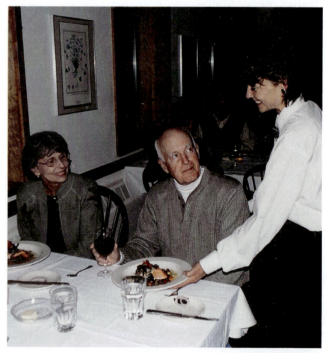

Figure 4-2. Standards not only apply to materials and products, but extend to service industries that use standards to assure their customers a satisfactory level of service. (The Creamery)

An example of a *de facto standard* that affects a large part of our population is the use of the VCR format for consumer video recording. The VHS format was gradually accepted by users of video equipment over other formats that existed in the field. Currently, this format is used by most video equipment manufacturers, even though no regulatory group specifies its use. A similar situation is developing with DVD recorders. The first DVD recorders used either the "plus" or "minus" format, but it has now become standard that a recording device should accept either format. This approach to standardization is relatively common in international business operations today.

A *general consent standard* is one voluntarily followed by a certain group. For example, several standardizing groups provide a means by which manufacturers and users can develop standards to provide maximum interchange of parts and fittings. In the fluid power field, examples of this type of standardization are thread size, seal designs, and mounting flange designs.

A *legal regulation standard* is one that must be followed due to laws that have been enacted. These standards often are related to safety issues. The governmental, safety based regulations on automobile manufacturers, such as those covering seat belts and other restraints, are good examples of this type standard, **Figure 4-3.**

Figure 4-3. Many of our standards are set by governmental agencies in an effort to protect the health and safety of citizens. The safety restraints in vehicles, including the air bag system, are examples of components that are influenced by governmental agencies.

Standardizing Organizations

Over 400 organizations are involved in the development of standards for use in this country. The standards that these groups develop cover a wide range of fields, with fluid power and its associated standards only a small part of the total picture. Some standards, such as those for safe drinking water, may have far-reaching public health implications. Others have a much narrower influence on the health, safety, or economic well being of the general population. Standardizing groups must keep in mind that the standards they develop can have a far-reaching influence on the economy of a nation, including the cost, quality, and availability of goods and the well being of individuals, including their health, safety, and quality of life.

Types of organizations

In addition to government regulatory agencies, several types of privately funded organizations are involved in the standardization process in the United States:

- Trade and business associations.
- Scientific and professional societies.
- General membership associations.
- Testing and certifying groups.
- Consortia formed of diverse groups.

These groups function as independent organizations and follow development procedures that have been independently designed and implemented. The standards are usually developed by each organization based on the needs and views of their membership following a process that allows adequate input from the membership.

Standards developed by trade and business associations are often created to promote the safety, technical, and economic interests of the sponsoring industry and its customers. Because these standards are developed by individuals working in the field, it is often felt they best reflect the needs of the field. The National Fluid Power Association (NFPA) is an example of this type standardizing group. The American Petroleum Institute (API) is another organization of this type. It has standards that indirectly affect large segments of the general population of this country. See **Figure 4-4.**

A

A goal of most scientific and professional societies is to advance theory and practice in their specific field. This goal is often reflected in the standards they develop, which tend to have a strong engineering emphasis. At times, some may feel these standards do not reflect current trends in the marketplace. The ASME, formerly known as The American Society of Mechanical Engineers, is an example of a professional society that develops and maintains standards. Its standards affect the fluid power industry in areas such as hardware, which includes bolts and other common components. See **Figure 4-5.**

The most broad based of the standardizing groups are those involving a general membership from a wide group of technical backgrounds, work experience, and interests. Generally, these groups use a standards-development procedure that assures broad-based input and a well-defined due process. This is designed to produce standards that are appropriate for the large audience the organization represents. The ASTM International is an example of this type standardizing group. The organization provides a format for the testing that is done on materials and components. The testing covered by this standardizing group is often used in the manufacture of fluid power equipment.

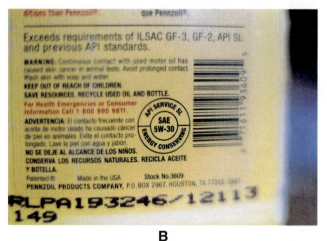

B

Figure 4-4. Some of the standardizing groups affect very large segments of our population. An example is the American Petroleum Institute, which not only affects the hydraulic area of fluid power, but also every automobile owner that purchases motor oil for their vehicle. A—Common motor oil available for purchase. B—The API mark on an oil container.

Figure 4-5. Standards affect many of the basic component parts of fluid power equipment. The bolts used in the assembly of components are manufactured to meet thread and strength standards developed by a standardizing group. This close-up of a bolt head shows the standardized marks for indicating strength.

Testing and certifying groups are usually not thought of as organizations that develop standards. Most people think of these groups as organizations employed by a manufacturer to test a product and certify that it meets an existing standard. In some situations, that is the process followed. However, in other situations, the groups are also heavily involved in the development of standards. Underwriters Laboratories, Inc. (UL) is a familiar example of this type organization that not only tests and certifies, but also has over 800 standards they have developed or modified for their use. The UL tests and certifies the electrical equipment used in many fluid power systems.

The rapid changes that are occurring in many industries have resulted in the need for standards for technologies that combine several different fields. Consortia made up of diverse groups have been formed to meet the need for guidelines and standards in these new fields. For instance, these types of groups have been formed in the volatile information and communications fields. The rapidly changing field of materials handling, where fluid power, electronics, and mechanical devices are being combined to produce the desired results, is another area where a consortium could produce the needed standards.

Standards-coordinating organizations

Although most standardizing groups function independently, they coordinate their efforts through national or even international organizations. In the United States, the primary coordinating activity is done through the *American National Standards Institute (ANSI).* This group has no legal claim to this position, but has designated itself as the national coordinating organization for standards development organizations. Most national standardizing organizations function through ANSI, although several major groups continue to function as independent bodies.

ANSI was founded in 1918. It is a federation that represents more than 120,000 companies and well over 3.5 million professionals. Membership categories include companies, government agencies, not-for-profit organizations, international groups, and individuals. The organization offers several thousand standards developed by themselves and other qualified organizations following ANSI's due process procedures. These standards are designated as American National Standards only after careful review and consensus for approval has been found among those groups that are directly and materially affected. ANSI is also the U.S. member of the International Organization for Standardization (ISO) and International Electrotechnical Commission (IEC).

The *International Organization for Standardization (ISO)* is an international agency for standardization located in Geneva, Switzerland. It was founded in 1947 and includes the recognized national standards bodies of approximately 155 of the countries of the world. The basic objective of the organization is to obtain international agreement on standards in all fields other than electrical and electronic engineering, which are the responsibility of the IEC. The organization encourages the application of standards to increase quality and productivity while promoting international trade. This is achieved through the use of over 3000 technical committees and subcommittees. The organization primarily functions in English and French, but does produce materials in other languages, **Figure 4-6.**

The *International Electrotechnical Commission (IEC)* is another international agency for standardization located in Geneva. This group has concentrated on electrical and electronic standards since its founding in 1906. Its membership includes manufacturers, users, trade associations, government agencies, scholars, and engineers. The group operates over 200 committees and subcommittees to achieve their objectives. The IEC and the ISO have established a joint Technical Programming Committee to ensure satisfactory coordination of standards involving technical areas that overlap their designated technical fields or when an issue is unclear.

Groups providing related standards, technical information, and certification programs

A large number of different groups provide services that have a direct influence on the fluid power industry, as well as a wide variety of other industries. These organizations provide information and standards that relate to such things as:

■ Composition of materials.

■ Generic parts like bolts and other fasteners.

■ Commonly used materials.

■ Component parts.

ISO 1219-2:1995(E/F)

5 Rules for identification of equipment in fluid power circuits	**5 Règles d'identification des appareils dans les circuits hydrauliques et pneumatiques**
5.1 General An identification code for the equipment shall be provided on the circuit diagram next to the respective symbol. It shall be used in all related documents.	**5.1 Généralités** Un code d'identification pour les appareils doit être utilisé sur le schéma de circuit à côté de leur symbole respectif. Cette identification doit être utilisée sur tous les documents connexes.
5.2 Identification code of components (excluding piping) The following identification code for components shall be used if no other code is stipulated. The identification code shall contain the following elements and should be enclosed in a box:	**5.2 Code d'identification des composants** (à l'exception des tuyauteries et raccords) Le code d'identification suivant doit être utilisé pour les composants si aucun autre code n'est stipulé. Le code d'identification doit comporter les éléments encardrés suivants:

NOTE 1   See annex A for information.

NOTE 1   Voir annexe A pour information.

Figure 4-6. The International Organization for Standardization (ISO) is one of the groups involved in coordinating standards development throughout the world. The group communicates in English, French, and other languages. (ANSI, see copyright statement on page 2 of this text)

■ Lubricants.

■ Seals.

■ Electrical solenoids.

Many of the organizations also provide training seminars designed to keep their members technically current in their field. Many operate testing and certification programs to verify the competence of individual workers. The discussion that follows does not provide an all-inclusive list, but the organizations included illustrate typical groups and the diverse topics with which they deal.

The *American Petroleum Institute (API)* is a group made up of corporations in the petroleum and allied industries. It includes producers, refiners, marketers, and transporters of crude oil, lubricating oil, gasoline, and natural gas. The group provides advocacy, research, and technical services to increase the ability of the petroleum industry to meet the needs of its customers. These needs include the development of energy sources, enhancement of the environment, increasing the health and safety performance of the industry, and conducting research to promote petroleum technology, drilling and refining equipment, and standards. The institute also has an extensive information service, providing publications to the public about all aspects of the petroleum industry.

The *ASTM International* is a group that establishes standards for materials, products, systems, and services. It includes members representing educational institutions, research laboratories, government agencies, and businesses. Membership is available on an individual or organizational basis. Individual membership includes engineers, scientists, skilled technicians, teachers, managers, and general consumers. The ASTM International uses over 130 committees with numerous subcommittees to assure that an inventory of over 12,000 standards, specifications, definitions, and recommended practices remains current.

The *American Society of Mechanical Engineers (ASME)* is a technical society of mechanical engineers and students. It is an educational and scientific group organized to promote mechanical engineering and its associated arts and sciences. Standards development is an important part of the society's activities. The group actively participates in international standardization, working both with the ISO and the IEC. Many fields have benefited from the ASME standards and codes. These include drafting practices and symbols, screw threads, pipe fittings, and numerous equipment safety codes, to mention only a few. Nearly all of the standards have been processed through the American National Standards Institute.

Underwriters Laboratories, Inc. (UL) is an independent, nonprofit corporation that both develops standards and evaluates the performance of products. The organization was founded in 1894 and has been developing standards since 1903. Over 800 UL standards are currently published. Many of them are also approved by ANSI. The organization evaluates products and systems to determine the safety of various materials, devices, construction designs, and operational methods. These evaluations identify hazards that may cause fire, injury, or death. The standards are developed using procedures that provide for participation, review, and comment from industry, government, insurance companies, consumers, and the general public.

The UL stamp is often found on electrical components and appliances. This indicates the device has been tested by the Underwriters Laboratories and meets the standards set by the organization for safe operation.

Organizations Specifically Related to the Fluid Power Field

Several organizations deal primarily with the fluid power industry. These groups are supported by memberships from individuals working in the field and companies producing supplies, components, and equipment or providing fluid power–related services. The organizations provide a variety of services to individual and company members. They also promote the fluid power industry to education, the general public, and other areas of industry.

National Fluid Power Association

The *National Fluid Power Association (NFPA)* is a nonprofit trade association and standards developer serving manufacturers in the fluid power field (both hydraulic and pneumatic). See **Figure 4-7**. The association serves approximately 325 members.

Figure 4-7. The National Fluid Power Association is very active in the standards area. Most fluid power systems produced in this country have been manufactured following NFPA standards in their manufacture and assembly, including the systems on this hydraulic hoist.

Membership is limited to organizations associated with the fluid power industry. To become a member, a group must be a fluid power manufacturer, distributor, or supplier; related service provider; or offer training and education programs. An international classification allows overseas groups in these areas to also join the association.

The NFPA is a member-driven organization with most of the members classified as small- or medium-sized companies. The members use the association to stay aware of the changes in the industry and how they can effectively deal with the changes. The organization is very active in the standards area.

The NFPA uses a board of directors and committees to plan and implement a variety of projects and programs. These groups work in the areas of building awareness of hydraulic and pneumatic technology, helping members know and reach the marketplace, and providing education for and about the industry. They also administer standards development on the industry, national, and international levels. Activities also include cosponsorship of the International Fluid Power Exposition, as well as numerous other publications, workshops, and presentations.

Fluid Power Distributors Association

The *Fluid Power Distributors Association (FPDA)* is made up of over 300 member companies. These companies consist of distributors and manufacturers. The organization promotes the fluid power industry through representation at industry trade shows and exhibitions, educational and marketing brochures, and web-based seminars. The organization supports numerous educational programs, meetings, and workshops. Courses are offered to provide training in topics such as distribution management, quality processes, and sales management. In addition, the FPDA sponsors the nationally recognized Young Executives program (YES) for individuals who work for member companies. This program is designed to develop the skills of management-level executives who are under 40 years of age.

The FPDA also conducts a variety of studies analyzing the fluid power industry. Many detailed documents are produced, including reports such as industry productivity, company profiles, monthly sales, sales forecasts, and compensation of industry personnel. The association provides members a wealth of technical information related to the fluid power industry on items such as telecommunication, electronic data interchange, and bar coding. Member companies are also eligible to borrow, for a nominal fee, training programs and audio and video recordings for use in internal training programs.

International Fluid Power Society

The *International Fluid Power Society (IFPS)* is a group of approximately 5000 individuals interested in the use of all phases of fluid power and related motion control. Members are concerned with research, development, design, installation, operation, and application in a wide variety of technical fields. Society members work in every facet of the fluid power industry including manufacturing, aviation, agriculture, marine, mining, mobile, materials handling, printing, education, and others.

Membership in the IFPS makes it possible for individuals to participate in regularly scheduled meetings with other fluid power professionals to advance personal growth and develop networks in the field. Involvement in the group makes it easier to remain technically current in the field through attendance at conferences and workshops sponsored by local chapters as well as the national organization.

The IFPS also operates a program where individuals may become certified in the field. Mechanic-, technician-, specialist-, and engineer-level certifications are available. The certification process consists of an optional educational review session followed by a three hour written test. The mechanic-level certification also requires an additional hands-on test that is three hours in length. Certification is valid for five years with recertification based on continued work experience and involvement in the fluid power field. See **Figure 4-8.**

Fluid Power Educational Foundation

The *Fluid Power Educational Foundation (FPEF)* is an organization that coordinates, supports, and promotes education in fluid power. The FPEF, with the support of the fluid power industry, strives to raise the level of competence of young people entering the industry. It also aims to improve the visibility and image of the fluid power technology field. The organization funds several programs

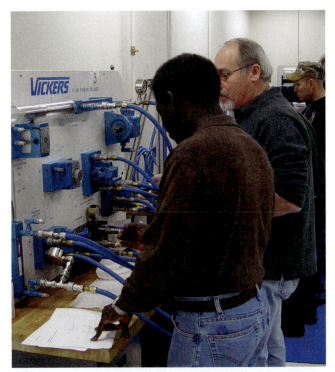

Figure 4-8. Professional organizations associated with the fluid power industry are very supportive of programs providing education and training. Support ranges from middle school to engineering curricula. The group shown here is receiving hands-on training experience at an Eaton/Vickers Training Center. (Eaton Fluid Power Training)

to meet these overall objectives. Programs include support of fluid power instructional programs in key schools located throughout the country, providing educational resources to schools in general, conducting summer teacher-training programs, providing assistance to students through scholarships, and matching students with employers. The FPEF works closely with the NFPA, FPDA, and IFPS to promote the fluid power industry.

Developing a Standard

All organizations that develop standards follow a carefully outlined set of procedures. This assures that individuals and organizations affected by the standard have input. These procedures are used to develop a consensus of the interested parties during approval, revision, reaffirmation, or withdrawal phases of standard development. The following discussions are based on the procedures handbook of the American National Standards Institute. The process may vary somewhat with individual organizations, but the concepts can be considered typical as many of the groups operate in cooperation with ANSI.

Development process

Due process is a very important part of the standardization development process. This means that any organization, company, government agency, or individual with a "direct and material interest" in the standard has a right to participate. This right to participate includes expressing a position and its basis, having that position legitimately considered, and the right to appeal adverse decisions. Participation in the process does not require membership in any particular organization, excessive financial input, or unreasonably restrictive technical or educational qualifications. Timely and adequate notice of any actions to be taken regarding a standard must be given to interested parties. This must include a description of the purpose of the proposed standardization activity and a source for further information.

The standards development process should have a balance of interests and should not be dominated by any special interest groups. *Dominance* means a position of or exercise of dominant authority, leadership, or influence within the group. This is further defined as superior leverage, strength, or representation to the point where fair consideration is not given to all viewpoints brought before the group. An individual or group that is affected by a standard can call for a *dominance test* for the group and the process that was used to develop the standard.

Representation in the standardization group should be based on the nature of the standard under development. The group should at least include representation from product producers, users, and those interested in the product. User participation should come from both individuals and representatives of organized groups. The group of those interested in the product may include general consumers, distributors and retailers, labor groups, trade associations, insurance companies, and others.

Written procedures should govern the development, revision, reaffirmation, or withdrawal of a standard. Initial approval or change in a standard usually depends on meeting these criteria:

- The standard was developed in accordance with the procedures on which the developer was accredited.

- All developmental procedures and appeals to the developer concerning the standard were completed using due process procedures.

- Notice of the development process was properly provided to interested parties.

- Significant conflicts with other standards have been resolved.

- Other known national standards were examined with regard to harmonization and duplication of content.

- The proposed standard was provided to the administrators of the appropriate technical advisory groups to assure appropriate international standards consideration.

- Patent policy and commercial terms and conditions were met, where applicable.

A standard has a good chance for approval if ANSI's Board of Standards Review (BSR) or the equivalent in another standardizing body determines that the above criteria have been satisfied. The board of review should deny approval if it is determined the proposed standard is not in the public interest, contains unfair procedures, is not appropriate for national application, or conflicts with an existing, approved standard. Standards that duplicate existing standards also should not be approved.

Standards should be made available in hardcopy or electronic form as soon as possible after approval. The developer should publish the standard no later than six months following approval.

Format of standards

The content of a standard can vary considerably in both technical complexity and length. However, standardizing organizations attempt to have the final form of their standards follow a format that is easy to interpret and allows accurate and precise display of text, data, and drawings in one or more languages.

The typical standard includes sections such as scope, purpose, references, definitions, identification statements, and general rules of the standard followed by a group of appropriate appendices.

The exact number and title of these sections will vary somewhat between standardizing groups, but should be recognizable among groups.

The scope section of a standard typically states the range of influence on the field. This may be a brief sentence or two or, if the standard is extensive, it may be in outline form. The purpose indicates the end use and reason for the development of the standard, such as "to provide drafting practices for drawings that depict fluid power systems." The references section lists other standards that form a basis for the standard or relate to it.

The definitions section defines terms that are required to understand the standard. These definitions may be found within the standard or the section may be a reference to terms listed in other standards. The identification statements section is simply a listing of the format that should be followed when referring to the standard in test reports, catalogs, and sales literature for a product that meets the standard.

The general rules section of a standard contains the key technical elements of the standard. These elements are usually concise with tables, charts, and drawings used to illustrate a concept whenever possible. Appendices are attached whenever needed to support the information supplied in the general rules section.

Fluid Power Symbols and Circuit Diagrams

Symbols and circuit diagrams are areas that have received a great deal of attention in relation to standardization in the fluid power field. They are also areas where considerable variation has been displayed over the years. Symbols and diagrams have been developed within companies to meet the needs at a specific time. Some of these symbols and diagrams have been adopted or adapted by other users. This section of the chapter examines several of the more common systems that have been in use for several years, as well as ISO graphic symbols and diagrams that are the latest versions of an internationally accepted system.

Types of Symbols

Symbols are used throughout the world in all areas and levels of the fluid power field. A knowledge of symbols is important as they serve as a special technical language. The *symbols* are used to communicate component design features, system construction, and system operation when constructing, operating, or troubleshooting circuits and systems. A knowledge of the common symbols is a valuable tool for any individual working in the hydraulic and pneumatic areas. Four different types of symbols are encountered in fluid power documents:

- Graphic symbols.
- Pictorial symbols.
- Cutaway symbols.
- Combination symbols.

Graphic symbols

Graphic symbols are the most common. They are made up of a series of lines and outlines of standard graphic figures such as squares, rectangles, triangles, and circles. Symbols of this type can be relatively simple to draw as they do not show the actual appearance of the component. They are suited to freehand sketching and both manual and computerized drafting procedures. The symbols for individual components can be easily connected by lines to represent a complex fluid power system. Circuit analysis is possible as internal operating characteristics of components can be shown in the graphic symbol. See **Figure 4-9.** Standardized forms

of graphic symbols can be easily developed. These are especially suitable for international applications involving a variety of nations and languages. Graphic symbols are the most standardized form of fluid power symbols.

Pictorial symbols

Pictorial symbols are made up of line drawings of the exterior shape of a component and show its major details. See **Figure 4-10.** All external interconnecting lines are shown when these symbols are used to illustrate a fluid power system. Pictorial symbols provide a simplified method of showing the location of components in a circuit or system and the placement of all connecting piping. Circuit analysis is not easily accomplished when these symbols are used as only external features of the components are shown. Standardized forms of these symbols are limited as the external shape and placement of connections on components vary between manufacturers and product models.

Single-acting cylinder   Three-way directional control valve

Figure 4-10. Pictorial symbols are line drawings showing the exterior shape and characteristics of a component. The appearance of these symbols will vary as the construction of the components vary among manufacturers.

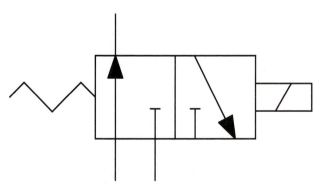

Figure 4-9. The method that is most frequently used to represent components in fluid power systems involves graphic symbols. Standard graphic figures, such as lines, squares, triangles, and circles, are combined into a symbol to represent a component, such as the directional control valve shown in this symbol.

Cutaway symbols

Cutaway symbols are really miniature section drawings of component parts, **Figure 4-11.** The symbols provide an excellent means of studying component and circuit design, examining system operation, and troubleshooting. Standard symbols do not exist for this classification of symbols. However, the following guidelines can assist a technical illustrator in preparing a drawing:

- Symbols should show all internal and external ports and clearly indicate flow paths through components.

- Actuator elements need to be shown to illustrate component operation.

- Component construction may be modified to clarify function or permit the shifting of movable parts to show flow paths during various phases of operation.

- The cutaway symbol should represent the relative size of the component compared to the other components shown in the circuit or system.

Single-acting cylinder   Three-way directional control valve

Figure 4-11. Cutaway symbols are section drawings showing the basic internal parts of a component. The drawings show the internal construction of the components. These symbols are very helpful when studying the operation of a valve or system.

Combination symbols

Graphic, pictorial, or cutaway symbols may be used in combination in a circuit diagram. This may be done to highlight a portion of a component or system. No particular standard exists for the methods to use when developing combination symbols. The combination should produce a diagram that results in a clear understanding of the design or operation of the highlighted portion. An example is shown in **Figure 4-12.** In that figure, cutaway symbols are used for the directional control valve and pump, pictorial symbols are used for the two cylinders, and graphic symbols are used for the remainder of the circuit diagram.

Structure and Use of Basic Graphic Symbols

Graphic symbols are commonly encountered in fluid power applications. The symbols lend themselves to standardization. Several organizations publish versions of graphic symbols that are very similar. This section follows ISO 1219-2 international recommendations and shows the basic elements of that standard. This discussion does not present all of the possible combinations, but it shows how standard graphic figures can be combined to represent a hydraulic or pneumatic circuit. The standard should be consulted when complete or composite functional symbols are required.

General rules

Symbols represent component type, function, method of operation, and external connections. They are *not* intended to show the actual construction of a component. If a symbol is not part of a circuit diagram showing a sequence of operations, it should be shown in the at-rest or neutral position of operation.

Basic symbols and functional elements may be combined to represent complex functions. When two or more symbols are contained in one unit, they should be drawn in an enclosure that consists of a thin chain line.

External ports should be shown in the symbol, but they need not represent the actual location of the ports on the component. Ports are indicated by the junction of fluid lines with the basic symbol of the component or the enclosure symbol that surrounds complex symbols. Only the connections that are functional need to be shown in the symbol.

Figure 4-12. A combination of symbol types may be used in a circuit diagram. For example, cutaway symbols may be used in combination with graphic and pictorial symbols to make component and system operation easier to understand.

Unless specifically stated in the standard, symbols may be drawn in any orientation without affecting their meaning. However, increments of 90° are preferred by most users in the fluid power field.

Symbols are *not* used to indicate quantities such as pressure, flow rate, or the setting of a component. Letters, when used, are merely labels and do not describe parameters or values of parameters.

Basic symbols

A variety of basic figures serve as the components of graphic symbols. Lines, circles, squares, triangles, dots, and arrows are commonly found as parts of a typical symbol. These figures are combined to produce the desired symbol, which is combined with other symbols to produce a diagram of a fluid power circuit.

A variety of graphic lines are used to represent hydraulic and pneumatic lines in fluid power diagrams. Refer to **Figure 4-13.** Working and return lines use solid, continuous lines as a symbol. Electrical lines shown in a fluid power circuit are also

Working or return lines

Drain or pilot lines

Component enclosure

Figure 4-13. The graphics used to show hydraulic and pneumatic lines in fluid power graphic symbol diagrams.

represented by this type of solid line. Pilot control, drain, bleed, and filter lines use a dashed line as the symbol. Enclosures for units that involve two or more functions are formed using a chain line, which is a broken line with alternate short and long dashes. The final line symbol is used to identify mechanical connections, such as shafts, levers, and piston rods. These are represented by double, parallel solid lines.

Circles are used to represent a number of different components. The size of the circle and its location in the diagram can be used to help identify the type of unit the symbol represents. The largest-diameter circles designate energy conversion units such as pumps, compressors, and motors. Medium-size circles, about 3/4 of the size of the conversion unit symbols, are used to show measuring units such as pressure gauges. Smaller circles are used to indicate a number of different components including nonreturn valves, rotary connections, and mechanical link rollers. Semicircles are used to designate motors or pumps with a limited angle of rotation.

A square resting on its side is used to indicate control components or a prime mover other than an electric motor. See **Figure 4-14.** Multiple positions of a control component, such as a multiple position directional control valve, are shown with multiple squares side-by-side to indicate the functioning of the valve in each of its positions.

A square resting on its corner represents conditioning apparatus. These includes filters, separators, lubricators, and heat exchanges. An additional symbol is shown within the square to indicate specific information about the apparatus.

Cylinders and valves have a rectangle as their basic symbol shape. No recommendations are included in the standard concerning the size of these symbols in relation to the graphic shapes representing other components. Rectangles are also used to represent the piston of cylinders, control mechanisms for valves, and cushioning devices for actuators.

Two additional basic graphic shapes are often used as component outlines. The first of these shapes is one-half of a rectangle that is open at the top with the height being one-half of the length. This shape serves as the symbol for a reservoir in a hydraulic system. The second form is a capsule. This looks like a rectangle with a semicircle on each end, as shown in **Figure 4-14.** The capsule may be placed in either a horizontal or vertical position, depending

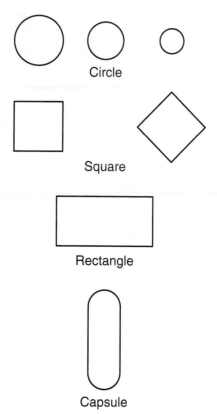

Figure 4-14. A variety of basic shapes serve as the base for fluid power graphic symbols. Circles represent energy conversion units, such as pumps, compressors, and motors. Squares resting on their sides indicate control components or prime movers, while squares resting on their corners represent components involved in fluid conditioning. Rectangles are used for work devices and valves. Capsules represent a pressurized device, such as an accumulator.

on the component it represents. The shape is used to designate a pressurized reservoir, air receiver, accumulator, or auxiliary gas bottle.

Functional elements

A variety of graphic elements are used to designate the operating medium, direction of movement, source of energy, and the way a part of a component is operated. These elements are added to the basic graphic shapes to provide more detail about the characteristics of a system component.

A small, equilateral triangle is used in conjunction with the previously discussed basic graphic shapes to show the direction of fluid energy movement through a component. This symbol is also used to show the type of fluid used in a system or component element. An open triangle indicates a pneumatic power source, while a solid triangle represents hydraulic operation. See **Figure 4-15.**

A variety of arrows are used as symbols to indicate factors such as fluid flow direction, equipment motion, heat flow, and the adjustability of pumps and other components. See **Figure 4-16.** Straight-line and sloping arrows indicate the direction of fluid flow, rectilinear equipment movement, and heat movement. Longer straight-line arrows are used to indicate variability or adjustability of equipment or component parts. These arrows are normally shown in a sloping position, rather than vertical or horizontal. Curved-line arrows are used to show shaft and equipment rotation.

Symbols for a number of additional, miscellaneous functions are shown in **Figure 4-17.** The closed path or port, spring, and restriction symbols are very commonly found in system diagrams.

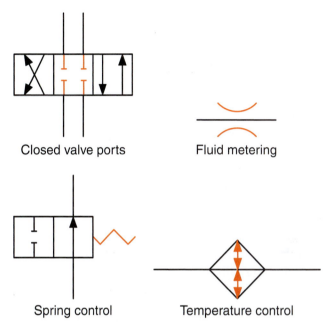

Figure 4-17. Symbols for a variety of functions are shown in this figure. The functions represented include closed ports to block fluid flow, restrictions for the metering of fluid, springs for control and activation of components, and temperature indication or control.

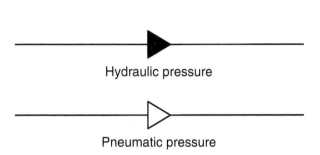

Figure 4-15. The source of energy for a system is indicated by a small, equilateral triangle. An open triangle indicates a pneumatic system, while a solid triangle indicates a hydraulic system.

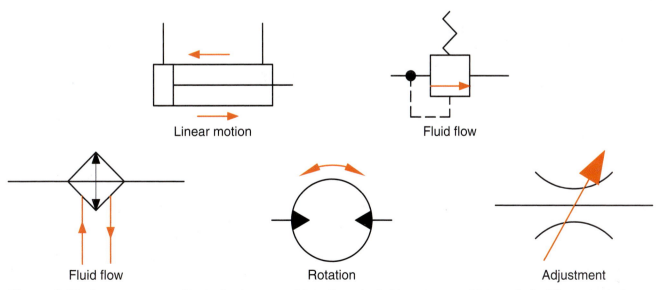

Figure 4-16. Arrows are used to indicate several functions in fluid power graphic symbols. These include rectilinear motion, direction of fluid flow, direction of heat flow, direction of rotation, and the adjustability of components.

These symbols may be either attached to or enclosed in one of the basic graphic shapes, as shown in the figure.

Flow lines and connections

Fluid power circuits are made up of a number of individual components. Standardized flow line and connection symbols provide a means to show how the components are interconnected to form the desired circuit.

The junction and crossing of lines is common in the structure of a fluid power circuit. The joining of lines in a circuit is indicated by a dot that covers the point where the lines connect. Care should be taken to avoid having this junction symbol indicate the connection of four lines at a single intersection point. The preferred form to follow is shown in **Figure 4-18** where two lines connecting to a third line are shifted slightly to eliminate the possibility of confusing them with lines that are simply crossing. Lines that cross each other, but do not join, are shown as crossing without the junction symbol dot. Hose, which is usually used to connect moving parts, is shown by a curved line connecting junction symbol dots at the end of fixed lines.

Symbols are not used to show the plumbing involved in the various lines of a fluid power system. Separate detailed drawings usually provide the information needed to assemble this aspect of the equipment. However, symbols are used for a number of connections that are important in the structure of a system.

Quick disconnect couplings are connection devices represented by a graphic symbol. See **Figure 4-19.** These devices are especially common in mobile applications where implements are attached and disconnected from tractors and other power sources. On mobile equipment, these connectors commonly include a mechanically opened, nonreturn check valve to allow easy connections with a minimum loss of fluid during attachment and detachment. Air bleeds and air exhaust ports are other connection devices that have symbols, as shown in **Figure 4-19.**

Rotary connections allow limited movement between sections of the plumbing in a circuit. These connections are used when a fluid power system is on equipment that swivels or turns in its operation. These devices are used in both hydraulic and pneumatic systems.

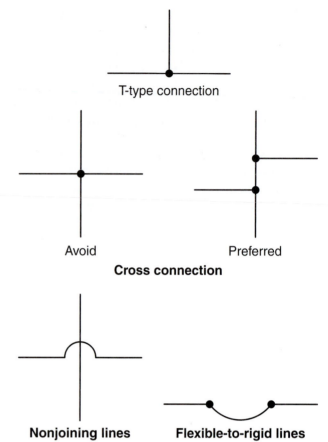

Figure 4-18. Junction points in lines are indicated by a dot that covers the points where the lines contact. Lines that are not connected simply cross each other. The correct method to use when showing two lines that cross without joining is with a semicircle. A hose connecting to two rigid lines is shown with an arc and two dots.

Control mechanisms

A number of different *control mechanisms* are used to control the valves found in fluid power systems. Several basic symbols are used, separately or in combination, to indicate the control method. The valve control symbols are drawn across the end of the basic valve symbol. Some of the control symbol information is enclosed in standard basic control rectangle symbols, while other symbols are stand-alone figures.

Several basic symbols are used to show control-related elements of fluid power valves. See **Figure 4-20.** These symbols show a number of mechanical components and control methods. The mechanical components are the physical devices

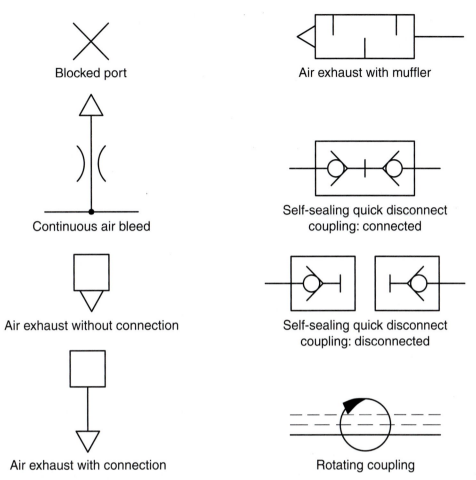

Figure 4-19. A number of different connections are used in hydraulic and pneumatic systems. Shown here are symbols for typical types of connections.

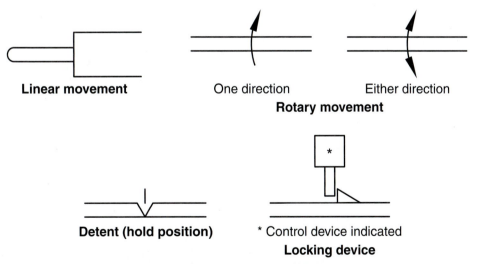

Figure 4-20. Four common mechanical devices used to direct the movement of control devices in components are shown here. These include a rod to cause linear movement, shaft to cause rotary movement, detent to hold a position against a limited force, and latch to lock a device into a selected position.

that shift the parts of a component to obtain the desired operation. The control method is the means used to move the mechanical controls in the valve.

The mechanical components of control mechanisms include rods, shafts, detents, latches, and over-center devices. Rods and shafts are common devices that allow a control device to be pushed, pulled, rotated, or moved to obtain the desired reaction. Detents, latches, and over-center devices are used to assure the control mechanism holds the position where it has been set.

Many different methods are used to control the operation of fluid power valves. However, the symbols for these devices are classified as either:

- Muscular.
- Mechanical.
- Electrical.
- Pressure.
- Feedback.

The symbols for these devices are shown in **Figure 4-21.** *Muscular controls* are those involving the direct effort of a human operator, such as a hand-operated push button or a foot pedal. *Mechanical controls* are operated by machine elements and include devices such as plungers and rollers. *Electrical controls* primarily involve solenoids and torque motors. *Pressure controls* can be either hydraulic, pneumatic, or a combination of both. They can involve either pressurizing the control, releasing pressure, or using opposing

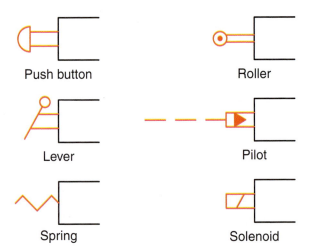

Figure 4-21. A wide variety of control methods are used with fluid power valves. The six symbols shown here represent methods commonly found in hydraulic and pneumatic systems.

differential pressures. *Feedback controls* involve sensing values internal and external to the valve, with the control device responding to these values and accordingly adjusting the operation of the valve.

Energy conversion

The *energy conversion devices* are hydraulic pumps, air compressors, motors, semirotary actuators, and cylinders. These devices either convert the energy of the prime mover into system pressure and flow or use the energy of the system fluid to do work. The symbols for these devices must be able to represent components from very basic pumps, motors, and cylinders to complex units that are pressure compensated and provide fully variable displacement.

The basic graphic shape for a pump or motor is a circle. The basic triangle symbol is used to indicate a hydraulic or pneumatic application. A pump is designated when a corner of the triangle points to the circumference of the circle. A motor is indicated when a corner of the triangle points to the center of the circle and one side of the triangle is in contact with the circumference. The direction of rotation for both pumps and motors is indicated by a curved arrow shown over a shaft symbol protruding from the side of the circle. Example pump and motor symbols are shown in **Figure 4-22.**

The basic graphic shape for a cylinder is a rectangle. Double-acting cylinders are drawn with a closed rectangle and ports at either end. The piston is a small rectangle drawn across the short dimension of the cylinder rectangle, while the cylinder rod is a double line extending from the piston through one end of the basic cylinder rectangle. A single-acting cylinder is drawn with the rod end of the cylinder exhausting to atmosphere in a pneumatic system and to a drain in a hydraulic installation. There are simplified versions of the symbols as well. See **Figure 4-23.**

Energy storage

In a pneumatic system, the principle energy storage device is the *air receiver.* The symbol is a capsule drawn horizontally with lines extending from both ends. See **Figure 4-24.** No additional internal graphic figures are normally used to designate this component.

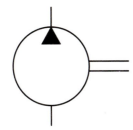

Hydraulic pump, one flow direction, fixed
displacement, one rotation direction

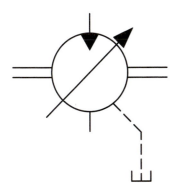

Hydraulic motor, one flow direction, variable
displacement, undefined control mechanism,
external drain, two shaft ends

Figure 4-22. Rotary energy conversion devices are
represented by circles. Many variations of these
units are found in circuits. The two symbols shown
here represent two of the possible variations.

Accumulators serve as energy storage devices
in specialized hydraulic circuits. The basic symbol
for these components is also a capsule. However,
in this situation, the shape is drawn vertically with
a single line extending from one end of the unit.
Internal graphic figures are used to indicate the
energy-storage method used in the device.

Energy control and regulation

Directional, pressure, and flow control devices
are grouped in the energy control and regulation
category of graphic symbols. These symbols are
made up of one or more boxes. When the symbol
is made up of more than one box, the boxes are
drawn with contiguous sides. These boxes can be
either rectangles or squares, depending on the
valve depicted. The number of boxes usually cor-
responds to the number of operating positions of
the valve. Two parallel lines along the length of a
valve symbol indicates that it has two or more distinct operating positions and a number of intermediate positions. See **Figure 4-25.**

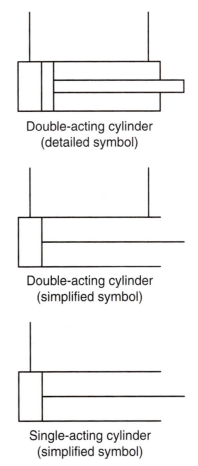

Double-acting cylinder
(detailed symbol)

Double-acting cylinder
(simplified symbol)

Single-acting cylinder
(simplified symbol)

Figure 4-23. Linear energy conversion devices
(cylinders) are represented by rectangles. As shown
here, the symbols can be detailed or simplified.

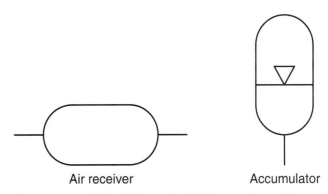

Air receiver   Accumulator

Figure 4-24. The left symbol is for an air receiver,
which is the energy-storage device for a pneumatic
system. The right symbol is for a gas-loaded accumulator, which is a common energy-storage unit in
a hydraulic system. This symbol can be modified
to represent a spring- or weight-loaded device by
simply changing the internal gas charged triangle to
show a spring or weight symbol.

pressure control valve. Simplified symbols for several types of basic pressure control valves are shown in **Figure 4-31.**

Flow control valves are devices used to control fluid flow in a circuit that, in turn, controls the speed of actuators. The preferred basic symbol for these valves includes the features of the valve in a square or rectangular box. The box is often dropped in simplified versions. **Figure 4-32** shows a number of variations that may be used to show the capability of a flow control valve. The curved lines represent the orifice (restriction) that controls fluid flow. The arrow passing through the orifice at a 45° angle indicates that the orifice is adjustable. The dimension shown in **Figure 4-32** indicates the physical diameter of the orifice. All of the symbols shown are acceptable. The level of information needed and company preference are the final determining factors in which symbol to use.

The orifices in flow control valves normally restrict flow in both directions through a valve. Often, it is desirable to have controlled flow in one direction and free flow in the opposite direction. **Figure 4-33** shows the symbol for a typical one-way restrictor valve. This valve combines flow control elements and a check valve to allow nonrestricted flow when the fluid flow reverses.

Pressure and temperature changes in a fluid power system affect the flow rate through the system. Flow control valves use compensation devices in an attempt to maintain accurate flow. Pressure compensation is shown in a simplified symbol by

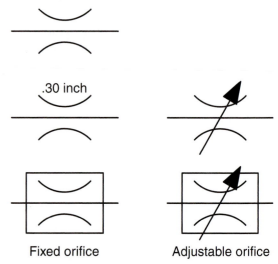

Fixed orifice   Adjustable orifice

Figure 4-32. This illustration shows several versions of symbols that may be used for flow control valves. An adjustable valve is shown with a slanted arrow passing through the orifice. The size of a fixed orifice can also be indicated with a dimension.

Single–stage relief valve   Sequence valve

Pressure–reducing valve   Counterbalance valve   Unloading valve

Figure 4-31. Simplified symbols are shown in this illustration for several common pressure control valves.

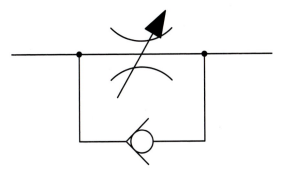

Figure 4-33. This illustration shows the symbol for a flow control valve that allows control of flow in one direction and free flow (through a check valve) in the opposite direction.

placing an arrow across the flow line of the symbol. Temperature compensation is shown by placing a thermometer symbol across the symbol flow line. **Figure 4-34** shows simplified symbols for pressure-compensated and temperature-compensated, adjustable flow control valves. If a single valve has both of these features, they can be combined into a single symbol, as shown in the figure. This is an indication of how easy a symbol may be structured to match the features of the component.

Fluid storage and conditioning

A number of different devices are used to condition and store both liquids and gases for fluid power systems. Simple symbols indicate these storage devices, which keep a ready supply of hydraulic fluid and compressed air. Somewhat more-complex symbols show the components used for fluid conditioning, which involves removal of dust and dirt, maintenance of proper

temperatures, and separation of moisture (from pneumatic system air).

Hydraulic reservoirs are shown as an open rectangle with the cut made along the long dimension of the figure. The open top indicates that the reservoir is not a sealed unit. Pickup, return, and drain lines touch the bottom of the open figure to indicate that they are below the surface of the hydraulic fluid. See **Figure 4-35.**

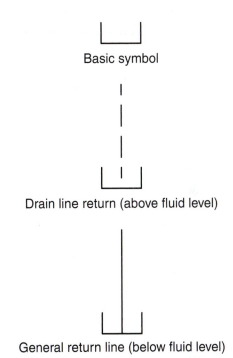

Figure 4-35. The hydraulic reservoir is drawn as an open rectangle, as shown in the top illustration. The middle illustration shows the reservoir with a drain line return that is above the fluid level. The bottom illustration shows the reservoir with a general return line from the system that is below the fluid level.

Figure 4-34. The left symbol is a simplified symbol for a flow control valve that is pressure compensated to help maintain a uniform flow rate through the valve. Pressure compensation is indicated by the arrow perpendicular to the flow line through the symbol. The middle symbol is for a temperature-compensated valve. The temperature-compensation feature is shown in the symbol by the temperature indicator symbol positioned across the flow line. The right symbol is for a valve that is both pressure and temperature compensated.

A number of reservoir symbols are shown in a hydraulic circuit, even though only a single reservoir is used in the system. This situation exists as return and drain lines from complete component symbols in a circuit may each terminate locally with a small reservoir symbol. This practice simplifies the diagram by eliminating a number of lines that otherwise would need to be shown returning to the main system reservoir.

A sealed or pressurized hydraulic reservoir is designated by a horizontally oriented capsule symbol. System line connections to this symbol are perpendicular to the long sides of the graphic. Below-fluid-level connections to the unit are indicated by lines that pass through the figure and contact the line representing the bottom side of the component.

A number of different fluid conditioners are used in hydraulic and pneumatic systems. These components include filters, separators, air dryers, lubricators, and heat exchangers. The basic graphic for these units is a square shown resting on one corner. The preferred orientation of the inlet and outlet lines is horizontal for all of these components. Symbols of separators or assemblies with separators are drawn *only* in the horizontal position. Symbols representing a number of the common conditioning components are shown in **Figure 4-36.**

A simplified symbol is often used for a group of common conditioning components used in many pneumatic systems. This group usually consists of a filter, separator, pressure regulator, pressure gauge, and lubricator. **Figure 4-37** shows both the detailed and simplified symbols for this common pneumatic combination unit.

Supplementary equipment

The supplementary equipment category of symbols primarily contains measuring instruments and specialized electrical switches. The basic symbol for measuring instruments is a circle, while a square is used for switches. The symbols in this category are used to indicate instruments that measure fluid pressure, flow rate, temperature, and level. Symbols also exist for a tachometer (used to measure the rotational speed) and torque meter (used to establish the torque output of motors and other rotational devices). **Figure 4-38** shows the symbols for common measuring instruments.

Filter

Manually drained separator

Lubricator

Heat exchanger

Figure 4-36. These symbols represent common fluid-conditioning devices found in hydraulic and pneumatic systems.

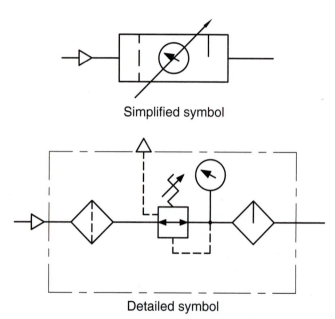

Simplified symbol

Detailed symbol

Figure 4-37. The top symbol is a simplified symbol for a common grouping of pneumatic system components used for the conditioning of air at the workstation. This grouping often includes a filter, separator, pressure regulator, pressure gauge, and lubricator. The bottom symbol is the detailed symbol for this functional group of components. Notice the complexity of the symbol.

Two types of switches that often appear in fluid power circuits are the pressure switch and limit switch. Both are used in circuits to provide control of system operation by sensing system pressure and actuator movement. The basic symbols for these components are shown in **Figure 4-39.**

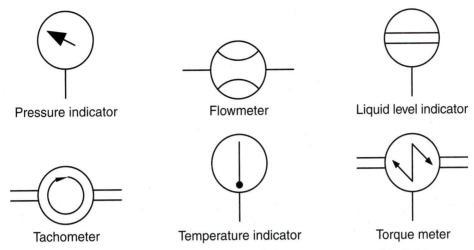

Figure 4-38. A wide variety of measuring instruments are used in fluid power systems. The six symbols shown here represent instruments that may be found in system diagrams.

Figure 4-39. The electrical components represented by the these symbols are often found in fluid power systems.

Circuit Diagrams

Circuit diagrams give very complete information on a fluid power system. This information is useful for system assembly, operation, and testing. Circuit diagrams include schematics to illustrate the structure of a system, showing each component and interconnecting line. Diagrams also contain information on conductor sizes and a list of components. Data are provided on component sizes, flow capacity, pressure settings, and speed. In addition, information is provided on the sequence of operation for the circuit. The ISO standard 1219-2 indicates that these items should be included in circuit diagrams:

- Schematic diagram.
- List of components.
- Sequence of operation.
- Conductor size.
- Component sizes.
- Flow capacity.
- Pressure settings.
- Component speed.

Other standards use a slightly different format, but include very similar content.

General rules

The schematics used in circuit diagrams should be clear. They must include all components and connections so it is possible to follow circuit movement and commands throughout the entire sequence of the system's operating cycle. The schematics need not take into account the exact physical arrangement of the equipment, but must include adequate drawings and related details to form a comprehensive document detailing system construction and operation. See **Figure 4-40.**

The lines or connections between the different components of the system should be drawn with a minimum of crossing points. The position of codes and indices should not overlap the space reserved for equipment parts and lines. In complex systems, the schematic should be divided into sections based on related control functions. When multiple pages are required, line connections between sheets should be carefully identified.

Figure 4-40. This is a hydraulic system circuit diagram based on the ISO standard 1219-2 for symbol and diagram formats. (ANSI, see copyright statement on page 2 of this text)

The symbols for the component parts in the fluid power system should be logically arranged from the bottom to the top and from left to right on the schematic. This places the energy source on the lower-left side of the drawing and the actuators on the upper-right side. The control components of the system should be in sequential order, moving upward from the power source and right toward the actuators. The symbols should show a powered-up system in the ready-to-start position.

Component identification

Identification should be provided for all equipment component parts in the circuit diagram. A code number should be located next to the symbols on the schematic. The ISO standard 1219-2 recommends a four-element code. The first position in the code is reserved for a number to identify the installation in which the component is located. The second position in the code should refer to the number of the circuit in the installation. The normal procedure is to start circuit numbering with 0 and proceed with numbers in sequential order. The third position is reserved for identification of the type of component. The component is normally identified by a letter using a code such as:

- P for pumps and compressors.

- A for actuators.

- M for prime movers.

- S for sensors.
- V for valves.
- All other equipment uses any letter not shown above.

The fourth code position is used to indicate the number of similar components. For example, four limit switches in a circuit are consecutively numbered 1 through 4. See **Figure 4-41.**

In addition to the coding system for components, several additional items of information should be provided about system piping. Lines should be labeled to indicate their function, such as P for pressure supply lines in both hydraulic and pneumatic systems. In hydraulic systems, tank return (reservoir) lines should be labeled T and leakage drain lines indicated with an L. Connections for the piping of circuit subassemblies should also be identified on the diagrams.

Technical information

A wide variety of information can be included in a circuit diagram. ISO standards suggest a complete list of technical items that can be included.

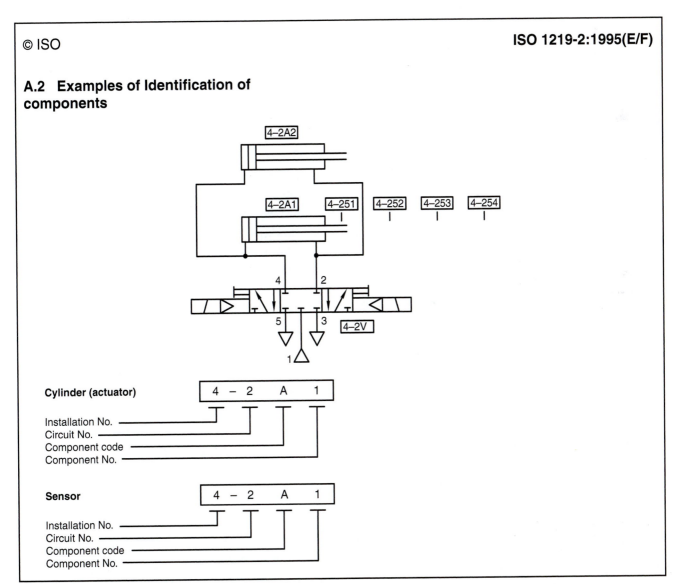

Figure 4-41. This is an example of the circuit diagram component identification system used by the ISO standard 1219-2. Each component in the circuit is labeled using the numbering system outlined in the standard. The table shows what each element of the number identifies. (ANSI, see copyright statement on page 2 of this text)

The following items should be considered basic. They provide valuable information about a circuit or system.

- Prime mover rated power, direction of rotation, and speed.
- Flow and pressure ratings of the pump or compressor in the system.
- Fluid capacity recommendations for the hydraulic reservoir.
- Type and rating of hydraulic fluid used in the system.
- Level of filtration required in the system.
- Capacity and maximum pressure rating of the pneumatic receiver.
- Nominal sizes of pipes, tubing, and hoses used in the system.
- Details of the actuators, including type, size, pressure rating, and function.
- Pressure control valve information, including function and pressure settings.
- Flow control valve information, including type and flow rate.
- Directional control valve information, including activation method and center position configuration.
- Accumulator volume, precharged pressure, and working pressure range.

Additional support information also can be included to assist in circuit installation, operation, and maintenance. For example, a component list, sequence of operation, function chart, and suggested circuit or system arrangement may be included.

Chapter Test

Answer the following questions. Write your answers on a separate sheet of paper.

1. Standards are established using one of three processes. List these processes.

2. A(n) _____ standard is one that must be followed due to laws that have been enacted.

3. List the five types of privately funded organizations involved in the standardization process in the United States.

4. The primary coordinating work with standardizing groups in the United States is done by the _____.

5. A number of organizations from outside the fluid power industry provide information and standards on items that support the hydraulic and pneumatic fields. List three items that this type of information or standardization may relate to in the fluid power field.

6. The nonprofit trade association that concentrates its efforts in the fluid power products manufacturing area and limits its memberships to companies is the _____.

7. The fluid power group that limits membership to individuals interested in the field and provides certification for individuals is the _____.

8. The procedure that assures adequate input during standards development from individuals and organization that have direct and material interest in the standard is known as _____.

9. A key factor in the process of standards development is that input be balanced and the process not be dominated by any _____.

10. Standardization groups (committees) should include members of at least three groups. List these three groups.

11. The key technical elements of a standard are found in the _____ section of the document.

12. List the three items that fluid power symbols are used to communicate.

13. What are the symbols called that are really miniature section drawings of component parts?

14. Graphic symbols are *not* intended to show the:
    A. Function of the component.
    B. Location of the component.
    C. Appearance of the component.
    D. Type of fluid.

15. The preferred orientation of graphic symbols is in increments of _____.

16. The symbol for a component involving two or more functions is surrounded by a chain line that represents a(n) _____.

17. The symbol for a control component or a prime mover appears in a graphic diagram as a(n) _____ resting on its side.

18. The type of fluid used in a system and fluid direction are indicated by a small equilateral triangle. A solid triangle indicates a(n) _____ operation and an open triangle indicates a(n) _____ operation.

19. The symbols for the devices used to operate directional control valves are drawn on the end of the basic valve symbol. These devices are referred to as _____.

20. Describe the basic symbol for a hydraulic pump.

21. The modes of operation of multiple-position _____ are shown by using a box for each of the operating positions.

22. In a simplified symbol for a flow control valve, an arrow that passes through the orifice in the symbol indicates the valve is _____.

23. In a hydraulic system drawing, multiple _____ symbols are used to terminate fluid return lines to reduce the number of lines returning to the main reservoir.

24. According to ISO standard 1219-2, the symbols used to depict fluid power systems should be arranged on the schematic from _____ to _____ and from left to right.

25. List the four elements of the component identification code recommended by ISO standard 1219.

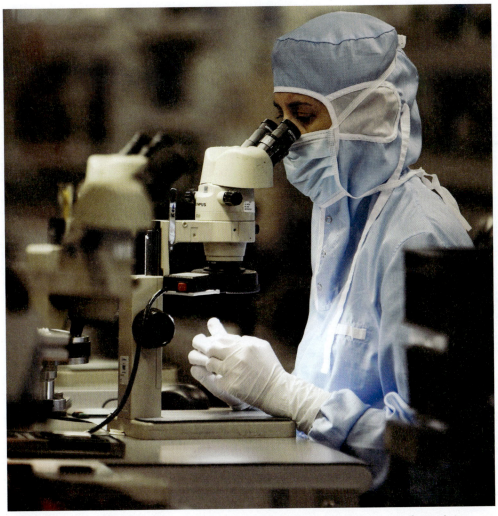

As fluid power standards are revised to reflect new materials, manufacturing techniques, and filtration standards, extra care must be taken to assure quality production. (Atlas Copco)

The use of equipment is an everyday activity in our industrialized society. Technology has produced a vast number of machines that range from simple hand tools to huge, complex machines. A large number of these devices have the potential to cause injuries to the operator or individuals working in the area. The number of workers injured on the job each year indicates that many people do not understand the environment in which they work or respect the equipment they operate or work near. Individuals working in the fluid power field will be exposed to a wide variety of work environments. They must be aware of safety and health issues related to those areas, as well as the specific issues dealing with fluid power. This chapter discusses items that influence general safety and specific issues that involve hydraulics and pneumatics.

Objectives

After completing this chapter, you will be able to:

- Describe how the workplace environment influences the safety of individuals working in an industrial setting.
- Explain how awareness of safety issues in a situation influences the safety of an individual.
- Identify personal traits that may make an individual more prone to violate safety procedures.
- Identify general equipment considerations that influence safe working conditions in the fluid power industry.
- Describe safety rules that need to be followed when working with individual hydraulic and pneumatic components.
- Identify government regulations that influence safety regulations and practice in the fluid power industry.

Chapter 5

Safety and Health
Promoting Proper Practices

Selected Key Terms

The following terms will be used in this chapter. As you read the text, record the meaning and importance of each. Additionally, you may use other sources, such as manufacturer literature, an encyclopedia, or the Internet, to obtain more information.

air-line respirators
back injuries
cumulative injuries
dust mask
earmuffs
earplugs
emergency showers
equipment maintenance
eye protection devices
eyewash stations
first aid kit
hearing-protection devices
housekeeping
lockout devices
Occupational Safety and Health Administration (OSHA)
pressure safety valve
respirator face masks
safety helmets
safety shoes
self-contained breathing devices
training

Internet Resources

www.cdc.gov/workplace/
Centers for Disease Control and Prevention
Presents various topics and resources on workplace safety and health, including chemical safety, eye safety, machine safety, and protective clothing.

www.nsc.org
National Safety Council
A nonprofit, nongovernmental public service membership organization that provides safety and health information and training programs.

www.osha.gov
Occupational Safety and Health Administration (OSHA)
OSHA provides extensive information on workplace health and safety, as well as training materials and other services. OSHA is a part of the United States Department of Labor.

www.hallite.com
Hallite Seals International—Company Overview
Choose the "Health & Safety Policy" link on the "About Us" page to review the health and safety/corporate responsibility policies of an international company.

http://commerce.rosscontrols.com/
Ross Controls
From the "Industry Solutions" menu, choose "Safety Industry" to return an overview of fluid power machine safety, including areas of concern for safety and risk reduction.

http://wyomingworkforce.org/employers-and-businesses/osha/Pages/safety-and-health-programs.aspx
Wyoming Department of Workforce Services
Provides several resources for complying with OSHA workplace health and safety guidelines. Choose the "Personal Protective Equipment Program" document to view a sample of a hazard assessment, equipment selection, and training document for personal protective equipment.

Adequate illumination must be provided in all work areas. The required light intensity depends on the work being performed in the area. Applicable local, state, and federal standards should be obtained and closely followed. Regulations vary on light intensity, but care should be taken to provide adequate general lighting in all areas and supplementary lights at workstations for detailed work, such as machine or test bench locations. Higher light levels are also required in classrooms and training facilities.

Appropriate fire-fighting equipment must be available in each work, training, or classroom area, **Figure 5-4.** Carbon dioxide, foam, or dry powder fire extinguishers need to be placed in the facility following the recommendations of local fire officials. Fire extinguishers have a class rating, which is the types of fire on which the extinguisher can be used. See **Figure 5-5.** A fire blanket, which can be used to put out a clothing fire, should be readily available. All flammable liquids, such as solvents and fuels, should be stored in metal containers in an area away from a heat source. These materials should be carefully handled, with any spillage immediately wiped up. Rags soiled with oils and solvents should be removed from the work area at the end of the day and stored in fireproof containers. Each area of the facility should have fire-emergency procedures posted and all individuals working in an area should be familiar with those procedures.

All of these factors contribute to a safe working environment. Failure to provide the basic elements of efficient facility layout, cleanliness, lighting, noise control, and fire prevention can considerably reduce individual safety. Care should be taken to provide an environment that meets at least the base requirements in each of these areas.

Employee Factors

Although often blamed, unsafe working conditions are *not* the chief cause of industrial accidents. Employees are really the chief cause. Statistics show that up to 85% of on-the-job accidents can be traced to unsafe acts by an individual. A wide variety of reasons are responsible for the acts that lead to these injuries. Four general areas of knowledge and activity that contribute to the safety record of employees are discussed in this section.

Figure 5-4. Fire extinguishers of the appropriate type need to be located throughout a facility following regulations issued by the local fire department.

Knowledge

A primary reason for performing unsafely is a basic lack of knowledge about the job. It is the responsibility of the employer to provide employees with the information needed to safely perform the job. This information should include:

- Steps necessary to perform the job.
- Hazards associated with the job.
- Need for and use of personal protective equipment during the performance of the job.

The employer must provide training sessions so an employee has the opportunity to practice the procedures before actually attempting to do the job. However, employees must assume the responsibility of *learning* the procedures during the training sessions, as knowing the job is a critical part of job safety, **Figure 5-6.**

Fire Extinguishers and Fire Classifications

Fires	Type	Use	Operation
Class A Fires Ordinary Combustibles (Materials such as wood, paper, textiles.) *Requires... cooling-quenching* Old / New	**Soda-acid** Bicarbonate of soda solution and sulfuric acid	Okay for use on A Not for use on B C D	Direct stream at base of flame.
Class B Fires Flammable Liquids (Liquids such as grease, gasoline, oils, and paints.) *Requires...blanketing or smothering.* Old / New	**Pressurized Water** Water under pressure	Okay for use on A Not for use on B C D	Direct stream at base of flame.
Class C Fires Electrical Equipment (Motors, switches, etc.) *Requires... a nonconducting agent.* Old / New	**Carbon Dioxide (CO₂)** Carbon dioxide (CO_2) gas under pressure	Okay for use on B C Not for use on A D	Direct discharge as close to fire as possible, first at edge of flames and gradually forward and upward.
Class D Fires Combustible Metals (Flammable metals such as magnesium and lithium.) *Requires...blanketing or smothering.* D	**Foam** Solution of aluminum sulfate and bicarbonate of soda	Okay for use on A B Not for use on C D	Direct stream into the burning material or liquid. Allow foam to fall lightly on fire.

		Multi-purpose type	Ordinary BC type	
	Dry Chemical	Okay for A B C	Okay for B C	Direct stream at base of flames. Use rapid left-to-right motion toward flames.
		Not okay for D	Not okay for A D	

	Dry Chemical Granular type material	Okay for use on D Not for use on A B C	Smother flames by scooping granular material from bucket onto burning metal.

Figure 5-5. Fire extinguishers are rated for the type of fire on which they can be used. This chart shows the identification symbols that appear on the fire extinguisher and explains where each type may be used.

increase in benefit and insurance fees, as well as the loss of valuable employee work time. In addition, the morale of sick or injured employees is often low.

A good safety and health program must start at the top management level. The involvement of management must be visible, continuous, and credible. Four areas that management can promote to help establish long-term employee support for an overall program are:

- Safety and health training programs.
- Protective clothing and equipment programs.
- Realistic emergency care and medical services.
- Established plan for disasters.

Safety and Health Training

Training can serve as an important influence and a substantial motivator in relation to workplace safety and health issues. Care must be taken to be certain the training is directed at a defined need of employees, rather than applied in an overall blanket approach used only to meet the requirements of an outside agency. As in other areas of employee development, general training should be used to solve problems dealing with lack of knowledge, skill, or motivation. A considerable amount of helpful and interesting safety and health-related content can be presented under these general classifications.

Training can be presented using a variety of formats, ranging from informal, one-on-one supervisor-employee contact to extensive, formal classroom training sessions. The style of the presentation is usually controlled by the complexity and extent of the topic. The first essential step in any training, however, is the need to define the objectives of the training.

Objectives should always be spelled out before content and methods are selected. These objectives should state how the training participants should perform on the job when the training is completed. For example, an objective might be stated as:

> At the conclusion of this training, the participants will be able to select the correct type of fire extinguisher for use on specific classes of fires as defined by the National Fire Protection Association.

A wide variety of commercial training materials are available in both the safety and health fields. These materials provide a wide choice in the methods that may be used for instruction.

Safety and health-related training should be provided to all employees that could be adversely affected by the jobs they perform. Safety and health issues critical to the completion of these jobs should be considered as possible training topics. After appropriate topics have been selected, objectives for each of them must be developed and instructional content structured. The last step in the process should be the selection of the presentation method.

The selection of employees for safety and health-related training should be based on the needs of the group or individual. One obvious group is new employees, who need to be introduced to the general policies and procedures of the organization. In addition, new employees need to receive specific training related to their individual position. Experienced workers must receive training when new equipment or procedures are introduced or when they are transferred into a new work situation. See **Figure 5-9**. Fluid power specialists, because their work assignments cross into a number of different departments in an organization, need to receive broad-based training so they can safely function in their job.

Protective Clothing and Equipment

The need for protective clothing and equipment implies that hazards have not been totally eliminated in some occupations. The nature of some jobs is such that the probability of accidents and long-range health hazards are higher. Total elimination of risk in these jobs is difficult. Protective gear is, therefore, needed to assure the safety of individuals performing these jobs. Fluid power specialists are exposed to most of the work environments in a factory. They need to be aware of what gear is needed to decrease the likelihood of an accident or a job-related health problem.

Ear protection

Noise may be considered a health issue. Long-term exposure is typically the reason for hearing loss due to noise. However, a single, high-intensity incident may cause damage. Protecting employees

Figure 5-9. A—Training programs need to be well planned and delivered to maintain employee interest. B—Training participants should be provided materials with well-defined objectives so they are aware of what content will be presented. (Eaton Fluid Power Training, Lab-Volt Systems, Inc.)

from these noise hazards is a very common use for personal protective equipment.

A variety of *hearing-protection devices* are available. These devices reduce the decibel level of workplace noise reaching a worker's ears. Earplugs, earmuffs, and helmets are three devices that can be used for hearing protection.

Earplugs are the most popular and very commonly used by workers, **Figure 5-10.** Rubber, plastic, or foam plugs are readily available in most work situations. Earplugs are generally inexpensive and inconspicuous, which is an important factor for many individuals that must wear noise control devices.

Earmuffs are larger, more expensive, and more conspicuous. However, they can be more effective than earplugs. Also, some workers feel earmuffs are more comfortable than earplugs.

Helmets are the most expensive form of hearing protection. They are usually used only in the most severe noise conditions or where a combination of a hardhat and sound protection is required.

Eye protection

Eye protection programs are widespread in industry. Many safety programs require that safety glasses be worn by all employees and visitors in any manufacturing area. *Eye protection devices* come in a wide variety of styles, ranging from plastic goggles to framed prescription eyeglasses, **Figure 5-11.** Face shields are also available for eye and face protection when working with dangerous liquids and chemicals.

Figure 5-10. Earplugs and earmuffs are two popular forms of noise protection for workers. They are available in a variety of materials and designs.

Figure 5-15. Safety helmets are primarily used to protect the head from impacts. A variety of designs are available to meet the special needs of particular industries. (Photo: Atlas Copco)

Safety shoes are another common piece of protective equipment for the fluid power specialist. Specially designed shoes or shoes with built-in foot protection are needed in many industrial operations. See **Figure 5-16.** Wearing proper shoes can significantly reduce the number or severity of foot injuries. Manufacturing firms often provide proper safety footwear for their employees or assist in the purchase of such gear.

Steel-toed shoes provide protection from heavy weights. However, safety shoes are available in many styles and forms far beyond the traditional steel-toed work shoe. For example, metal-free footwear is available for use where there are severe electrical, fire, or explosion hazards. Wood-soled shoes are designed for working on wet floors or for jobs that require walking or standing on hot floors.

Figure 5-16. Steel-toed shoes are a common safety shoe found in manufacturing situations. Many other types of safety shoes are available.

Metal safety guards worn over shoes to protect the toe and arch area of the foot are also available. These are worn in areas where heavy objects, such as metal castings and timbers, are routinely handled.

Other protection

A variety of other personal protective clothing and equipment is manufactured for use in specialized situations. See **Figure 5-17.** These include complete garments, aprons, gloves, headgear, eye and face protection, and many more. The development of many new materials has allowed many of these items to provide substantially more protection than in the past. Fluid power specialists need to be aware of what clothing and equipment is available to provide a safer work situation and to diligently use the items recommended by their employer.

Emergency Care and Disaster Planning

Accidents where individuals are injured happen even in organizations that have good safety programs. Additionally, workers can suffer sudden and serious illnesses on the job and disasters, such as fires and storms, can occur. Rapid response is a primary consideration in these situations. Effective first aid and the time saved because disaster plans exist, are understood, and are followed can mean the difference between life and death.

Figure 5-17. A wide variety of specialized clothing is available to promote safety in various industries. Welders wear leather garments to protect themselves from hot slag and metal splatter encountered during the welding process. (Miller Electric Mfg. Co.)

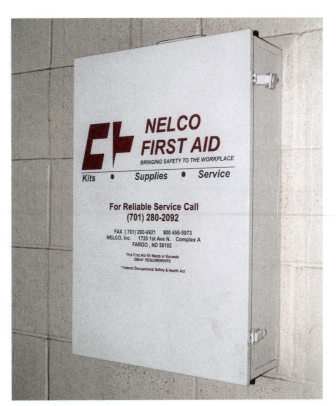

Figure 5-18. A first aid kit is a basic tool that should be part of any emergency care and disaster plan.

The simplest of the tools needed is a *first aid kit,* **Figure 5-18.** These kits are commercially available or may be assembled following the advice of company medical staff. The kits need to be readily available and restocked on a regular basis. At least one person adequately trained in first aid is required in a facility if an on-site infirmary does not exist or a hospital is not close to the facility. *Emergency showers* and *eyewash stations* are required in facilities where exposure to corrosive material is a possibility, **Figure 5-19.**

Each supervisor and worker should know how to rapidly obtain medical assistance in routine first aid, emergency, and disaster situations. When serious injury or illness occurs, emergency medical assistance should be requested as soon as possible. First aid procedures should start immediately if the situation is life threatening. The sick or injured worker should not be moved unless the location itself threatens further injury. Care of the sick or injured individual should be turned over to the emergency medical personnel as soon as they arrive on the scene, **Figure 5-20.** Be prepared to provide information on any first aid administered to the victim.

The facilities, equipment, and procedures for first aid and medical treatment should be a part

Figure 5-19. Eyewash stations are required in areas where exposure to corrosive fluids is a possibility. (Used with permission of CNH America LLC)

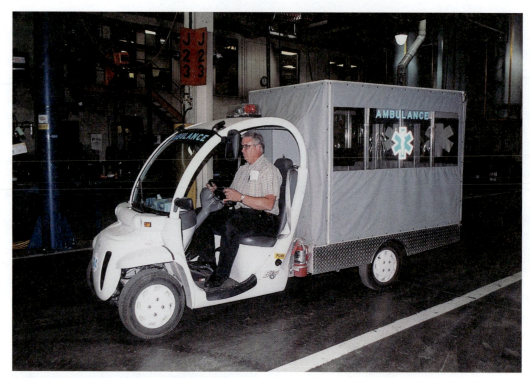

Figure 5-20. The need for an on-site infirmary is dependent on the number of employees, the type of work carried on in the facility, and the proximity to a hospital. (Used with permission of CNH America LLC)

of the design and operating plan of all facilities. Also included in this is planning for the personnel required to care for employees in first aid, emergency, and disaster situations. The plan should indicate the position that has the primary responsibility for directing the operation of the team involved. Company personnel that work in several areas of a facility, such as the fluid power specialist, need to be familiar with the emergency plans for all the area in which they work, **Figure 5-21.** This knowledge may prove to be extremely beneficial to both themselves and their fellow workers in any number of situations.

General Industrial Hazard Control Considerations

There are many different reasons for injuries in industrial settings. These include factors such as inattentive machine operators, inappropriate equipment or processes, defective or poorly designed equipment, and a multitude of other causes. Safety personnel generally feel that every effort should be made to promote safety by first removing as many hazards as possible in facilities and equipment through design before using employee safety promotion to reduce the remaining hazards. It must be kept in mind that there are situations in which hazards cannot be totally eliminated by design and the safe performance of workers must be relied on to eliminate accidents.

Because fluid power specialists work in such a variety of locations and situations in a factory, they should be aware of factors that can affect overall worker safety and moral. Problems in these areas can contribute to increased accidents and injuries in a facility:

- Housekeeping.
- Machine guards.
- Equipment maintenance.
- Materials handling.

Fluid power personnel can promote safety by being aware of these factors and providing information to the appropriate individuals so corrective steps may be taken to improve or eliminate the problem.

Figure 5-21 The procedures to follow in case of a fire or a natural disaster should be readily available and understood by the employees. Often, evacuation routes like this one are posted throughout the facility.

Housekeeping

General housekeeping can have a great influence on the number of accidents and injuries in manufacturing facilities. It should be ranked extremely high on the list of factors that can prevent accidents and increase productivity. *Housekeeping* not only includes keeping floors and work surfaces clean and free from dirt and debris, but also includes the proper storage of raw materials, partly finished products, and tools, **Figure 5-22.** Aisles and walkways need to be clean and clear of raw materials and products. Doing so prevents injuries from employees tripping over or running into these items or having the items fall on them. The idea that a busy and profitable manufacturing plant will be cluttered and dirty is *not* correct.

Figure 5-22. Keeping a facility clean, with aisles and walkways clear, promotes good general housekeeping and also helps prevent accidents. (Vladimir Melnik/Shutterstock)

Machine Guards

All mechanical motion in a machine is, to some degree, hazardous. The *Occupational Safety and Health Administration (OSHA)* spells out when operators must be protected. If the hazard points in a machine cannot be eliminated, guards must be installed. These guards can take many forms, including:

- Simple, fixed enclosures that cover hazardous areas to prevent exposure to moving machine parts, **Figure 5-23.**

- Interlocking guards that shut off the power to the machine when they are opened.

- Automatic guards that pull or push an operator's hands, arms, or body from the hazardous area as the machine performs its work.

- Specialized machine designs that require two-handed operation or keep the operator away from hazardous areas by automatic placement, feeding, or ejecting of parts, **Figure 5-24.**

Figure 5-23. A multitude of fixed-enclosure guards are used to cover hazard points in machines in industry today. This guard covers the belt drive between the prime mover and compressor.

Figure 5-24. Several types of specialized guard designs are used on equipment to promote safe operation. One specialized design requires two-hand operation of the equipment. Both hands must be on control switches, which are placed well out of any danger, in order for the equipment to operate. (Used with permission of CNH America LLC)

A multitude of devices must be considered for guarding, including items such as rotating shafts, keys and setscrews on shafts, gears, chain drives, belt drives, flywheels, and so on. Maintenance staff, including fluid power specialists, should be very familiar with machine guarding, including lockout devices. *Lockout devices* prevent a machine from being accidentally turned on while maintenance work is underway. When machine service is complete, all guards must be replaced to the manufacturer's specifications.

Equipment Maintenance

Proper *equipment maintenance* is a key factor in a good safety program. This not only means the service work done on large stationary pieces of equipment, but also maintenance of portable tools and hand tools. A systematic program of preventive maintenance should be set up to keep machines, tools, and equipment safe and operating at their maximum efficiency. Some equipment may need daily maintenance, while others can be serviced after a specified number of hours, days, or weeks.

The fluid power specialist should be heavily involved in developing and operating a preventive maintenance program. Equipment that could cause an accident if it fails during operation should be inspected and maintained on a closely controlled schedule. Parts should be carefully adjusted and replacement of key parts should be done on a definite schedule.

Care of portable tools is especially important, as many workers use some form of this equipment. Portable tools account for a high proportion of injuries, especially injuries to the hands. However, eye injuries and electric shock are also very commonly associated with operation of these tools. Extreme care should be taken while operating this type of equipment.

Materials Handling

The handling of materials in an industrial setting can be as hazardous as the manufacturing process itself. These hazards range from injuries caused by a worker applying improper lifting techniques to accidents involving mechanical material handling equipment, such as forklift trucks and cranes.

Back injuries represent the largest single group of injuries for which compensation is awarded. These injuries are often related to lifting. Care should be taken in all situations in which lifting must be done. It is possible to injure the back even with light loads if care is not taken to position the back and legs to obtain appropriate leverage. A conscious effort to position the legs and back comfortably while using the legs to do the lifting should reduce the number of back injuries. In addition, back supports are available to help reduce injuries, **Figure 5-25.** However, lifting injuries can occur even under the best of conditions.

The amount of materials physically lifted by humans in an industrial plant is usually tiny by comparison to the total weight involved in the manufacturing process. These materials are designated by tons, rather than in pounds and ounces common to the loads individuals lift. Hand- or electric-operated dollies, forklift trucks, conveyors, and cranes are commonly used to move the vast majority of materials in a factory. The selection of the system used to move the material is based on

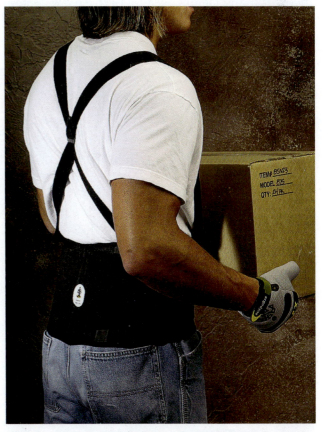

Figure 5-25. Back injuries are common when work involves the handling and lifting of materials. Special care should be taken to correctly position the back and legs before performing the lift. Use the legs rather than the back for the actual lifting. Back supports such as the one shown here also help prevent back injuries. (Ergodyne)

the weight and size of the material, type of operation being performed on the material, and configuration of the facility in which it must be moved. The equipment operator's skill, attitude, and awareness of hazards is critical to the safety performance of the equipment. Safety performance is also influenced by the awareness of workers to the factors involved in the operation of the systems. The more conscious the workers are of the operating characteristics of the systems, the more efficiently the systems can operate and the better the safety record.

Specific Fluid Power Safety Factors

Working safely as a fluid power specialist involves many factors. Several of these are general industrial safety items that have been mentioned

earlier in this chapter. The importance of these safety issues cannot be overemphasized. A number of additional detailed factors relating to hydraulics and pneumatics need to be emphasized. These include specific hazards related to pressurized fluids and system components. The following sections discuss factors that are especially important to any individual working with fluid power.

Factors Regarding Pressurized Fluids

Individuals working with equipment that uses fluid power systems must constantly be aware of the fact that they are working with pressurized liquids and gases. Being aware of this fact and at all times treating the system as if it is pressurized and ready to function can eliminate many accidents, **Figure 5-26.**

Pressurized fluids must be treated with respect. Compressed gases in pneumatic systems and in

Figure 5-26. Operators of fluid power systems, from simple presses to large complex machines, will have fewer accidents if they treat the equipment as if it is pressurized at all times. (Yates Industries, Inc.)

accumulators in hydraulic systems can cause actuators to move even though the system appears to be turned off and out of service. Liquids compress very little, so fluid compressibility is usually not critical in hydraulic systems without accumulators. However, in hydraulic systems using large-volume cylinders, movement is possible. This is especially true if small quantities of air have become entrained in the system oil.

Pneumatic systems, of course, use compressed air for their operation. The systems should always be approached as if they are pressurized and ready for operation. Shutting off the system and cycling it to eliminate compression should be a standard step before repair work begins.

Compressed air is commonly used for cleaning parts and equipment, **Figure 5-27.** Blowguns or air nozzles attached to air hoses can be used to remove dust, dirt, and liquid. Extreme care needs to be taken when using these devices to prevent eye and skin damage from flying liquid and dirt particles produced by the airstream. When using these devices, workers should avoid near or direct contact between the nozzle tip and the skin. It is possible for the air to penetrate the skin, causing

Figure 5-27. Using compressed air for cleaning parts is a common, but very dangerous practice. Use adequate eye protection and treat the air nozzle with the respect of a tool that can easily produce injuries.

severe medical problems. Pressurized air must be treated as a potentially dangerous tool. It is used for cleanup in a shop situation, not as a toy for any type of horseplay.

Unprotected, direct physical contact with pinhole leaks in hydraulic systems should also be avoided. The velocity of the liquid may cause penetration problems in cuts and abrasions on the hands. The pressure of the liquid is not a factor, as the pressure of the escaping liquid drops to atmospheric pressure as soon as it leaves the system.

Factors Regarding System Components

Fluid power specialists and the operators of equipment involving fluid power are working with controlled forces. These forces can reduce the effort of the operator and produce tremendous gains in production. They can also produce situations in which the safety of individuals is compromised and equipment is damaged. Understanding equipment operation and knowing the basic safety guidelines for the system involved will greatly reduce the chances for accidents. A specific equipment training program is beyond the scope of this book and usually considered the responsibility of the employer. However, this section of the chapter discusses hazards associated with different types of fluid power components. Procedures and safety rules are suggested to reduce or eliminate problems.

Actuators

Actuators are generally considered the most dangerous of the fluid power components because they involve considerable linear or rotary motion, **Figure 5-28.** Cylinders and motors are the two most common actuators in both hydraulic and pneumatic systems. Various other actuator designs are used in special circuit applications. The safety hazards associated with actuators are usually related to high-speed movement or high-force output. Generally, pneumatic systems produce the highest speeds, while hydraulic applications yield the greatest forces.

The safe operation of actuators in an application begins with the proper selection of the component, its correct installation in the system, and its protection from heat, corrosion, abrasive dirt, and shock as it functions. Each actuator

Figure 5-28. Care must be taken when working with actuators, as they are probably the most dangerous components in fluid power circuits. The large hydraulic cylinders on this backhoe generate both dangerous movement and extremely high forces that can cause serious injuries.

has manufacturer specifications that should be matched to circuit needs before installation. These specifications include factors such as pressure rating, cylinder bore and stroke, and motor torque output. Recommended mounting methods must be followed for the component as it is installed in the system.

Once the system is in operation, regular inspections should be performed during its service life. These inspections should include checking for items such as:

- Excessive system pressure.
- Excessive system operating temperature.
- Internal shock caused by the cylinder piston striking the cylinder end caps at the end of the stroke.
- External damage to the actuator caused by other machine members striking the component.

Pumps and compressors

Pumps and compressors are generally considered the second most dangerous group of components in fluid power systems. As with actuators, this designation is based on the fact that movement is involved in the operation of the units. Another factor that contributes to this designation is the coupling device used to connect the unit to the system prime mover. Coupling devices can range from simple, direct-drive shaft couplings or belt drives to complex, variable-speed couplings found on larger, stationary compressors. Often, coupling devices are exposed, allowing a machine operator or maintenance person to make direct contact with rotating parts, which can cause serious injury.

Proper sizing, alignment, and guarding of the couplings are important to both the efficient and safe operation of pumps and compressors. Correctly sized and installed couplings will provide many hours of safe power-unit operation. Guarding for the coupling may be a shield specifically designed for just the coupling or belt drive. A guard may also be a larger enclosure that covers or restricts access to the whole unit, including the pump or compressor and the prime mover. Whatever the guarding method, it is essential that the guard keeps hands and clothing away from the moving parts of the coupling or drive unit.

The prime mover and pump or compressor need to be properly sized for the application if the system is to operate both efficiently and safely. Units that are too large waste energy, while units that are too small may fail because they are operating at inappropriate pressures or temperatures.

Be certain a *pressure safety valve* is built into both hydraulic and pneumatic systems close to the pump or compressor. This valve is placed in the system to open if the normal maximum-system-pressure control device fails. If the normal pressure-control method fails, system pressure could reach an unsafe level. The prime mover could then stall or the pressure could damage system components and possibly injure machine operators or other workers in the area.

Air receivers

An air receiver is the device in a pneumatic system that stores an adequate volume of compressed air and assures a steady supply of pulse-free air to the system. This device needs to be carefully sized and located in a system to promote effective system operation. The receiver must be large enough to supply the air demanded by the system while reducing air pulsations that affect tool operation. It also must be appropriately located in the system to reduce pressure drops in the distribution line. Both of these situations contribute to a safer operating system, as they contribute to an air supply of uniform pressure and volume.

The air receiver also needs to be equipped with a valve that allows the release of condensed water vapor from the tank. These valves are available in both automatic and manual designs. However, under all circumstances, a procedure should be in place to routinely check the amount of condensed water vapor in the tank and drain it. The operating efficiency and service life of circuits and tools are affected by the condensed moisture circulated in the system. Both of these factors contribute to a safer operating system.

Air receivers are commonly used as the location of the system safety valve, an overall system pressure gauge, and a manually operated valve that can be used to relieve the pressure in the system. All three of these devices allow an operator to better control the system and provide a clean, dry, and pulse-free air supply that contributes to a safer pneumatic system operating environment.

Pressure, flow, and directional control

Pressure, flow, and directional control are three basic areas of fluid power. A distinct group of components supplies the control aspect for each of these areas. Terminology varies slightly between the hydraulic and pneumatic fields, but the concepts are very similar. Understanding these key concept areas can have a definite influence on how the operation of fluid power components and circuits is approached and handled. Understanding a system can result in the system being handled more safely.

Pressure control valves are used in both hydraulics and pneumatics systems to control the force output of actuators and to block their movement until a desired system pressure is achieved. Flow control valves are used to control the speed of actuators. Actuator directional control is achieved by changing the fluid flow routes through directional control valves so fluid flows to the desired section of the actuator to cause the unit to move in the desired direction. When initially setting up a circuit, it is safest to set pressure and flow control valves at their lowest setting and directional valves at their center or normal position.

Most pressure and flow control valves have few moving parts and seldom malfunction. However, they must be properly mounted and initially adjusted to the correct pressure and flow rate. It is considered good practice to install tamper-resistant devices on pressure and flow control valves once the machine has been set up. Operators should be advised not to change valve settings as the valves have been set at their safe operating levels. If the operator encounters reduced pressure or actuator speed, it indicates a problem somewhere else in the circuit. Changing the valve setting will not solve the problem and may produce a major safety issue.

Considerable care should be taken when locating and setting up directional control valves or other machine controls. These are general, safety-related rules to follow when locating and installing controls:

- Machine operators should be able to easily reach the controls without leaving their normal workstation.

- Controls should not interfere with the operator's normal work space or work surface.

- Controls should be located away from all moving machine components.

- Controls must be designed to prevent the accidental activation of the machine.

- The control mechanisms should be protected against accidental damage.

- Easy access for maintenance should be part of the control design.

Accumulators

An accumulator stores pressure in a hydraulic system through the use of weights, compressed springs, or compressed gases. This pressure can be maintained even after the system has shut down. Circuits that contain accumulators should be clearly marked to help prevent accidents from unexpected actuator movement resulting from the stored energy.

Circuits that contain accumulators should provide a means to automatically unload the accumulators when the system is shut down. Several different designs are available to provide this important safety feature. Fluid power specialists and maintenance staff should always be alert to

15. When working with a blowgun for clean-up, compressed air should be considered a potentially dangerous tool. The air can penetrate _____ and cause severe medical problems.

16. Name three factors that are critical to the safe installation and operation of actuators.

17. Excessive system pressure in a fluid power system can be avoided by having a(n) _____ valve built into the system close to the location of the pump or compressor.

18. It is considered good practice to install _____ devices on pressure and flow control valves to prevent operators from resetting the system to unsafe pressure and flow levels.

19. Machine operators and maintenance workers must always be alert to the possibility of _____ in the system when an accumulator is included in the circuit.

20. A consideration when selecting pneumatic tools is the safety factor involved in the storage and movement of the _____.

Casting can be dangerous and requires special safety equipment. Notice all of the different safety equipment used by these workers. (Joe Rossi, Saint Paul Pioneer Press)

In hydraulic systems, energy is transmitted through a liquid, rather than through shafts, gears, cables, and belts, as in a mechanical power transmission system, or the wires and complex control components used in electrical systems. Although hydraulic systems may sound less complicated, they are still complex devices requiring consideration of many elements. One of the fundamental elements that must be considered is the liquid to be used in the system. This chapter deals with the many factors relating to liquids that must be considered when selecting a hydraulic fluid for a particular application.

Objectives

After completing this chapter, you will be able to:

■ Describe the various functions a hydraulic fluid performs in a fluid power system.

■ Identify and explain the general properties of a liquid that would make it suitable as a hydraulic fluid.

■ Name and describe the general categories of materials that are commonly used as hydraulic fluids.

■ Explain the terms used to describe the basic characteristics of hydraulic fluids.

■ Explain procedures to follow for the selection and performance monitoring of hydraulic fluids.

■ Describe appropriate procedures for handling, storage, and disposal of hydraulic fluid.

Chapter 6

Hydraulic Fluid
The Energy Transmitting Medium

Selected Key Terms

The following terms will be used in this chapter. As you read the text, record the meaning and importance of each. Additionally, you may use other sources, such as manufacturer literature, an encyclopedia, or the Internet, to obtain more information.

additives
anti-wear agents
biodegradable fluids
capillary viscometer
catalysts
emulsion
extreme-pressure agents
fire point
fluid
Four-Ball Method
friction
HWCF
kinematic viscosity
lubricity
oxidation
oxidation inhibitors
phosphate ester
polyglycol
pour point
rust inhibitors
Saybolt viscometer
sludge
specific gravity
spontaneous ignition
viscosity
viscosity index number

Internet Resources

www.azom.com/details.asp?articleid=2235
Azom: The A to Z of Materials—Fire Resistant Hydraulic Fluids
An article that addresses the three main types of fire-resistant hydraulic fluids.

www.castrol.com
Castrol USA
Castrol is a major supplier of hydraulic fluids. Their website provides extensive details about the various fluids available from the company, as well as international information.

www.globalspec.com
GlobalSpec: The Engineering Search Engine
Provides details about companies that manufacture hydraulic fluids, allowing a comparison of various products. Links allow easy comparison of characteristics of various fluid types and brand names.

http://news.thomasnet.com
ThomasNet News
In the "Search News" box, enter "hydraulic fluids". This returns several articles from a variety of hydraulic fluid producers.

http://www.ars.usda.gov/News/News.htm
United States Department of Agriculture— News & Events
From the News & Events page, search "biodegradable hydraulic fluid". This search returns several articles on the development of commercial-grade, biodegradable hydraulic fluid for use in heavy equipment.

http://en.wikipedia.org/wiki/hydraulic_fluid
Wikipedia: The Free Encyclopedia—Hydraulic Fluid
Provides a basic review of the purpose of hydraulic fluids, and the various types and trade names used in industry. Several additional sites are suggested, as well.

Functions of a Hydraulic Fluid

The primary function of the liquid in a hydraulic system is to transmit the energy to do the work the system is designed to complete. The *fluid* is the medium by which the energy is transmitted from the pump to the cylinders and motors that do the actual work. To do the best possible job, the fluid needs to easily flow and quickly respond to pressure and directional variations. The fluid in a system should be considered just as important to the system as any of the hardware components, **Figure 6-1.** In addition to transmitting energy, hydraulic fluids act in a number of other roles to assure the effective operation of a hydraulic system:

- Lubricating component parts.
- Helping maintain proper system temperature.
- Sealing clearances between internal parts.
- Helping prevent rust and corrosion of parts.
- Cleaning the internal parts of the system.

A key function is the lubrication of the internal parts of the various components. The amount of movement of internal parts differs between components. Some valves involve only slight movement for pressure or flow control, while pumps and actuators can involve both high-speed rotation and sliding movement. The hydraulic fluid must provide lubrication to reduce friction and minimize wear of parts in each of these situations.

Assisting in the maintenance of proper system temperature is another basic function of the fluid. Heat generated by friction, pressure drops across valves, or external heat sources can raise the temperature of system components to a level where system efficiency is lost or damage occurs to parts. The fluid must be capable of readily absorbing the heat and transferring it to other sections of the system where it can be dispersed. Heat can be transferred from the fluid to the atmosphere or other medium, such as the water in a heat exchanger.

Sealing clearances between the internal parts of components is another role a hydraulic fluid must perform. Many valve designs depend on close manufacturing tolerances to control internal fluid leakage. An example is directional control valves, where fluid-film strength and small clearances, rather than separate seals, control leakage between the spools and the valve body. See **Figure 6-2.**

Another important function of a hydraulic fluid is to protect the internal surfaces of component parts from rust and corrosion. Rust and corrosion may occur when water enters the system through condensation or directly during machine cleaning.

Figure 6-1. All hydraulic systems use some type of fluid to transmit power within system circuits and perform a variety of additional functions. This fluid should be considered as important as any of the mechanical components of the system.

Close tolerances and the fluid form the seal

Figure 6-2. The hydraulic fluid acts as a seal when close machining tolerances, such as in this spool design, are used to control internal leakage.

Other contaminants can also enter systems during machine operation. Combinations of these materials often result in the formation of chemicals that can cause rapid deterioration of component parts. Fluids must be formulated to protect systems against these elements.

A hydraulic fluid must also function as a basic system cleaner. Fluid movement through a hydraulic system constantly flushes the system. Insoluble contaminants that enter the system, materials that are present from system assembly, or other materials that have formed because of chemical changes slowly move through a system. A quality fluid allows rapid settling and separation of these contaminants once they have reached the relative calm fluid found in the system reservoir.

Performance Characteristics of a Hydraulic Fluid

The numerous functions required of a hydraulic fluid mean it must be carefully selected if it is to perform satisfactorily in a system. A variety of characteristics must be considered when choosing a fluid for a particular system, including:

■ Lubricating power.

■ Resistance to flow.

■ Viscosity stability.

■ Low-temperature-operating ability.

■ Resistance to oxidation.

■ Reaction to condensation and other water.

■ Resistance to foaming.

■ Resistance to fire.

These factors are discussed in the following sections.

Lubricating Power

Lubrication deals with the reduction of friction between the surfaces of two bodies. *Friction* is the resistance to movement between two surfaces in contact. It is evident when you attempt to slide one body of material that is resting directly on the surface of a second body. Friction occurs in this situation because of the roughness of the surfaces in contact. This roughness exists even

in highly polished surfaces. The amount of friction varies, depending on the force pushing the surfaces together and the surface roughness. A smooth surface or light load produces less friction than a rough surface or heavy load.

The resistance to movement between two surfaces can be greatly reduced by placing a thin layer of liquid between them. This layer of liquid, called a *film*, separates the surfaces of the materials. The film is made up of molecules that are strongly attracted to each other, but can freely move past each other. The molecules in the film fill the irregularities in the surfaces of a bearing, for example, and form a layer to separate the parts of the bearing. See **Figure 6-3.** When the liquid film is in place, the two surfaces of a bearing are separated from each other. The molecules of the liquid slide over each other and allow the parts of the bearing to move with little or no actual contact between their surfaces. The resistance to movement of the bearing is then determined by the resistance of the molecular movement within the film, rather than the rough surfaces of the bearing material.

The ability of a liquid to form a strong film between two bearing surfaces and adhere to the surface of the material is known as *lubricity*. This characteristic is very important in a hydraulic

Figure 6-3. A fluid fills the irregularities on the surfaces of bearings and provides a film that separates the surfaces. The molecules of the fluid slide over each other providing smooth operation.

fluid, as the fluid must have the ability to minimize the wear of component parts under a variety of conditions. Wear in pumps, cylinders, motors, and control valves causes leakage, pressure loss, and reduced accuracy of control. Hydraulic fluids of all types must have the ability to adequately lubricate these basic system components in order to minimize friction and provide smooth, efficient component operation with a minimum of wear.

Resistance to Flow

The internal resistance to flow of the molecules of a liquid is known as *viscosity*. A fluid with high viscosity resists flow, while one with low viscosity flows more easily under the same conditions. Viscosity is one of the more important properties of the fluid used in hydraulic equipment, **Figure 6-4.** A fluid with the proper viscosity forms a strong film to provide an internal seal in components and reduce wear by protecting parts from direct contact. However, it must not be excessively viscous, requiring excessive energy output to move the parts against the resistance of the fluid.

Changing the temperature or pressure of a hydraulic fluid also changes its viscosity. Warming a fluid causes it to flow more easily, while cooling it causes the fluid to become more viscous. Increasing the operating pressure of a system causes the viscosity of the system fluid to increase. Care must be taken to select a fluid based on the expected normal operating temperature and pressure of the system, **Figure 6-5.**

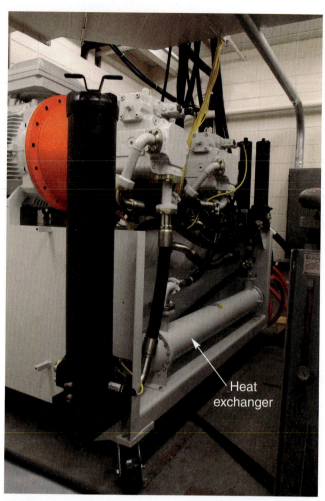

Figure 6-5. The viscosity of a fluid changes with its temperature. It is important to monitor the operating temperature of a system. If correct temperature levels cannot be maintained, a heat exchanger may be required in the system. (MTS Systems Corporation)

Figure 6-4. Using the proper viscosity fluid in a hydraulic system is important, as excessively low viscosity will not provide adequate lubrication and too high of a viscosity will require excessive energy output to operate. The oil on the right is more viscous than the oil on the left.

If a fluid is selected with a viscosity that is too low for system operating conditions, a number of problems can result:

- Increased wear.
- Increased internal leakage.
- Decreased pump efficiency and control accuracy.
- Increased system temperature.

There can be an increase in the overall wear in the system, as the fluid will not provide a sufficiently strong film to protect parts from friction. There can also be an increase in internal leakage in components because the fluid cannot provide an adequate seal of the clearances between close-fitting, moving parts. This leakage will cause less-efficient operation of pumps and motors, less-accurate control and slower response time for all actuators, and an increase in system operating temperature.

Selecting a fluid with a viscosity that is too high for system operating conditions may also cause a number of problems. A majority of these problems are caused by the higher internal friction of a high-viscosity fluid. Increased friction may result in increased power consumption, higher pressure drops across components, overall sluggish system operation, and an increase in system operating temperature. The higher viscosity fluid can contribute to pump cavitation problems because of increased resistance to flow through the inlet line of the pump. Separation of air and dirt particles from fluid returned to the reservoir can also be inhibited by an excessively high viscosity.

Viscosity Stability

It would be very desirable if the viscosity of a hydraulic fluid was fixed and did not vary during the changing conditions encountered throughout the operation of a fluid power system. However, fluid viscosity changes as system temperature and pressure vary. A measure of the relative change in viscosity that can be expected for a given change in temperature is provided by a viscosity index number. A fluid with a high index number has less viscosity change than a fluid with a low number. Viscosity index numbers are discussed later in this chapter.

The viscosity index of a hydraulic fluid is especially important when the operating temperature range is great. Special attention to viscosity index is wise when selecting fluid for mobile equipment operating with minimum system temperature control and under varying ambient temperatures. For example, equipment that operates in both a warm manufacturing area and a cold warehouse must have fluid with the proper viscosity index. The importance of the viscosity index becomes less in a system with carefully designed temperature controls and a relative constant ambient temperature, **Figure 6-6.**

Low-Temperature-Operating Ability

When hydraulic systems operate in a cold environment, the fluid must be able to flow under atmospheric pressure to the intake side of the pump. The characteristic that indicates the ability of a liquid to flow under these conditions is called

Figure 6-6. A fluid that maintains a uniform viscosity is especially important in hydraulic systems that experience a wide range of ambient temperatures. Viscosity stability is indicated by a viscosity index number, with a higher number indicating greater stability. (Atlas Copco)

pour point. This fluid property must be carefully considered when extremely cold system-start-up temperatures are a possibility. It is an especially important consideration for equipment stored outside or in unheated facilities in cold climates. Generally, the pour point of the fluid used in a system should be 20° Fahrenheit below the lowest-expected ambient temperature.

Resistance to Oxidation

Oxidation is a chemical reaction that increases the oxygen content of a compound. Many hydraulic fluids are susceptible to oxidation. System operating conditions that promote oxidation and can lead to general deterioration of hydraulic fluids include:

- High temperatures.
- Entrainment of air in the fluid.
- Reactions caused by contact with different metals in the system.
- Contaminants in the system, such as water and dirt.

All of these conditions tend to reduce the chemical stability of system fluid and shorten its service life.

High operating temperatures in any area of a system increase the oxidation rate of the fluid in the system. The fluid temperature in the reservoir of an operating hydraulic system should ideally be between 110° and 140° Fahrenheit, **Figure 6-7.** In many systems, varying load conditions cause the temperature to sporadically exceed this level. In some systems developed for special applications, design limitations may require higher general operating temperatures. Higher temperatures may also occur in localized areas of a system where design features do not allow the fluid to freely circulate to dispense heat.

Hydraulic fluid normally contains small quantities of air that have been dissolved as the fluid passes through the system. The higher the system pressure, the greater the quantity of air that can be dissolved. When the fluid moves to a low-pressure portion of the system, the dissolved air forms bubbles that then move with the fluid through the system. Air bubbles can also be mixed into the fluid by improperly designed return lines in the reservoir. A reservoir with inadequate capacity or leaks in the pump intake

Figure 6-7. The oxidation rate of hydraulic fluids increases as the temperature of the fluid increases. Most systems recommend a reservoir fluid temperature range of 110°–140° Fahrenheit.

line or any other line or component that operates below atmospheric pressure will also increase oxidation. These *entrained* air bubbles increase the surface area of the contact between the fluid and the air and tend to increase the rate of fluid oxidation.

A number of metallic materials commonly used in the construction of hydraulic systems can act as **catalysts,** promoting the oxidation of the system fluid. Copper is generally recognized as the greatest promoter of chemical reaction in systems. Iron, aluminum, and some other materials act to a lesser degree as oxidation promoters. Care must be taken to minimize the amount of these materials in system construction.

Reaction to Condensation and Other Water

Hydraulic systems are usually vented to the atmosphere at the reservoir. When this is the case, the system fluid and atmosphere have direct contact. As a result, moisture in the air can condense and mix with the hydraulic fluid. Water can also enter a system during equipment clean-up or exposure to weather in exterior installations. Regardless of the type of fluid used in the system, this water can cause operational and service problems.

In a system using petroleum-based fluid, part of this moisture remains separate from the hydraulic fluid and can be drained from the bottom of the reservoir. However, the water and hydraulic fluid can form an *emulsion,* which is a mixture of fluids that usually cannot be mixed. This is difficult to remove and can promote corrosion, decrease heat dispersion, and generally shorten the service life of the fluid. In situations where exposure to water is possible, selecting a fluid that resists emulsification is important to assure a properly operating system.

In systems using water-based fluids, additional water modifies the fluid. The fluid characteristics can be changed to the point where system performance, component service life, and fluid service life are negatively affected. Viscosity and lubricating properties are primary among these changes, although many of the other basic properties are reduced as the water content increases in the fluid.

Resistance to Foaming

Fluid foaming and air entrainment can cause a variety of problems in a hydraulic system. Air can be drawn into a system through leaks in any line that operates below atmospheric pressure. Air can also enter at a pump inlet line that is partially above fluid level because of low system fluid. Return lines that end above fluid level because of low fluid levels can also cause air to be churned into the fluid.

In addition to increased oxidation, as discussed earlier, the foam that forms can, in extreme cases, actually overflow through the vent of the reservoir. The resulting fluid loss creates a messy work environment. Additionally, a serious system problem results when the foam is drawn into the pump and forced out into other system components. The air in the foam/fluid mixture is compressible and can cause slow, spongy, and erratic system operation. Foam combined with other system contaminants, such as water and dirt, can also form *sludge* that reduces the performance and service life of both the fluid and the system components.

Resistance to Fire

The danger of fire exists, to some extent, in any situation where hydraulic fluid power is used to operate equipment. These potential fire situations range from typical pieces of equipment on the floor of a manufacturing plant or assembly operation to coal mining equipment and aircraft installations. The level of fire resistance required of a fluid varies considerably by application. In many situations, petroleum-based fluids can provide the required safety level, while in other situations water-based or synthetic liquids must be used.

Three terms are used to indicate a fluid's susceptibility to heat and fire. The first of these is *flash point.* This is the lowest temperature at which a liquid gives off sufficient vapors to ignite when an open flame is applied under specific test conditions. The second term is *fire point.* This refers to the temperature at which a substance is vaporizing rapidly enough to ignite and then continuously burn once an open flame or spark has been applied. Flash and fire points are discussed in more detail later in the chapter. The third term is *spontaneous ignition.* This is the temperature at which a fluid will ignite and burn without the application of an external flame or spark. One or more of these terms is often used to designate the required fire resistance of a fluid for a particular operation.

Commonly Used Hydraulic Fluids

Many liquids could be used in hydraulic equipment to transmit power, thus performing the primary function of a hydraulic fluid. However, many liquids do not have essential properties that allow them to provide the other functions key to the effective operation of a typical hydraulic system. For example, water is readily available and inexpensive for many installations. However, water alone is a poor lubricant, promotes rust, freezes, and rapidly evaporates at temperatures within the operating range of many systems. Petroleum-based liquids are very suitable for hydraulic fluids as they typically possess many characteristics that are desirable for efficient system operation. However, under certain operating conditions, fire resistance greater than that of petroleum-based fluids is essential. Fire-resistant fluids may be water based or synthetic.

Petroleum-Based Fluids

The most common hydraulic fluid consists of a blend of petroleum oils and additives that produces a desired set of properties, **Figure 6-8.** This blend produces a combination of properties that allow the fluid to perform satisfactorily in an operating hydraulic system. A quality petroleum oil blended for use as a hydraulic fluid should:

■ Readily transmit pressure and energy.

■ Provide lubrication for the bearing surfaces in system component parts.

■ Seal close-fitting parts against leakage.

■ Act as a cooling agent for the system.

■ Remove dirt and wear particles from internal parts.

■ Protect component parts from rust and other corrosion.

In addition, petroleum-based fluids can also withstand the effects of oxidation, resist foaming, and readily separate from entrained air, dirt, and water particles once the fluid has been returned to the system reservoir.

Biodegradable Fluids

Biodegradable fluids have been developed to meet the need for environmentally safe fluids for use in hydraulic systems in mining applications, offshore platforms, national parks, wildlife refuges, ski resorts, and farming operations, **Figure 6-9.** Small spills of these fluids will not harm soil or waterways. The fluids are broken down by organisms that normally exist in nature.

Figure 6-9. Biodegradable fluids have been developed for use in situations where environmentally safe fluids are required. Parks, nature trails, and resorts are typical examples of these situations. (Grayling Recreation Authority, Hanson Hills Recreation Area)

Element	Function
Refined mineral oil	Primary carrier that provides the base of the fluid
Viscosity-index improver	Reduces viscosity changes due to temperature variation
Anti-wear additive	Reduces bearing surface wear under light to normal loads
Extreme-pressure additive	Protects bearing surfaces from severe-loading conditions
Corrosion inhibitor	Prevents corrosion of components from oxygen and water
Oxidation inhibitor	Prevents oxidation of hydraulic fluid
Pour-point improver	Lowers the pour point of the hydraulic fluid
Defoamer	Prevents formation of foam in the fluid
Demulsifier	Allows rapid separation of oil and water
Dispersant	Reduces deposit buildup by holding contributing elements in suspension

Figure 6-8. Elements of a typical petroleum-based hydraulic fluid.

However, larger spills should be treated the same as for a regular petroleum-based product. Used oil should also be disposed of according to federal, state, and local regulatory codes.

Biodegradable fluids are vegetable– or synthetic ester–based oils that provide good service in moderate to severe conditions. Many of the fluids have a tendency to darken during normal service. Therefore, viscosity change is a better indicator of oxidation changes than fluid color. Care needs to be taken not to contaminate these fluids with mineral-based oils as this will adversely affect the biodegradable characteristics of the products.

Soluble-Oil Emulsions

Some fluid power systems that involve huge component parts, such as major forging and extrusion presses, use water as the primary hydraulic fluid, **Figure 6-10.** However, water by itself cannot provide sufficient lubrication, sealing, and rust

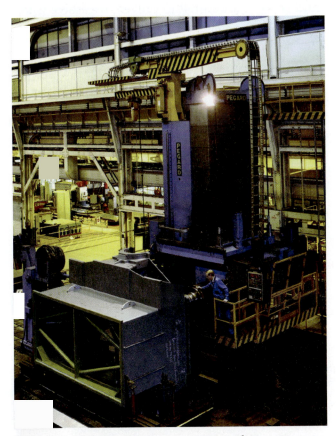

Figure 6-10. Large forming presses often use soluble-oil emulsions as a hydraulic fluid. These fluids are susceptible to both evaporation and freezing because of their high water content. (Brand X)

prevention characteristics to assure effective system operation and a reasonable service life for the components parts. The water is, therefore, mixed with 1% to 5% soluble oil to form an *oil-in-water emulsion.* This combination can function well with specialized pumps at pressures in the range of 2000 to 3000 psi. The high water content of this fluid requires that it not be subjected to freezing temperatures. Also, the system operating temperature should not exceed 140° Fahrenheit. Higher operating temperatures produce an excessive rate of evaporation, requiring considerable maintenance time.

High-Water-Content Fluids

High-water-content fluids are primarily water to which 2% to 5% soluble chemicals have been added. These chemical additives provide the lubrication, rust prevention, and wear resistance needed to protect the various parts of system components. Because of the high water content and the water-to-additive ratio, the fluids are common known as *HWCF* or *95/5 fluids.*

These fluids can function in specially designed reciprocating pumps at pressures as high as 10,000 psi. The limited anti-wear characteristics of most of these fluids restricts the operating pressure of systems using typical piston and vane pumps to 1000 psi. They are not recommended for use in systems that use gear pumps. The high water content of these fluids requires that freezing temperatures be avoided and the maximum system operating temperature be limited to 140° Fahrenheit to avoid excessive water loss by evaporation.

Fire-Resistant Fluids

Fluid power systems operate under diverse conditions with many situations involving high ambient temperatures and, in some cases, actual exposure to open flames. To reduce the possibility of fires in those applications, a fire-resistant fluid is typically used in the system. The fluids are not generally fireproof. However, they will not burn without sustained exposure to an ignition source.

Certain properties of fire-resistant fluids differ considerably from petroleum-based fluids. This results in the need to reduce system operational specifications, such as pressures and operating speeds. The expected service life of components is also reduced in many cases. Lower lubricity;

chemical reactions with seals, gaskets, and non-ferrous metals; and environmental toxicity are several of the problems that may be encountered when using fire-resistant fluids.

Fire-resistant fluids include water-oil emulsions, water-glycol fluids, and synthetic fluids, **Figure 6-11.** The following sections include information on fluid formulation, properties, and areas of application for these types of fluids.

Water-in-oil emulsions

Water-in-oil emulsions contain approximately 40% water in an oil base to provide the desired fire resistance. The oil surrounds the finely divided water droplets to form what is often called an *inverted emulsion* because the water is suspended in oil instead of the oil in water. These fluids should not be confused with the soluble oil and high-water-content fluids, which have a much higher water content and the oil is suspended in the water.

The initial viscosity of water-in-oil emulsions is usually higher than the viscosity recommendations for most fluid power systems. However, the viscosity of the fluid tends to decrease during system service to the point where it is approximately the same as a comparable petroleum product.

Pumps need to be specially selected for systems that use water-in-oil emulsions. Also, system operating temperatures should be held below 140° Fahrenheit and water evaporation must be carefully monitored. A laboratory test is required to determine the amount of water to add to the system in order to maintain correct fluid viscosity.

This test is essential as the viscosity of an inverted emulsion is higher than the viscosity of either of its primary fluids.

Water-glycol fluids

Water-glycol fluids are made up of 35% to 50% water with the remainder being some type of polyglycol. *Polyglycol* has a chemical structure similar to automotive antifreeze. The glycol and water mix to form a homogeneous, or uniform, fluid, rather than a suspension of one fluid within the other, as is the case with an emulsion. The viscosity of water-glycol fluids is comparable to petroleum-based fluids. In addition, they have better stability, are less expensive, and provide better fire protection than inverted emulsions.

The water-glycol fluids adversely affect some types of seal material and paint. Also, system pressure and load durations need to be reduced to ensure adequate service life of pumps using these fluids. The maximum recommended operating temperature for systems using this type fluid is 140° Fahrenheit to prevent excessive water evaporation.

Synthetic fluids

Synthetics have been used in a variety of fluid power applications, including engine oils and hydraulic fluids. The term *synthetic* has been defined in many ways. For purposes of this text, synthetic fluids are considered to be refined products containing additives that produce high levels of certain factors desired by end users. The

Fluid Type	Composition	Comments
Water Content	Oil-in-water (5% oil, 95% water)	• Extremely high fire resistance • Requires typical hydraulic fluid additives • Has strengths and weaknesses of plain water
	Water-oil (40% water, 60% oil)	• High fire resistance • Requires typical hydraulic fluid additives • Has inherent instability and maintenance needs
	Water-glycol (35-50% water, remainder glycol)	• Most popular of the fire resistant fluids • Can be used with most equipment designed for oil • Must be closely monitored for evaporation
Synthetic	Phosphate esters, chlorinated hydrocarbons, or various combinations	• Contain no water or petroleum oil • Some available brands are biodegradable • Cost is a selection factor

Figure 6-11. Characteristics of common fire-resistant hydraulic fluids.

synthetic fluids may contain highly refined petroleum as a carrier for the formulated elements.

Phosphate ester is the most common of the synthetic fluids manufactured for use in hydraulic systems, although other synthetic fire-resistant fluids are produced. These other fluids include chlorinated hydrocarbons, blends of various phosphate-ester-based fluids, and other materials such as mineral oil. All of the synthetic products have very high flash points and provide excellent fire resistance. They function well as hydraulic fluids and provide viscosities similar to petroleum-based products, satisfactory performance at high system pressures, and good component lubrication.

Synthetic fire-resistant fluids are very expensive in relation to other fluid types. Several other factors must also be considered before installing these fluids in a system. Special seal and gasket materials are required in all system components, as conventional sealing materials are not compatible with the chemical makeup of the synthetic fluids. The fluids also tend to dissolve most paints. Environmental factors must also be considered as the fluids are often toxic and may require special handling for disposal.

Hydraulic Fluid Additives

A number of *additives* are used to increase the chemical stability and overall performance of hydraulic system fluids. Typically, these materials are used to enhance most of the performance characteristics of the fluids. A wide variety of chemicals are used in these additives to improve the performance and extend the life of system components, as well as increase the service life of the fluid, **Figure 6-12.**

Extreme-pressure agents are used in fluids to help prevent metal-to-metal contact between the surfaces of the moving parts of components. These additives react with the metal surfaces to form a protective coating on the part. The agents are generally organic compounds containing one or more of elements such as sulfur or phosphorus.

Another group of additives that also prevent metal-to-metal wear are the *anti-wear agents.* These additives function at lower operating pressures and temperatures than the extreme-pressure agents. These materials are also typically formulated from organic compounds, including materials such as phosphates and ammonia-based salts.

Vane tip and
cam ring contact

Figure 6-12. Extreme-pressure and anti-wear agents help protect bearings and parts, such as pistons and vanes in pumps, from high loads and surface speeds that can cause rapid wear.

Viscosity-index-improver additives are blended into hydraulic fluids to reduce the rate of viscosity change caused by system temperature variations. These additives allow the fluid to have the flow characteristics of a low-viscosity fluid when cold and a higher-viscosity fluid at operating temperature. A number of organic chemicals are used to provide this quality.

Low-temperature performance of fluids is increased by the use of *pour-point depressants.* These additives allow a hydraulic fluid to flow more freely at low temperatures, permitting cold-weather operation of equipment. Most of these additives use chemicals that prevent the formation of crystals in fluids. The crystals reduce the fluid's ability to freely flow. The ability to flow is especially important in pump intake lines. If a fluid becomes so thick it cannot be drawn into the system, noisy system operation and pump damage will result.

Oxidation inhibitors are used to reduce the tendency of hydraulic fluids to oxidize. These materials are formulated to interfere with the complex chemical reactions that continually occur in an operating hydraulic system. The rate at which these reactions occur is influenced by system operating temperature, the amount of contact the fluid has with air, and the catalytic effects of the various metals used in the system. The service life of hydraulic fluids can be extended by controlling these factors, as well as selecting a fluid that contains appropriate oxidation inhibitors.

at 40° and 100° Celsius. The final viscosity rating is calculated by multiplying the flow time by a calibration constant established for each individual viscometer. The unit of measure used is the stoke (St) or centistoke (cSt). ASTM standard D-445 outlines the procedure commonly used in the United States for conducting a kinematic viscosity test.

ISO viscosity grades

The need for a wide range of liquid lubricant viscosities becomes more critical as new materials, designs, and applications develop for industrial equipment. Internationalization of trade also requires a system accepted by business and industry around the world. These criteria have been meet by ISO standard 3448, which includes twenty *viscosity grades*, **Figure 6-16.** This system has been gaining acceptance in world markets for lubricants, hydraulic fluids, electrical oils, and other applications. The standard defines each of the grades in kinematic viscosity at 40° Celsius. These ISO grades are defined in this country by ASTM standard D-2422. Each of the grades includes the prefix ISO VG followed by a nominal viscosity number. These grade numbers range from a low of 2 to a high of 3200. The range of viscosity within each grade can vary plus or minus 10%. For example, an ISO VG 10 fluid could range from 9 to 11 cSt. Hydraulic fluid viscosity typically ranges from 22 to 150 on this viscosity grading scale.

ISO Viscosity Grade	Nominal Viscosity at 40°C	Minimum Viscosity at 40°C	Maximum Viscosity at 40°C
ISO VG 2	2 cSt	1.8 cSt	2.2 cSt
ISO VG 3	3 cSt	2.7 cSt	3.3 cSt
ISO VG 5	5 cSt	4.5 cSt	5.5 cSt
ISO VG 7	7 cSt	6.3 cSt	7.7 cSt
ISO VG 10	10 cSt	9 cSt	11 cSt
ISO VG 15	15 cSt	13.5 cSt	16.5 cSt
ISO VG 22	22 cSt	19.8 cSt	24.2 cSt
ISO VG 32	32 cSt	28.8 cSt	35.2 cSt
ISO VG 46	46 cSt	41.4 cSt	50.6 cSt
ISO VG 68	68 cSt	61.2 cSt	74.8 cSt
ISO VG 100	100 cSt	90 cSt	110 cSt
ISO VG 150	150 cSt	135 cSt	165 cSt
ISO VG 220	220 cSt	198 cSt	242 cSt
ISO VG 320	320 cSt	288 cSt	352 cSt
ISO VG 460	460 cSt	414 cSt	506 cSt
ISO VG 680	680 cSt	612 cSt	748 cSt
ISO VG 1000	1000 cSt	900 cSt	1100 cSt
ISO VG 1500	1500 cSt	1350 cSt	1650 cSt
ISO VG 2200	2200 cSt	1980 cSt	2420 cSt
ISO VG 3200	3200 cSt	2880 cSt	3520 cSt

Typical hydraulic fluid (spans ISO VG 22 through ISO VG 150)

Figure 6-16. ISO viscosity grades indicating nominal, minimum, and maximum kinematic viscosity for each of the twenty classifications used in the system.

SAE viscosity grades

The Society of Automotive Engineers (SAE) grading system is extensively used in the automotive field to rate engine oils, gear lubricants, and axle lubricants. Through the years, the system has also been used to rate hydraulic fluids. Although the system is not used as much in the fluid power field as it once was, the ratings are still found on some fluids.

This system defines grades ranging from a low of 0 to a high of 60, **Figure 6-17.** Each of the grades designates a defined kinematic viscosity (in cSt) at 100° Celsius. The grades are defined by SAE standard J300.

The system also uses a W suffix. This indicates the ability of that fluid to perform at specified cold temperatures. These cold temperatures range from –5° to –35° Celsius.

Fluids graded by this system may be designated by two numbers, such as 5W-20. This grade indicates the material can perform as a 5W-grade fluid under cold conditions and a regular 20-grade fluid under normal working temperatures. Fluids with two grade numbers are called multiviscosity fluids.

Saybolt universal viscosity

The Saybolt universal viscosity system has seen widespread use in the fluid power field. The *Saybolt viscometer* is an apparatus containing a cup with a calibrated orifice in the bottom. See **Figure 6-18.** Ratings are established by heating the test fluid in the cup to a standard temperature of 100° or 212° Fahrenheit and then measuring the number of seconds needed to fill a 60 mL sample flask with fluid flowing through the calibrated orifice. The resulting time measurement, called Saybolt Seconds Universal (SSU), serves as a means to compare fluids, but does not relate to the standard definition of viscosity.

Until recently, the Saybolt test procedure was listed as ASTM standard D88. However, that standard is no longer being supported by ASTM. Refiners and manufacturers are being encouraged to use more accurate means of measuring viscosity, such as those covered in ASTM standards D-445 and D-446. These standards outline test procedures and equipment for determining kinematic viscosity. ASTM also provides standard D2161 that lists the formula and tables needed to convert SSU to cSt. Even though the Saybolt test standard is no longer supported by a standardizing group, the procedure continues to be used in the profession. Most hydraulic fluid specification sheets list SSU information. Some data sheets carry the listing as Saybolt Universal Seconds (SUS).

Viscosity Index

A viscosity index number appears as part of the specifications for most hydraulic fluids. The *viscosity index number* is widely accepted as an

SAE Viscosity Grade	Minimum Viscosity at 100°C
0W	3.8 cSt
5W	3.8 cSt
10W	4.1 cSt
15W	5.6 cSt
20W	5.6 cSt
25W	9.3 cSt
30	9.3 cSt
40	12.5 cSt
50	16.3 cSt
60	21.9 cSt

Figure 6-17. Ten SAE viscosity grades that may be used with hydraulic oils in some applications.

Temperature control bath

HEAT TIME TEMP

Figure 6-18. Saybolt universal viscosity is based on the time required for 60 mL of fluid to flow through a calibrated orifice at a specified temperature.

A
ity of
rous
wate
proc
hydr
in-se
new

T
heat
test
stirr
desi
beak
test
amo
fluic
for 2
end
and
fluic
repc
AS1
cedi

Co

of tl
oils
can
sys1
par
bee
anc
nes

effe
sist
qua
ten
is t
ph

mc
the
cor
sha

Figure 6-23. The wear characteristics of hydraulic fluids can be established by using a pump test circuit made up of standard fluid power components operating under carefully controlled temperature, load, and time conditions.

Handling and Maintaining Hydraulic Fluids

Hydraulic fluids of all types are expensive. They can also be costly to maintain, if care is not exercised in their handling and upkeep. Proper care not only reduces costs of new replacement fluids, but reduces the time maintenance personnel spends in flushing systems and changing fluids. In many cases, proper fluid care also extends the service life of the total fluid power system, further

reducing costs. Awareness of a number of simple rules and other factors can reduce the costs related to the fluid used in a system and help extend the service life of equipment.

Storage and Handling of Fluids

Storing new, unused hydraulic fluids is an important consideration in any organization that makes use of fluid power systems. Improperly stored fluid can become contaminated. Contaminated fluid not only results in the loss of the fluid, but the loss associated with increased maintenance time if contaminated fluid is installed in a system. Considerable time can be lost due to system flushing that may be required or the downtime of equipment idled waiting for the arrival of a new supply of fluid. In general:

- Store drums in a cool, clean, dry place.
- Place drums on their sides to reduce chances of contamination.
- Carefully clean drum tops before removing bungs.
- Use clean fluid-transfer equipment.

Drums used for fluid storage should be stored indoors in a cool, clean, dry place. Storing the drums on their sides reduces the accumulation of dirt around the bung, **Figure 6-24.** It also eliminates any seepage of water or other liquids into

Figure 6-24. Fluid storage drums should be kept in a cool, clean, dry place. Tops and bungs should be carefully cleaned before opening the drums.

the barrels from spills in the storage area. In all situations, to reduce contamination, the barrel tops and bungs should be carefully cleaned before the bung is opened. General rules of cleanliness should be followed when transferring fluid from the barrel to the system reservoir. The use of clean containers, transfer pumps, and hoses is a must to assure no contamination occurs while changing or replenishing fluid.

System Cleanliness

The cleanliness of the fluid in a fluid power system involves factors related to:

- System design.
- System construction.
- Housekeeping practices of a facility.

Contaminants can enter a system through many routes. A poorly designed or constructed system can allow dirt to enter even when reasonable housekeeping is performed by maintenance personnel. A new fluid power system that has not been thoroughly cleaned after construction can also contain many contaminants. Cleaning solvents, core sand, metal chips, paint flakes, and rust scale are just a few of the foreign materials that can appear in the fluid of a newly constructed system if not properly cleaned.

Foreign materials may enter a fluid power system through a number of locations. Attention to construction details can reduce dirt entry at most of these locations. The reservoir is one of the major components in a system where special attention must be directed if fluids are expected to have a reasonable service life. For example, the cover of the reservoir should fit well with a gasket sealing the space between the cover and the tank. Gaskets are also needed to seal the space around the pump inlet, system return, and drain lines where they pass through the cover.

The reservoir filler hole should be constructed with both a screen and a tight-fitting cover. The reservoir also needs to be fitted with a breather including an air filter that can be easily cleaned. See **Figure 6-25.**

Cylinder piston rods are also an entry point for contaminants that can shorten the service life of hydraulic fluids and system components. Contaminants can enter the system during cylinder retraction. These contaminants vary with the type of installation, but can include dust, soil, water, cutting oil, metal chips, and a variety of other materials. Equipment should be designed to protect piston rods from excessive contact with these materials. In situations where cylinders are constantly exposed to quantities of foreign materials, *piston rod wiper rings* can be used to prevent entry of the materials. These devices wipe contaminants from the rod as it is retracted. Boots that cover the rods can also be used. See **Figure 6-26.**

The hydraulic systems of equipment should be regularly inspected to ascertain the condition of system fluid, as well as the overall condition of mechanical components. The frequency of these inspections depends on the type of equipment and the intensity of operation. A sample of fluid should be taken from the upper portion of the fluid in the reservoir after the system has reached operating temperature. The appearance of the sample should be visually checked with that of an

Figure 6-25. Many contaminants that are troublesome in a hydraulic system enter through the reservoir. Be certain that all reservoir fittings and openings are sealed or filtered to prevent the entrance of water, dirt, cutting oils, and other foreign materials.

Figure 6-26. Dirt and fluids adhering to the rod of a cylinder can enter a hydraulic system when the cylinder is retracted. Piston rod wiper rings or boots (shown here) can reduce this contamination. (A & A Manufacturing Co., Inc., Grotite)

unused sample of the fluid. If, by comparison, the used fluid is cloudy, muddy, or an inappropriate color or contains suspended particles, a closer examination of the fluid is justified. If the reason for the contamination cannot be identified by a closer visual examination of the fluid and system, an analysis by a qualified laboratory is required and legitimately justified.

Fluid Operating Temperature

Operating temperatures have considerable influence on the service life of hydraulic fluids and often reflect the efficiency of systems. The heat in the fluid of an operating fluid power system comes from the energy lost by forcing oil to flow through the system. Heat is generated when energy is used to move the fluid against a resistance within the system without completing work outside of the system. If a system is not properly designed, this heat may increase fluid temperatures above recommended levels. Higher-than-recommended temperatures cause fluid oxidation and general fluid deterioration. Another factor directly related to temperature is evaporation. This must be considered when fluids containing water are used in a system.

Most system designers indicate the ideal fluid temperature in the reservoir is between 110° and 140° Fahrenheit. This operating temperature may be slightly higher if a quality fluid is used that contains additional oxidation inhibitor additives. However, reservoir temperature does not tell the whole story regarding fluid temperatures in a system. Hot spots in system control or actuator components can raise the temperature of system fluid over the recommended level in localized areas. Fluid temperatures in these areas may be at a level that causes rapid fluid deterioration.

If the system reservoir fluid temperature is above the recommended level, several factors need to be examined:

- Ambient temperature in the area where the system is operating.
- Clearance for air circulation around the reservoir.
- General cleanliness of the unit.

The system may need to be shielded from external heat sources. It is also possible that the reservoir needs to be moved away from equipment to allow unrestricted air contact with all surfaces of the tank. Any accumulation of dust and dirt should be removed from the exterior surfaces of the unit. If these factors do not appear to be the source of the excessive temperature, several reservoir design elements should be examined:

- Reservoir fluid capacity.
- Relationship of the pump inlet line to the system return line.
- Physical shape of the reservoir.

A reservoir that is too small recirculates the fluid into the system too quickly. This does not allow adequate heat dispersal through the reservoir walls. As a result, the desired fluid temperature cannot be maintained. A general rule for reservoir sizing is a fluid capacity of two to three times the gallons per minute flow rate of the system pump.

The pump inlet and system return lines should be positioned as far apart as possible or separated by a baffle. This design element forces fluid returning from the operating system to traverse through the reservoir. Doing so facilitates cooling and the separation of dirt and other foreign materials.

The physical shape of the reservoir needs to be considered. See **Figure 6-27.** Much of the heat transfer that keeps system fluid within the proper operating temperature range occurs through the reservoir sidewalls. A shallow, broad-based tank provides a proportionally smaller sidewall area than a deeper, narrow-based unit. However, a tank that is too narrow and deep restricts the separation of entrained air and encourages fluid foaming.

Additional factors that can contribute to high fluid temperatures relate to improper system circuit design or operation. A larger-than-necessary pump, higher-than-required relief valve setting, smaller-than-necessary lines or components, and slower-than-necessary sequencing of the circuit operation are all factors that produce unnecessary heat in a system. A careful analysis of these factors plus an awareness of ambient temperature conditions and system reservoir design should lead to identification of most excessive fluid temperature problems, **Figure 6-28.**

Oil level/temperature indicator with glass tube and protective aluminum housing

Combination filler/air breather (10 Micron)

Inlet flange sized for easy strainer removal during routine cleaning

TEFC or ODP C-face electric motors

Dependable continental pressure-compensated, variable-volumn vane or axial piston pump with startup information

Coolant-resistant paint

Return couplings with reservoir downpipes cut at 45° angles

Removable top plate (3/8″ HRPO 1020 steel)

Channel gasket keeps out airborne particles; reduces vibration and noise

Baffle sized for 2 ft/sec fluid velocity or less

Single bolt cleanout covers with drains

Magnetic drain plugs

Tanks feature steel and rounded-corner, two-seam construction

Sloping botton to magnetic drain plugs

6″ High legs provide excellent air circulation for maximum heat dissipation

Strainer sized to pass two to three times the pump's maximum flow at minimum pressure drop

Figure 6-27. The reservoir is a very important component in relation to system temperature control. Fluid capacity should be two to three times the pump output (in gallons per minute) and have adequate sidewall surface area for cooling. (Continental Hydraulics)

Figure 6-28. Excessive heat may be generated if component selection and pressure settings are not appropriate for a system. For example, a pump with a capacity larger than required or a relief valve setting that is too high can cause system overheating.

Maintaining Fluid Cleanliness and Operating Temperature

During system operation, fluids will become contaminated to some degree regardless of the efforts to keep them clean. It is standard practice to include filtration devices in most systems to promote the cleanliness of both the internal parts of system components and the fluid used in the system. Specialized heat-exchange devices are also used in situations where fluid temperatures cannot be held within the recommended operating range. The design, construction, and operation of both of these types of devices and their related circuits are discussed in Chapter 12.

Chapter Test

Answer the following questions. Write your answers on a separate sheet of paper.

1. The primary function of the liquid in a hydraulic system is to:
   A. Clean system parts.
   B. Remove heat from the system.
   C. Transmit energy to do the work the system is designed to complete.
   D. Prevent rust and corrosion of system components.

2. List the two primary factors that determine the amount of friction between two surfaces.

3. Define *lubricity.*

4. Name three major problems that may appear if a fluid with too low of a viscosity is used in a hydraulic system.

5. The amount that the viscosity of a hydraulic fluid will change as the temperature of the fluid changes is indicated by a(n) _____ number.

6. Name three conditions in a fluid power system that promote oxidation and general deterioration of a hydraulic fluid.

7. Why do air bubbles entrained in a hydraulic fluid increase fluid oxidation?

8. Name three common ways that water can enter a fluid power system and contaminate the system fluid.

9. The lowest temperature at which a liquid gives off sufficient vapors to ignite when a flame is applied is called the _____.

10. List the characteristics of water that make it a poor hydraulic fluid should it be used alone in a fluid power system.

11. Vegetable-based oils are classified as _____ fluids because they can be broken down by organisms that normally exist in nature.

12. Hydraulic fluids that are formulated by adding 2% to 5% soluble chemicals to water are called _____ or _____ fluids.

13. The water-in-oil type of fire-resistant fluids is often called an inverted emulsion because the _____ is suspended in the _____.

14. Typically, what is the initial viscosity of a water-in-oil emulsion compared to that recommended by system manufacturers?

15. The most common of the synthetic fire-resistant fluids is _____.

16. The hydraulic fluid additives that are used to prevent metal-to-metal contact in system components are called anti-wear or _____ agents.

17. The hydraulic fluid additive that reduces the rate of change of fluid viscosity is known as a(n) _____ additive.

18. Improving surface tension characteristics increases the rate at which water separates from hydraulic oils and collects in the _____.

19. Rust and corrosion inhibitors in hydraulic oils protect metal system parts by neutralizing _____ or by developing a(n) _____ on the parts.

20. Define *absolute viscosity*.

21. The most precise viscosity rating that is usually found in specification sheets for hydraulic fluids is _____ viscosity.

22. The kinematic viscosity variation of a fluid between 40° and 100° Celsius is indicated by the _____.

23. Corrosion tests used with petroleum-based hydraulic fluids determine the effect on metal parts of components from _____ compounds that have remained in the fluid after refining.

24. Name two general methods that are used to establish the lubrication and wear-prevention qualities of hydraulic fluids.

25. List four general rules that need to be followed when storing and handling replacement hydraulic fluids.

26. When cylinders are exposed to extreme conditions, contamination of system fluid can be reduced by using _____ or _____ to prevent the entry of dirt from the surfaces of the piston rods.

27. Fluid samples taken from hydraulic systems for routine testing purposes should be collected only from the _____ of the fluid in the reservoir and only when the system is at _____ temperature.

28. _____ in the fluid of a hydraulic system is caused by forcing fluid through the system without completing work outside of the system.

29. Name three factors external to the system that can cause a hydraulic system to operate at a higher-than-desired temperature.

30. The cleanliness of the fluid in most hydraulic systems is assured by the addition of specialized _____ devices.

This is a vane-type hydraulic motor that develops high torque. Hydraulic fluid acts on the vanes, which causes the output shaft to turn. There are many important factors related to hydraulic fluid that must be considered when designing or servicing a hydraulic system. (Rineer Hydraulics, Inc.)

The power unit is the source of power for the remainder of the hydraulic system. The devices that comprise this unit are fundamental to the operation of the system. These include the prime mover that supplies the energy to operate the pump, which creates fluid flow in the system. In addition, the unit usually includes components to store system fluid, control maximum system pressure, and clean the fluid. This chapter introduces each of these components and then discusses pump designs in detail.

Objectives

After completing this chapter, you will be able to:

- Describe the function of a hydraulic power unit and identify its primary components.
- Explain the purpose of a pump in a hydraulic system.
- Explain the operation of a basic hydraulic pump.
- Compare the operating characteristics of positive-displacement and non-positive-displacement hydraulic pumps.
- Compare the operating characteristics of rotary and reciprocating hydraulic pumps.
- Compare the operating characteristics of constant- and variable-volume hydraulic pumps.
- Explain the principles involved in the operation of a pressure-compensated hydraulic pump.
- Describe general construction for each of the various hydraulic pump designs.
- Explain cavitation and its effect on pump performance and service life.
- Interpret performance data supplied by a pump manufacturer.

Chapter 7

Source of Hydraulic Power
Power Units and Pumps

Selected Key Terms

The following terms will be used in this chapter. As you read the text, record the meaning and importance of each. Additionally, you may use other sources, such as manufacturer literature, an encyclopedia, or the Internet, to obtain more information.

application information
axial-piston pump
balanced-vane pump
cavitation
dual pump
fixed-delivery pumps
gear pumps
general hydraulic horsepower
gerotor design
impeller
mechanical efficiency
non-positive-displacement pumps
pintle
power unit
prime mover horsepower
radial-piston pumps
reciprocating pumps
screw pump
spur gear
stationary-cylinder design
swash plate
valve plate
vapor pressure
variable-delivery pumps
volumetric efficiency
wear plate

Internet Resources

www.answers.com/topic/cavitation
Answers.com
Considers cavitation as it occurs in a number of situations, ranging from industrial hydraulics to applications in the health care field.

www.continentalhydraulics.com/
pvx_vane_pumps.aspx
Continental Hydraulics
Page for a series of vane pumps. Provides typical information available from manufacturers. Links are provided to product documentation, including specifications, control options, and accessories.

www.engineersedge.com/pumps/
screw_type_pump.htm

Engineers Edge
Provides an explanation of the design and operation of two- and three-screw pump designs.

http://hydraulicspneumatics.com/
fluid-power-basics/pumps
Hydraulics & Pneumatics Magazine
Series of three articles (Part 1, Part 2, and Part 3) on various hydraulic pump designs with illustrations and explanations of operation.

www.en.wikipedia.org/wiki/gear_pump
Wikipedia: The Free Encyclopedia—Gear Pump
Provides detailed explanation of the design and operation of gear pumps. Additional sites are listed that provide considerable detail about various gear pump designs.

Power Unit

The design and appearance of the power unit of a hydraulic system varies considerably, depending on the equipment in which the system is operating, **Figure 7-1.** However, even though the appearance is different, the *power unit* must provide the energy for operation of the system, move the fluid through the system, and provide a safe maximum limit on system pressure while assisting in maintaining proper system operating temperature and fluid cleanliness.

The hydraulic power unit is often supplied by manufacturers as a package. The package includes all of the desired components and required plumbing. Also, system controls are often included. Standard units are available or components can be ordered to meet specific requirements. Each of the components of the power unit are discussed in detail in various sections of this book.

Prime Mover

In order to serve as the overall source of energy for a hydraulic system, the power unit must contain a prime mover. The *prime mover* converts energy into mechanical movement. This mechanical movement then acts as the carrier of the energy through the system to the point where it is needed. In most systems today, the prime mover is an electric motor or internal combustion engine. See **Figure 7-2.** These sources may be individual units dedicated to operating the system or they can be power take-off arrangements where the motor or

Figure 7-1. The source of power for the hydraulic system is the power unit, which is made up of a pump driven by a prime mover and includes a reservoir, filters, driveshaft coupler, and basic pressure control valves. (Kim Hotstart Manufacturing Company)

engine operates several different units. An example of a power take-off setup is found in mobile farm or earth-moving equipment. The engine supplies power to move the equipment from one location to another, but also operates auxiliary systems, such as leveling devices, digging buckets, plows, or other attachments.

Coupler

In most hydraulic power units, the prime mover is mechanically coupled directly to the input shaft of the system pump. See **Figure 7-3.** One of several different designs of *coupler* can be used to maintain proper alignment between the two units. Care must be exercised to assure the coupler is adequate in relation to rated speed, load, and allowable misalignment. Belt, chain, and gear drives are also used to connect some prime movers and pumps. These types of connections are generally less preferred than direct drives because they are less efficient, have a lower safety factor, and place higher stress on the equipment. However, these devices may be required in situations where changing to a higher or lower speed is required.

Pump

The *pump* is the device that causes the fluid to move through the system, carrying the energy to the point where it is needed. A wide variety of pump designs are available for use on power units. The design selected for use depends on the required system flow rate, system operating pressure, and a variety of other requirements.

A

B

Figure 7-2. The prime mover for hydraulic power units are typically electric motors in manufacturing installations (A) and internal combustion engines in mobile equipment (B). (Continental Hydraulics; Deere & Co.)

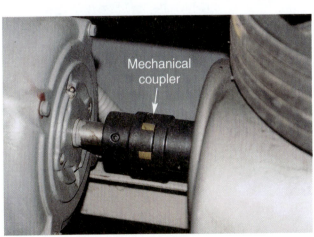

Figure 7-3. The coupler transfers power between the prime mover and the hydraulic system pump. Careful selection is important to provide quiet and trouble-free system operation.

Reservoir

Another major component of power unit is the *reservoir,* which serves multiple purposes in a power unit. First, it is a tank that holds system fluid. The pump draws fluid from the tank and circulates it through the various components to operate the system. Fluid is then returned to the reservoir to await continued system use. Second, while the fluid is being held in the reservoir, it cools while dirt and other unwanted materials settle out. Third, the top of the reservoir often serves as a mounting location for the prime mover and the pump.

Pressure Control Valve

The power unit normally includes a *pressure control valve* to establish a maximum operating pressure for the system. See **Figure 7-4.** The valve

may function as the *relief valve,* establishing the operating pressure for the system, or it may be set at a slightly higher pressure to act as a *safety valve* if the normal system relief valve fails. In both situations, the valve serves to protect system components from damage and the prime mover from being overloaded by high pressures.

Filters and Strainers

Fluid filters are often included as a part of the power unit of a hydraulic system. The importance of keeping system fluid free of contamination cannot be overemphasized. A *strainer* is often attached to the end of the intake line that is submerged in the reservoir fluid. This device is constructed of fine-wire screen and designed to remove larger dirt particles. Additional filtering devices are used in the pump intake or system return line to remove fine particles that have entered the fluid, **Figure 7-5.**

Basic Pump

A wide variety of pump designs are used to convert the mechanical power output of the prime mover into fluid flow. The appearance and structure of these pumps may be different, but the operating principles that cause the fluid to move

Figure 7-4. A basic pressure control valve in the form of a relief valve or a safety valve is a part of most power unit packages. These valves are included to establish the maximum operating pressure of a system and/or protect from system overload.

Pressure control valve

Figure 7-5. Protecting the system from dirt and other contaminant in the hydraulic fluid begins in the power unit. A strainer is often used on the intake end of the pump inlet to remove larger particles.

through the unit are very similar. The basic operating principles require that the pump have areas within its structure where less-than-atmospheric pressure, as well as above-atmospheric pressure, can be produced and maintained.

Structure of a Basic Pump

A hand-operated piston pump can be used to illustrate the basic concepts involved in pump operation. This pump is very similar to a simple tire pump used to inflate bicycle tires. The pump consists of a pumping chamber made up of a cylinder, closed at one end, in which a movable piston operates. Two one-way valves are located in the closed end of the chamber. One of these serves as the pump inlet valve and is attached to a line connected to a fluid reservoir. The second valve serves as the outlet valve and is connected to the hydraulic system. A handle is attached to the piston to allow an operator to easily slide the piston back and forth in the chamber, **Figure 7-6.**

Operation of a Basic Pump

The operation of a pump involves intake and pressurization phases. The discharge pressure is based on several factors. The next sections discuss these phases and the factors related to discharge pressure.

Intake phase

The inlet phase of basic pump operation is illustrated in **Figure 7-7.** As the piston is pulled toward the open end of the pumping chamber, the volume increases between the piston and the end of the chamber containing the valves. As the volume increases, pressure in the closed chamber drops below atmospheric pressure. This reduced pressure acts on the outlet valve, allowing it to be closed by atmospheric pressure if the pump is just beginning operation or by higher system pressure if the pump is attached to an operating system. The reduced pressure in the chamber also allows the inlet valve to be forced open by system fluid moved through the pump inlet line by atmospheric pressure pushing against the surface of the fluid in the reservoir. Fluid flow caused by this pressure difference fills the pump chamber.

A

B

Figure 7-6. The basic hydraulic pump can be compared to a simple, hand- or foot-operated pump. A piston moves within a cylinder to provide a variable-volume chamber. An inlet valve connects the chamber to a reservoir and an outlet valve connects it to the system. A—Hand-operated pumps. B—Foot- and hand-operated pumps. (Star Hydraulics, Inc.)

Inlet valve
open

Fluid
flow in
from the
reservoir

Outlet valve
closed

Figure 7-7. During the basic pump intake phase, the piston moves to enlarge the pumping chamber volume. This allows fluid to move from the reservoir, through the inlet, to the chamber.

Inlet valve
closed

Fluid
flow out
to the
system

Outlet valve
open

Figure 7-8. During the basic pump pressurization phase, the piston moves to decrease the volume of the pumping chamber, forcing the fluid to move through the discharge valve to the system.

Pressurization phase

The pressurization phase of basic pump operation is illustrated in **Figure 7-8.** When the piston is pushed back toward the closed end of the pumping chamber, the volume decreases between the piston and the chamber end. Force applied to the piston produces a higher-than-atmospheric pressure in the fluid trapped in the pump chamber. The pressure causes the inlet valve to close, blocking the return of fluid to the reservoir, and the outlet valve to open, allowing fluid to flow through it into the system.

Discharge characteristics

The maximum pressure developed in the system is determined by the resistance to the flow of fluid in the system and the force that can be applied to the piston. The output flow rate of the pump is determined by the size of the pumping chamber and the speed at which the cycle is repeated (the speed of the prime mover).

Basic Pump Classifications

All pumps used in hydraulic systems can be classified using three methods. The various categories involve the design, structure, and operation of pumps relating to:

- Displacement.
- Pumping motion.
- Fluid delivery characteristics.

Displacement

The displacement classification relates to the output of a pump per revolution under varying system operating conditions. The factor is dependent on the design characteristics of the pumping chamber of the pump and its ability to deliver fluid. Pumps are classified as either positive or non-positive displacement.

Positive-displacement pumps produce a constant output for each revolution of operation over a wide range of operating pressures. These pumps have very close tolerances between moving and stationary parts. The operating efficiency of these pumps causes some variation in output as system pressure changes. However, output remains relatively constant. Most hydraulic system pumps are positive displacement, **Figure 7-9.**

Non-positive-displacement pumps have larger clearances between moving and stationary parts. Because of this design factor, considerable slippage can occur in the pumping chamber, resulting in pump flow variations as system operating pressures change. Pumps classified in this category are usually used in low-pressure applications, such as fluid transfer, rather than for actuator operation, **Figure 7-10.**

Fundamental Pumping Motion

The pumping motion relates to the basic motion used to change the volume of the pumping chamber, which acts to move fluid through the pump.

Figure 7-9. The component parts of the pumping chamber of positive displacement pumps are machined with close tolerances, allowing uniform flow output per pump revolution.

Figure 7-11. The pumping action of a rotary pump is based on the continuous rotating movement of pumping-chamber components, such as these gears.

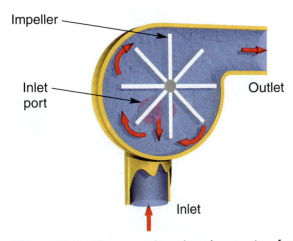

Impeller

Inlet port

Outlet

Inlet

Figure 7-10. The pumping chamber parts of non-positive-displacement have considerable clearance, allowing variation in pump flow output as system load conditions change.

The volume of the chamber increases during the intake phase of operation and decreases during the pressurization phase. Pumps are classified as either rotary or reciprocating.

There are several different rotary pump designs. *Rotary pumps* use rotating parts in the pumping chamber to produce the increase and decrease in volume needed to move fluid through the pump, **Figure 7-11.** Rotary motion may also be used to drive component parts that develop centrifugal force. This is another method that can be used to move fluid through a pump.

Reciprocating pumps use the up-and-down linear movement of a piston in a cylinder or similar structure to move fluid through the pump. As the piston moves downward, the volume of the chamber above the piston enlarges. This creates the low-pressure area needed to bring fluid into the pump from the reservoir. The upward movement of the piston reduces the volume of the chamber. This pressurizes the fluid and moves it out of the pump under pressure. A number of hydraulic pump manufacturers use variations of this concept to produce a wide variety of pump designs. See **Figure 7-12.**

Fluid Delivery Characteristics

Fluid delivery characteristics are associated with the fluid output of a pump. This is determined by the size of the pumping chamber of the unit. In some pumps, the volume of the chamber is fixed, **Figure 7-13A.** The fluid output of these *fixed-delivery pumps* is constant for each pump revolution. In other designs, the volume of the pumping chamber can be varied, **Figure 7-13B.** *Variable-delivery pumps* can produce fluid outputs ranging from zero, or no flow per revolution, to the maximum allowed by the size of the unit. Variable-delivery pumps include a mechanism that allows an operator to vary the internal pump geometry, thus changing the size of the pumping chamber.

Figure 7-41. The rotating-cylinder version of the radial-piston pump can be fitted with a mechanism to adjust the reaction ring to produce a variable-displacement pump. When the reaction ring (rotor) and cylinder block are concentric, no flow is produced. The greater the eccentricity, the greater he piston stroke and the higher the flow. (The Oilgear Company)

block and the input shaft are concentric, a no-stroke condition exists. This produces no pump flow. Tilting the cylinder block to the greatest angle allowed between the barrel and shaft results in the longest piston stroke. This creates the highest flow output.

The rotating-cylinder, radial-piston pump may also be a variable delivery. The feature is incorporated into this pump design with the use of a movable reaction ring, **Figure 7-41.** When the ring and fixed pintle are concentric, the cylinder block can rotate without the pistons moving. Moving the ring off center of the pintle results in piston movement and fluid flow from the pump. Like the other variable-displacement pumps, the flow output can vary between zero at the no-stroke position to

maximum flow when the reaction ring is offset to produce the longest piston stroke.

Pressure Compensation

The pressure-compensation feature of a pump allows the unit to sense system pressure and vary the volume of the pumping chamber. This allows the pump to provide only sufficient fluid to maintain a desired system operating pressure. When system actuator resistance is low and flow demand high, the volume of the pumping chamber is the largest and the pump produces maximum flow. As the load on the actuator increases, the higher resistance to fluid flow causes system pressure to climb. In a pressure-compensated pump, this increased pressure causes the geometry of the pumping chamber to change. As a result, only the amount of flow needed to maintain a preset system pressure is produced. When the load on the system stalls the actuator, the volume of the pumping element approaches zero with only sufficient flow produced to compensate for system leakage. See **Figure 7-42.** Pressure compensation is available only in unbalanced-vane and piston pumps. This is because a pump must be capable of variable-flow output to allow the compensating mechanism to operate.

In a basic, unbalanced-vane pump, the compensating mechanism involves a spring that forces the pumping chamber ring to the maximum offset. See **Figure 7-43.** The pump produces maximum flow at this setting. As resistance to fluid flow begins to increase, reflecting an increased system load, pressure on the pumping chamber ring gradually shifts the ring until it is centered on the rotor and vanes. When the ring is centered, flow output of the pump is near zero with system pressure maintained at the preset level.

In an axial-piston pump design, the compensating mechanism senses system pressure and alters the angle of an internal mechanism, **Figure 7-44.** In an inline pump, the angle of the swash plate is changed. In a bent-axis pump, the angle between the input shaft and the cylinder bore barrel is changed. These changes vary the stroke of the pump, which changes the pump displacement. Compensation is provided in the rotating-cylinder radial piston

Figure 7-42. The output flow of a pressure-compensated pump depends on the pressure in the system. When system pressure is below the compensator setting, the pump delivers maximum flow. As the pressure approaches the compensator setting, the pump output drops to only the flow needed to maintain system pressure.

Figure 7-44. Pressure compensation is achieved in variable-displacement piston pumps by automatically setting the stroke length of the pump. The compensator senses system pressure and mechanically changes the angle of the swash plate or cylinder barrel, or the eccentricity of the reaction ring, to produce the volume of fluid required to maintain pressure. (Continental Hydraulics)

Figure 7-43. The compensator in a pressure-compensated vane pump senses system pressure and automatically positions the cam ring to produce the fluid output required to maintain a desired pressure level. (Continental Hydraulics)

pump by using the compensating mechanism to move the reaction ring. Maximum flow output is produced when the ring is at the greatest offset. No flow is produced when the center points of the cylinder block and ring are concentric.

A major advantage of the pressure-compensated feature is the savings that result from reduced energy consumption. Basically, the energy consumed by the hydraulic system prime mover is determined by system load. System load is established by pressure and flow demands. The pressure-compensation feature reduces the amount of time that the prime mover must spend producing maximum flow against maximum system pressure.

As soon as the system actuator encounters a load condition that stops its movement, the compensator reduces flow output, which reduces energy consumption and operating cost. See **Figure 7-45.**

The compensating feature of these pumps serves to control maximum pressure in a system. However, compensated pumps should *not* be considered the safety device for limiting maximum system pressure. A relief valve *must* be included to protect the system.

Pressure Balancing

The pressure developed in a hydraulic system pump may range from several hundred to several thousand pounds per square inch. The pressure is used in system actuators to generate force, which is one of the concepts basic to the operation of fluid power systems. While the pressure is needed to develop actuator force, it can cause problems in

Figure 7-45. The power required to drive a hydraulic system pump depends on the flow rate and pressure of the fluid. With a pressure-compensated pump, the flow rate and, therefore, power requirements are greatly reduced when the required pressure is being maintained. The curves shown here represent three system pressures for the same pump.

the operation of units such as pumps. Examples of these less-than-desirable situations are:

- Side loading of pump bearings.
- Leakage of fluids on the end surfaces of gears and vane rotors.
- Wear from the heavy loads placed on piston pump components, such as the piston shoes and swash plates.
- Leakage between the vane tips and cam ring of vane pumps.

Pressure balancing is a design technique used to reduce or balance the pressure differences and overcome a number of problems. The following describes several design approaches used to overcome typical pressure-related problems.

Side loading of the main shaft bearings is inherent in the design of gear, vane, and radial-piston pumps. This loading is caused by the fact that 180° of shaft rotation is devoted to the inlet of fluid at less-than-atmospheric pressure, while the other 180° of the rotation is concerned with the discharge of the fluid at a higher pressure. As a result, the pressure difference may range from near 0 psi to the maximum system pressure. This difference in pressure generates unequal forces that place considerable side load on the bearings. The balanced-vane pump design solves this problem by the use of two crescent-shaped pumping chambers that allow two inlet and two discharge ports. See **Figure 7-46.** The inlets are located 180° from each other, as are the discharge port. This places an equal load on the shaft bearing under all system pressure conditions. Other pump designs compensate for uneven loading through the use of heavier bearings to withstand the increased load. In some cases, pressure lubricating of the shaft bearings is used to compensate for side loading.

The end surfaces of gears in gear pumps and the rotor and vanes in vane pumps cause special problems for the pump manufacturer. These surfaces are often referred to as *thrust* or *wear*

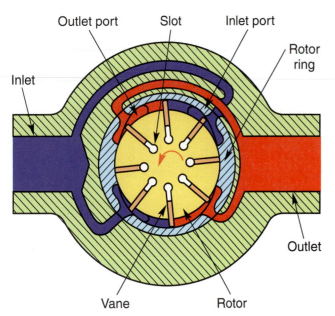

Figure 7-46. A balanced-vane pump has two inlet and two discharge areas that are opposite of each other. The pressures in the areas counteract each other, greatly reducing bearing loads. (Deere & Co.)

surfaces. The clearance between these surfaces and the pump housing must be small to reduce internal leakage. At the same time, the metal of the two surfaces must be compatible to provide a long service life. A common method used to promote both a good seal and long wear is including a **wear plate** on one or both ends of the gear or rotor. See **Figure 7-47.** This plate is fitted into the pump housing and held against the end surfaces of the gear or rotor by light spring pressure. In addition, a passageway from the discharge section of the pump is connected to the space between the wear plate and the pump housing. During pump startup, the spring pressure assures a good seal between the components. As actuator load causes the system pressure to increase, the wear plates are more-tightly forced against the end surfaces of the gear or rotor. Careful design assures that the force pushing the wear plates against the end surfaces does not become excessive, which could cause rapid wear and system power loss.

Heavy loads are placed on a number of pump components during operation. In several situations, pressure generated within the system can be used to balance forces and reduce the load on these components. An example is found in piston pumps. Piston shoes, slippers, or ball joints provide critical bearing surfaces connecting the

A   B

Figure 7-47. A—Light spring tension is often used to hold wear plates in contact with the end surfaces of the gears in gear pumps. B—Pressures exerted on the surfaces of the wear plates by the inlet and discharge areas are balanced through lines leading to chambers behind the plates. (Used with permission of CNH America LLC) pistons and the pump drive shaft. These bearing surfaces are heavily loaded when the pump is operating at maximum system pressure. Manufacturers incorporate features in the pump that deliver pressurized fluid from each associated piston chamber to these surfaces. That pressurized fluid is used at each bearing surface to counteract the load, as well as provide lubrication. See **Figure 7-26**.

In other situations, system pressure can be used to assure a positive seal between critical pump components. Vane pumps are usually designed to apply system pressure to the rotor end of the vanes to assure that a positive seal is maintained between the cam ring and vane. The slot in the rotor in which the vane slides contains a circular opening at the base. This section of the slot is exposed to pumping-chamber pressure through channels machined into the end surface plates of the pump. When the rotor is in the discharge phase of pump operation, system pressure forces the vanes out against the cam ring, assuring a positive pumping chamber seal. When the rotor is in the intake phase, the slot is exposed to lower pressure. This allows the vane to more easily move back into the rotor. See **Figure 7-48**.

Figure 7-48. The ends of the vanes inserted into the rotor are subject to pressure from the inlet and discharge chambers in many vane pump designs. This pressure and centrifugal force work together to hold the vanes against the cam ring during pump operation.

Dual-Pump Designs

A *dual pump* is a unit made up of two pumps housed in a single case and driven by a single prime mover, **Figure 7-49**. The units are usually

Figure 7-49. Dual pumps combine two pump units in a single case, usually for ease in system fabrication. In most of these units, one pump has high flow/low pressure ratings and the second has low flow/high pressure ratings. (Sauer-Danfoss, Ames, IA)

constructed using a basic gear- or vane-pump design. Models are available as basic, two-pump units or with integral valving. Those with built-in valving are typically fitted with relief and unloading valves. The pumps are produced as a single unit to meet the requirement of applications needing two fluid flow rates for operation.

A typical example of the use of dual pumps is a high-low circuit. See **Figure 7-50.** A high-flow output is required to extend a large cylinder a considerable distance under low pressure. After the initial movement, a much-higher pressure is required to complete the work. In this circuit, one of the pumps in the dual-pump design has a large flow capacity, while the second has a much lower flow capacity. The flow from both pumps is combined to provide the initial high input to quickly extend the cylinder. At the point where the operation requires high pressure, the high-volume pump is unloaded to the reservoir and the low-volume pump continues to operate the system as resistance increases.

Pump and Power Unit Design and Operating Considerations

The effectiveness of the hydraulic system depends on the careful selection and assembly of components that make up the system. This is especially true of the pump, prime mover, and plumbing of the power unit. Each component needs to be adequately sized to assure that the expected tasks can be performed, but not oversized to the point where initial component price and continuing operating costs are higher than necessary. Several concepts need to be understood to properly select components and then assemble, operate, and troubleshoot the power unit. These concepts include reading pump specifications, calculating system horsepower requirements, and designing pump inlet lines to eliminate pump cavitation.

Figure 7-50. A typical application for a dual pump is a high/low circuit, where the combined flow of the high- and low-volume pumps extend a large cylinder. Once the cylinder is extended, the low-volume/high-pressure pump is used to create the high system pressure required to finish the operation.

Interpreting Information Sheets of Pump Manufacturers

Most pump manufacturers supply considerable information about the pumps they produce. Typically this information is broken down into four categories:

■ General specifications.

■ Performance data.

■ Installation drawings.

■ Application information.

The amount of information and the presentation method varies between manufacturers. Often, this varies even between the models produced by the same company.

General specifications include operating information such as displacement per revolution, flow at rated speed and pressure, maximum rated speed, continuous rated operating pressure, and allowable inlet line vacuum. Specification tables may also list physical information, such as pump weight and type of mounting flange. Specifications are usually presented in table form. See **Figure 7-51.** Often, a table is available that presents information about several variations of the same pump. This provides an easy means of comparing models during initial pump selection or when identifying a pump during service.

Graphs are typically used to present pump *performance data.* Overall pump efficiency and volumetric efficiency are two performance data that are commonly charted on all pump designs. Pressure-compensated designs usually plot power requirements to show performance during standby operation. See **Figure 7-52.** Typically, two or more operating pressures are shown for each factor. Considerable variations will be found in the form used by companies to illustrate what they consider to be critical performance information about the operation of their pumps.

Installation drawings provide detailed information about the shape and size of a pump. See **Figure 7-53.** This information is very useful to system designers for placement and mounting of the pump. It is also used for inlet and discharge line routing. Often, these drawings include dimensions for the various input shaft configurations available for the pump.

The *application information* section of data sheets lists general information about pump installation and operation. Information typically included is specific directions on the positioning of the pump, location of oil drain lines, recommended filtration level for the system, and recommendations for the type and viscosity of hydraulic fluid. Cold weather startup is another factor that is commonly discussed in application information.

Prime Mover Power Requirements

The size of the prime mover required for the power unit needs to be established when developing the specifications for a hydraulic system. The size is primarily determined by the peak power needed during the operating cycle of the components in the system. This approach is necessary because:

■ System pressure changes as actuator loads fluctuate.

■ Pump efficiency varies as system pressure changes.

■ Varying system operating temperatures result in a range of hydraulic fluid viscosities that change fluid flow resistance.

Each of these changes results in a somewhat different power requirement. The following sections discuss the source of the basic horsepower formula, the involvement of pump efficiency in the calculation, and a general horsepower formula for the power unit.

General hydraulic horsepower formula

Horsepower is defined as 33,000 foot pounds of work per minute. Moving an object weighing 33,000 pounds one foot in one minute or one weighing 550 pounds one foot in one second requires one horsepower of effort. In a fluid power system, a pressurized fluid is moved instead of an object. The weight of the moved fluid functions the same as the weight of the object. The pressure of the fluid converted to

**4-Bolt
Flange Mounted**

PVR50 SERIES
VANE PUMPS
VARIABLE DISPLACEMENT PRESSURE COMPENSATED

TYPICAL PERFORMANCE SPECIFICATIONS

			PUMP SIZE			
			32A15	42A15	50B15	70B15
VOLUMETRIC		cu. in./rev.	7.7	9.9	7.7	9.9
DISPLACEMENT*		ml/rev.	126	162	126	162
	91.5 psi	gpm	39	182	223	284
PUMP DELIVERY	6.3 bar	lpm	148	26.5	34	41
AT 1750 RPM*	rated	gpm	32	42	50	70
	pressure	lpm	121	159	189	265
	Max.	psi	1500	1500	1500	1500
COMPENSATED		bar	103	103	103	103
PRESSURE	Rated	psi	1500	1500	1500	1500
RANGES		bar	103	103	103	103
	Min.	psi	350	400	350	400
		bar	24	27.6	24	27.6
OPERATING	Min. rpm		800			
SPEEDS**	Rated rpm		1200	1200	1800	1800
	Max. rpm		2200	1500	2200	1800
POWER INPUT AT RATED		hp	36	42	50	60
FLOW & PRESSURE (1750 rpm)		kW	27	31	37	45
MAXIMUM POWER	Max.	hp	100			
INPUT TO DRIVE SHAFT		KW	75			
SUCTION	Max.	in./Hg	5			3
		bar	-0.17			-0.10
	PRESSURE Min.	psi	20	10	20	10
	Specific Grav. < 1	bar	1.4	.07	1.4	0.7
	Min.	in./Hg	5			3
	Specific Grav. > 1	bar	-0.17			-0.10
FLUID	Max.	ft./sec.	5			
VELOCITY		m/sec.	1.5			
CASE CRAIN	NOMINAL FLOW AT DEADHEAD PRESSURE	Max. Pressure gpm	3			
		mlpm	11			
		Min. Pressure gpm	2.5			
		mlpm	9.5			
	MAXIMUM CASE PRESSURE	psi	10			
		bar	0.7			
WEIGHT		lbs.	119			
		kg	54			

NOTES:

*   Volumetric displacement is measured displacement at 91.5 psi (6.3 bar) and rated rpmper ANSI specification. Volumetric displacement varies with both pressure and rpm. Flow rates at any rpm other than the rated rpm may be approximated as follows:

$Q_2 = Q_1$ (N-142)/1667 where Q_1 = Flow (gpm) at rated rpm at 91.5 psi (6.3 bar).

Q_2 = Flow (gpm) at N rpm.

N = rpm at which Q_2 is to be determined.

**   When operating above 1500 psi (103 bar), it is recommended that a direct-acting differential relief valve be used at the pump to relieve pressure spikes and surges.

Maximum rpm at full displacement - 1900 rpm. For higher rpms up to 2000 rpm, pump displacement must be reduced to limit flow to 60 gpm (227 lpm) maximum.

OVERALL EFFICIENCY
At Maximum Displacement, Fluid Viscosity 130 SUS

TYPICAL SOUND LEVEL

PRESSURE and VOLUME ADJUSTMENT SENSITIVITY

	PUMP SIZE		32A15	42A15	50B15	70B15
PRESSURE ADJUSTMENT	Press Change/Turn	psi (bar)	115 (8.0)		135 (9.4)	
	Max. Torque	ft./lbs.(kg/m)	26.5 (13.7)			
VOLUME ADJUSTMENT	Flow Change/Turn	gpm (lpm)	14 (53.0)		22 (83.0)	
	Min. Flow Adjust.	gpm (lpm)	6.0 (22.7)	8.0 (30.3)	9.5 (36.0)	12.5 (47.0)
	Max. Torque	ft./lbs. (kg/m)	28 (3.9)	16 (2.2)	28 (3.9)	16 (2.2)

Figure 7-51. A general specification sheet for a hydraulic pump lists a wide variety of information about a specific pump model. Typical information on these sheets includes displacement per revolution, recommended operating speed, and allowable inlet line vacuum. (Continental Hydraulics)

PVR50 SERIES
VANE PUMPS
VARIABLE DISPLACEMENT PRESSURE COMPENSATED

NOTE: Typical performance curves are based on ISO VG46 oil at 120° F. (49° C.). Above 400 SUS (84 CS), add 2% hp/100 SUS.

NOTE: Deadhead horsepower is read from curves at 0 gpm flow and pressure compensator setting psi.

32A15 (at 1175 rpm)

42A15 (at 1175 rpm)

50B15 (at 1750 rpm)

50B3L (at 1750 rpm)

Figure 7-52. A performance data sheet for a hydraulic pump provides information about factors such as pump efficiency. This information is presented in chart and graph form, as well as in written descriptions. (Continental Hydraulics)

PVR50 SERIES
VANE PUMPS
VARIABLE DISPLACEMENT PRESSURE COMPENSATED

CONTINENTAL HYDRAULICS ®

PUMP DIMENSIONS
Right Hand Rotation (CW) - Code RF

Dimensions shown in: Inches
(millimeters)

Pressure Port
1-1/2 Inch SAE 4-Bolt Connection Pad

Inlet Port
2 Inch SAE 4-Bolt Connection Pad

Figure 7-53. Information concerning the physical size of a pump is listed on the installation drawings. This includes details on shaft sizes, keyways, port locations, and sizes, as well as all exterior dimensions. (Continental Hydraulics) fluid head serves as the equivalent to distance moved. The time factor is the same in either example. This is a basic formula calculating *general hydraulic horsepower:*

$$\text{general hydraulic horsepower} = \frac{\text{fluid flow weight in pounds per minute} \times \text{pressure in feet of head}}{33,000 \text{ foot pounds per minute}}$$

Fluid flow weight is determined for the basic formula as:

$$\text{fluid flow weight} = \frac{\text{pump flow rate (Fr)} \times \text{weight of water (Ww)} \times \text{specific gravity of fluid (Sg)}}{\text{gallons per cubic foot (Gf)}}$$

The pressure in feet of head is determined for the basic formula as:

$$\text{pressure in feet of head} = \frac{\text{system pressure (Sp)} \times \text{square inches in a square foot (Si)}}{\text{weight of water (Ww)} \times \text{specific gravity of fluid (Sg)}}$$

These symbols represent the various elements in the fluid flow weight and pressure in feet of head portions of the formula:

Fr = pump flow rate in gallons per minute

Ww = 62.4 pounds per cubic foot

Sg = specific gravity of the hydraulic fluid

Gf = 7.48 gallons per cubic foot

Sp = system pressure in pounds per square inch

Si = 144 square inches in 1 square foot (used as a conversion factor)

Substituting the fluid flow weight and pressure in feet of head elements into the basic formula produces this formula:

$$\text{general hydraulic horsepower} = \frac{\dfrac{(Fr)(Ww)(Sg)}{Gf} \times \dfrac{(Sp)(Si)}{(Ww)(Sg)}}{33,000 \text{ foot pounds per minute}}$$

Reducing the various elements of the equation produces this working formula:

$$\text{general hydraulic horsepower} = \frac{\text{pump flow rate (Fr)} \times \text{system pressure (Sp)}}{1714}$$

or

$$= \frac{\text{gpm} \times \text{psi}}{1714}$$

The formula can also be expressed using a decimal instead of the 1714 numerator. The fraction 1/1714 can be converted to .000583 and used in this format:

$$\text{general hydraulic horsepower} = \text{gpm} \times \text{psi} \times .000583$$

> **Note:** The horsepower formulas presented here provide the theoretical horsepower required to move the fluid in the system. They do not take into consideration losses that occur in the system.

Power unit efficiency

The horsepower actually required of the prime mover will be greater than that calculated with the formulas presented in the previous section. This is because energy is lost in the operation of both the pump and prime mover. These losses must be considered as power units are designed and assembled.

One of the energy losses in the power unit is caused by the *leakage* that takes place between the moving parts of the pump. This leakage is caused by basic pump design factors, manufacturing tolerances, and wear caused by operation. These losses expressed as a *volumetric efficiency* factor, which is less than 100%. Volumetric efficiency is indicated as a percentage and calculated using this formula:

$$\text{volumetric efficiency (Ve)} = \frac{\text{actual pump flow rate}}{\text{theoretical pump flow rate}} \times 100$$

A second energy loss also occurs in the pump and is caused by *internal friction*. This friction is caused by factors such as shaft bearings; other close-contact surfaces; the viscosity of the hydraulic fluid, which causes resistance to flow; and fluid turbulence. These losses, identified as *mechanical efficiency,* are expressed as a percentage less than 100% and calculated with this formula:

$$\text{mechanical efficiency (Me)} = \frac{\text{theoretical horsepower needed to operate pump}}{\text{actual horsepower required to operate pump}} \times 100$$

Measuring the actual horsepower needed to operate the pump requires the use of a dynamometer to measure the actual power required to operate the unit

The final efficiency calculation is for *overall efficiency.* This calculation considers *all* energy losses in the pump. This formula is used to determine the overall efficiency rating:

$$\text{overall efficiency (Oe)} = \frac{\text{volumetric efficiency} \times \text{mechanical efficiency}}{100}$$

Using pump efficiency information

Efficiency ratings contain a number of variables that often are unclear to the user of hydraulic systems. Data for each of these terms may be found for most hydraulic pumps, although some manufacturers supply only part of the information or use variations of the terminology. Becoming familiar with the terms and the information a manufacturer supplies is critical to the success of a fluid power technician.

The information supplied by volumetric efficiency tables or graphs allow the fluid power technician to calculate the actual output flow of pumps at different speeds and pressures. For example, the overall efficiency rating of a pump can be used in conjunction with the general hydraulic horsepower formula to establish the size of the prime mover needed for a hydraulic power unit.

This example shows how the general hydraulic horsepower formula can be used with the overall efficiency rating to determine the *prime mover horsepower* needed to operate the power unit of a hydraulic system:

$$\text{required prime mover input horsepower (Rhp)} = \frac{\text{gpm} \times \text{psi}}{1714\,(\text{Oe})}$$

or

$$= \frac{\text{gpm} \times \text{psi} \times .000538}{\text{Oe}}$$

where Oe = overall efficiency of the pump.

Effects of Inlet-Line Design on Pump Operation

All of the fluid used in the operation of a hydraulic system enters the system through the inlet line of the pump. As the pump rotates, a low-pressure area is formed within its pumping chamber. Atmospheric pressure acting on the surface of the fluid in the reservoir then pushes fluid into the pump through the inlet line. The relatively low pressure differential and the lower-than-atmospheric pressure (vacuum) that exists creates special problems for hydraulic system operation. System designers must be aware of the relatively low pressure differential that is available to move the fluid from the reservoir to the pump.

The inlet line must be carefully sized to eliminate excessive fluid velocities and pressure drop. Inlet lines should be no smaller than the diameter of the pump inlet with a fluid velocity calculated at no more than 4 feet per second.

The height of the pump above the level of the reservoir fluid must also be limited. A general rule is that the pump inlet should not be more than 3 feet higher than the fluid level of the reservoir.

Fluid viscosity must be carefully monitored so excessive pressure drops are not created. Strainers and inlet-line filters must be carefully selected to minimize pressure drop in the line.

Care must be taken with each of the points discussed above. Otherwise, it is possible that the pressure in the inlet line will drop to the point where adequate fluid cannot enter the pumping chamber of the pump.

When the pressure in the pump inlet section drops too far below atmospheric, a principle called liquid vapor pressure becomes a factor. *Vapor pressure* is the pressure at which a liquid begins to form a vapor. See **Figure 7-54.** The point at which this occurs is controlled by the liquid type, temperature, and pressure. When the vapor pressure is reached in the inlet of a hydraulic pump, a condition known as *cavitation* begins to appear. This is identified by bubbles of vapor that form in the hydraulic fluid in the pump inlet. These bubbles produce three conditions that affect pump operation:

- The bubbles reduce the ability of the fluid to lubricate, causing increased part wear.
- When the bubbles encounter the high-pressure section of the pump, they suddenly collapse and generate tons of force that can produce severe damage to the metal surfaces of the pump chamber.
- The collapse of bubbles also produces a high noise level in the pump.

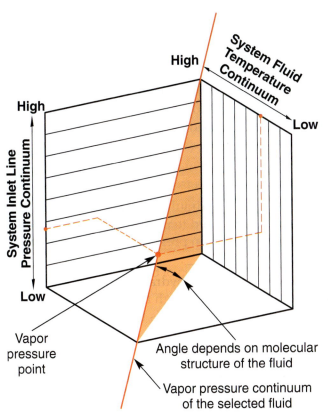

Figure 7-54. The point at which the vapor pressure of a liquid is reached and gas bubbles begin to form depends on the molecular structure of the material, its temperature, and the external pressure applied (such as that encountered in the inlet portion of a hydraulic pump).

Figure 7-55. Serious physical damage to pump components can occur if the reason for cavitation is not identified and eliminated from a hydraulic system. (Courtesy of Eaton Fluid Power Training)

Continued operation of a pump that is cavitating can severely damage or destroy the pump, **Figure 7-55.** Metal particles released from damaged pump parts may also cause excessive wear in other system components.

Another factor somewhat similar to cavitation is *entrained air.* This condition results from air that enters the inlet line, is drawn into the pump, and remains suspended in the fluid. The air may enter the line through leaks in fittings, lines, or accessories such as filters or due to a low reservoir level. A pump operating under this condition produces high noise levels similar to cavitation. Damage from this condition can be similar to that produced by cavitation. This condition can be differentiated from cavitation by determining inlet line pressure. Cavitation produces a high, negative inlet-line pressure (vacuum), while a slightly negative (closer to 0 psi) pressure may indicate a condition caused by entrained air.

Chapter Test

Answer the following questions. Write your answers on a separate sheet of paper.

1. Name the five functions of a hydraulic system power unit.
2. The device that is used to move fluid through a hydraulic system is called the _____.
3. List the three basic ways in which hydraulic pumps are classified.

The fluid in a hydraulic system serves several important functions, including energy transmission and component lubrication, cooling, and cleaning. These functions require that fluid be circulated through the system to remove excess heat, flush dirt from internal areas, and lubricate bearing surfaces while moving actuators to do the work for which the system was designed. These functions are facilitated by hardware such as the reservoir and fluid conductors. The reservoir acts as a storehouse for fluid while aiding in system cleaning and cooling. Conductors and associated connectors allow the movement of fluid between system components to transmit energy and also aid with cooling and cleaning. This chapter discusses each of these components and a number of design variations used by system component manufacturers.

Objectives

After completing this chapter, you will be able to:

- Describe the function of the hydraulic system reservoir and identify the primary design features involved in its construction.
- Explain the factors that must be considered when establishing the size of a reservoir for use in a system.
- Describe the factors that are basic to the design and structure of fluid conductors used in hydraulic systems.
- Compare and contrast the various types of pipe, tubing, and hose used in hydraulic systems and explain the ratings for each conductor.
- Identify and explain the factors that must be considered when selecting a conductor for use in a hydraulic system.
- Explain the information typically found on data sheets provided by the manufacturers of hydraulic system pipe, tubing, and hose.

Chapter 8

Fluid Storage and Distribution
Reservoirs, Conductors, and Connectors

Selected Key Terms

The following terms will be used in this chapter. As you read the text, record the meaning and importance of each. Additionally, you may use other sources, such as manufacturer literature, an encyclopedia, or the Internet, to obtain more information.

adapter fittings
baffle
braided wire
burst pressure
compression fittings
drain return line
Dryseal standard pipe thread
ferrule
flared fittings
fluid turbulence
hose
hose-end fitting
laminar flow
manifolds
nominal sizing
pipe
quick-disconnect coupling
reservoir
schedule number
shock pressures
spiral clearance
spiral-wound wire
tubing
vortex

Internet Resources

www.daycoproducts.com/daycoweb.nsf/
Hydraulics?OpenPage
Dayco Products
Provides links to information on Dayco's line of hose, couplings/adapters, and crimpers. Also provides a link to the Dayco Hydraulic Catalog. This site provides the opportunity to examine one manufacturer's product specifications and recommendations for selection and installation.

www.eaton.com/Eaton/ProductsServices/
ProductsbyCategory/Hydraulics/connectors/
Index.htm
Eaton Corporation
Provides details on one company's quick disconnect coupling products. Coupling construction, selection information, and safety information are included.

www.gates.com/index.cfm?location_id=5215
Gates Corporation
Provides links to hydraulic hose catalogs, part interchange information, selection wizard, and other engineering tools.

www.gates.com/index.cfm?location_id=3042
Gates Corporation
Provides links to calculators that determine pressure losses for user-input data. These calculators are based on a group of assumptions and average conditions. Results should only be used as a guide.

www.hydacusa.com/literature/accessories/
accessories.pdf
HYDAC Technology Corporation
Provides an overview of reservoir accessories. Includes descriptions of filters, coolers, fluid-level indicators, breathers, and temperature sensors.

www.jwftechnologies.com/manufacturers/
air-way-fittings
JWF Technologies
Provides an extensive listing of hydraulic pipe and tube fittings. Includes product data sheets and an index illustrating each of the fittings.

Reservoir

The *reservoir* is the component that holds the system fluid not currently in use in the system pump, control components, actuators, and lines. This reserve fluid supply is necessary to make up fluid lost in leakage during system operation. It is also needed to compensate for the varying volume between extension and retraction positions of cylinders. The availability of this reserve fluid supply is basic to the operation of any hydraulic system, even though the reserve quantities required will vary depending on circuit design and operation.

Functions of the Reservoir

The primary role of the reservoir is to provide storage for the hydraulic fluid required for system operation. However, a well-designed and well-constructed reservoir provides a number of important additional functions, as listed below. See **Figure 8-1.**

- Removal of heat.
- Separation of solid particles.
- Release of trapped air.
- Separation of water.

In even the least-complex designs, the reservoir dissipates system heat to the atmosphere through

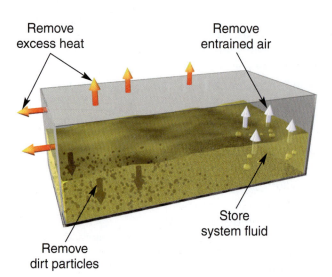

Figure 8-1. Besides storage, the reservoir helps cool and clean the hydraulic fluid. This helps to prolong the life of both the system fluid and components.

radiation. This radiation primarily occurs through the reservoir walls. The amount of heat transferred may be increased by incorporating simple devices, such as cooling fins placed on reservoir walls. Baffles in the interior of the reservoir may also be used to increase cooling by directing warm fluid returning from the system to the exterior walls of the tank to promote rapid heat transfer.

Fluid movement is slow in the reservoir when compared to the fluid velocity through system components and lines. This reduction in the rate of movement allows some of the heavier particles being carried by the fluid to settle to the bottom of the reservoir. This process helps clean the fluid, but it should *not* be considered a reliable fluid filtering method. Larger, heavier metal or dirt particles will settle out, but many of the smaller particles will not. This is especially true when the fluid has a high viscosity. Regular filters properly placed in the system provide much more reliable fluid cleaning.

Most reservoirs are vented to the atmosphere. Venting exposes the surface of the fluid in the reservoir to normal atmospheric pressure. This is important to the operation of a hydraulic system as it allows air that has become trapped in the fluid to escape. Typically, hydraulic fluid contains small quantities of air when it is being used in a system. The air may actually be dissolved in the fluid without a visible indication of its presence. This type of situation normally does not cause operational problems. However, when the fluid begins to appear cloudy, it is an indication that large quantities of small air bubbles have become entrained in the fluid. This can cause foaming, spongy system operation, and increased operating temperatures. A reservoir that is adequately sized and containing proper baffles will allow entrained air to form larger bubbles. These larger bubbles rise to the surface of the liquid and escape to the atmosphere, thus reducing or eliminating the problems associated with entrained air.

Water that may have become mixed with petroleum-based hydraulic fluids also has an opportunity to separate as the mixture passes through the reservoir. Water is heavier than the hydraulic fluid. This allows water to settle to the bottom of the reservoir. It can then be drained off, if a sufficient quantity exists. Separation of the two liquids is relatively slow, so it is important to have a reservoir of adequate size to allow the fluid sufficient time to separate as it moves through the tank. If the water and oil form an emulsion, this separation process is not possible.

Construction of a Reservoir

The construction details of a reservoir may vary considerably, depending on the application of the hydraulic system. In most manufacturing situations, the reservoir is a tank located in close proximity to the system pump. It is easily recognized as a tank. In other situations, the cavities in structural members of a machine may serve as the reservoir. However, each design must provide fluid storage and help condition system fluid by assisting in the removal of heat and contaminants.

The typical hydraulic system reservoir is a rectangular, covered tank made from welded or formed steel plate. The tank is often fitted with short legs to permit circulation of air for cooling of the bottom surface. Also, the bottom of the tank is often dished and includes a drain hole to allow easy drainage. The ends, sides, or top of the tank is often fitted with ports with removable covers to permit easy cleaning. The tank is also fitted with a vented breather that allows the tank to take in and release air as the fluid level changes during system operation. In many systems, the breather is incorporated into the filler cap.

An especially important part of the construction of a reservoir is the baffle. A *baffle* separates the pump inlet and system return lines. This baffle is often nothing more than a vertically positioned, flat plate that divides the tank lengthwise. See **Figure 8-2.** The baffle prevents the hydraulic oil from moving directly from the return line to the pump inlet. Forcing the oil to travel around the baffle provides longer contact with the sides of the tank. This allows better cooling as well as a longer time for separation of air and the settling of dirt and water. The reservoir is fitted with three lines connected to the system. These are the pump inlet line, system return line, and a drain return line.

The *inlet line* should be large enough to provide fluid at a velocity and pressure drop lower than that suggested by the pump manufacturer. The hole where the inlet line enters the reservoir should be fitted with a *packed slip joint* to prevent dirt and water from entering the tank through the opening while absorbing normal pump and power unit vibration. See **Figure 8-3.** The pickup end of the inlet line should be positioned well below the

Figure 8-2. Baffles are used inside reservoirs to direct the flow of hydraulic fluid. They control fluid movement in the tank to promote fluid cooling and cleaning.

normal fluid level in the tank to prevent the formation of a vortex. A *vortex* is a whirlpool that can form around the inlet line if the pickup is too close to the fluid surface. In extreme cases, a vortex can actually allow air to be drawn into the pump. Also, the inlet should be placed at least 1″ from the bottom to minimize recirculation of sediment from the bottom of the tank.

The *return line* should be large enough in diameter so fluid velocity is not excessive during the maximum-expected return flow. It must always be positioned on the opposite side of the baffle from the inlet line. A common practice is to cut the end of the return line at a 45° angle. The end of the line should be installed approximately two pipe diameter off of the bottom. The sloped end should be positioned to cause the fluid to circulate along the sides of the tank.

The *drain return line* must provide drainage with a minimum of back pressure. This is necessary as drainage from components needs to escape as easily as possible to allow effective operation and lubrication.

The top surface of the reservoir is often used as the mounting base for the system power unit, **Figure 8-4.** It is generally considered a location as the pump is placed close to the fluid supply. This

Packed slip joint
Pump inlet line
Drain line
Return line
Fluid level

Inlet line end:
1. Minimum of 1″ from reservoir bottom
2. May be fitted with a sump strainer

Reservoir baffle

Return line end:
1. Cut at 45° angle
2. Two pipe diameters from bottom of reservoir
3. Directs fluid flow along reservoir sides

Figure 8-3. Pump inlet, return, and drain return lines must be carefully positioned in a reservoir. The distance from the ends of these lines to the bottom of the reservoir will affect system performance.

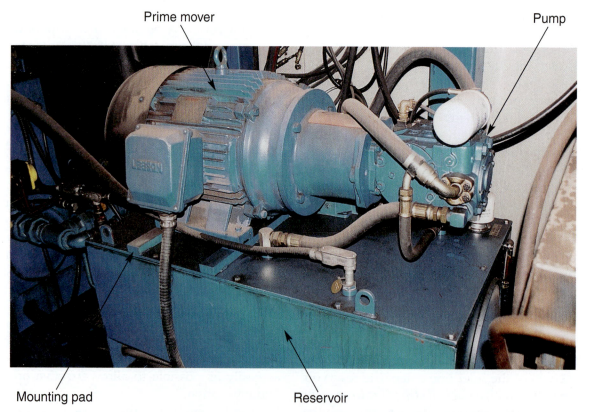

Prime mover
Pump
Mounting pad
Reservoir

Figure 8-4. The reservoir is often used as a mount for the system prime mover, pump, and other system components. The tank must be made from adequately strong material to support these units. (Used with permission of CNH America LLC) can minimize inlet line pressure drop. Locating the power unit on the reservoir requires the addition of a metal support plate and other fittings to facilitate attachment. In larger-capacity units, the weight of the pump and prime mover become important factors in determining the strength of the structural members of the reservoir and the mounting plate.

When placing a reservoir in use for the first time, a great deal of care should be taken to be certain the inside of the tank is clean. Any machining chips and welding slag obviously should be removed as well as any cutting oils. The inside of the tank should be finished with a material that will withstand the chemical characteristics of the fluid used in the system. It must also be able to withstand the chemical activity promoted by contaminants that enter the system during equipment operation or cleaning.

Reservoir Configurations

A rectangular tank with the pump located above the fluid level is probably the most commonly used form. This configuration is found in both stationary and mobile hydraulic-powered equipment. The requirements regarding reservoir construction can easily be met using this form. However, other system characteristics, such as limited physical space or the need for positive pump inlet pressure, may require reservoir variations.

L-shaped or overhead reservoirs may be used if a positive pressure (greater than atmospheric) is needed on the pump inlet line. Both of the designs place the pump below the level of the oil in the tank. See **Figure 8-5.** This difference in height may be only a few inches, but it allows gravity to assist in moving fluid into the pump. Of course, these systems must have inlet line shutoff valves to minimize fluid loss from the reservoir when the pump must be serviced.

Other shapes and forms may also be used for hydraulic reservoirs. The cavities of large machine tools can be used as the reservoir. For example, the hollow structural members of equipment, such as the arms of front end loaders, may serve for both storage and fluid transmission. In other mobile pieces of equipment, the oil used for the hydraulic system also functions as the lubricating oil for the transmission and gear train. In those situations, the transmission/differential housing is, in a sense, serving as the system reservoir. See **Figure 8-6.**

Figure 8-5. An L-shaped or overhead reservoir design is often used when the pump inlet line needs a slight positive pressure for more efficient operation. A—L-shaped reservoir. B—Overhead reservoir.

Figure 8-6. In some equipment using hydraulic power, machine cavities or transmission cases in conjunction with a constructed reservoir serve as the hydraulic system reservoir. (Used with permission of CNH America LLC)

Sizing the Reservoir

It is often stated that a hydraulic system reservoir should have a storage capacity three times the rated flow of the pump. Although this is a very general statement, it has proved to be appropriate, especially in stationary industrial systems with typical actuator sizes. In any case, it is a good starting point when determining the size of a reservoir in a new or modified system.

In actuality, many factors must be considered when determining reservoir size. First, the reservoir must be large enough to store more than the anticipated maximum volume of oil that will be returned to the reservoir during any part of the system cycle. An example is the fluid returned to the reservoir in a system containing a single-acting cylinder with a large bore and long stroke. Other special considerations should involve the basic reservoir functions discussed earlier in the chapter. For example, an increase in the capacity of the reservoir increases the amount of time the fluid is in the tank. This assures increased heat dissipation and separation of dirt or water.

Many times, the final selection of the reservoir size is a compromise influenced by space and weight limitation. Fixed installations, where weight is not a factor, often have a ratio of pump output

to reservoir capacity above 3:1. On the other hand, mobile installations, where increased weight can be a problem, have ratios below that guideline.

Conductors

The individual component parts of a hydraulic system cannot function alone. They must operate as an integral part of a system. This is accomplished by moving fluid from the reservoir to the pump and on to the control valves and actuators. In most system designs, it is not practical to directly connect these components. This is because many of the components must operate machine members as well as interact with other hydraulic components in the system. *Conductors* assure distribution of system fluid to all components, no matter how widely they are dispersed, **Figure 8-7.** The proper selection, installation, and maintenance of these conductors is of utmost importance to proper operation of a hydraulic system.

Inherent Design Requirements

A variety of different types of conductors and associated connectors and fittings are available for use in hydraulic systems. Pipe, tubing, and flexible hose are considered basic conductor categories. Of

Figure 8-7. A wide variety of conductors, connectors, and fittings are available for use in hydraulic systems, as can be seen on this piece of earthmoving equipment. Selecting appropriate components requires familiarization with each of the types available.

course, the items in each of these groups vary considerably in appearance, construction, and application. However, a number of basic factors are related to items in all of the categories. Part of these factors are concerned with safe system operation, such as physical strength and shock resistance. Other factors deal with the efficiency of system operation, such as resistance to fluid flow and achieving correct fluid velocity. Still other factors deal with ease of installation and service.

Adequate strength

Conductor strength must be considered a critical factor when selecting a conductor for a system. The most obvious reason for this concern is system operating pressure. Strength must also be considered from the standpoint of pressure surges or shock caused by sudden, unexpected load changes during the operating cycle. In addition, other strength aspects that must be considered are mechanical stresses produced by thermal expansion, physical abuse, and even the weight of other system components or machine members the conductor is expected to support.

A conductor must have a continuous operating pressure rating equal to the expected system operating pressure plus an adequate safety factor. A factor of 4:1 is often stated as a minimum. The actual factor selected is based on experience with specific applications.

Safety factor is a margin of error used to provide protection from hydraulic shock pressures.

Shock pressures result from machine-member or work-load kinetic energy when a directional control valve is centered or reversed and the load suddenly stops or reverses. Shock pressures must be considered when selecting a conductor, as they are often higher than most circuit designers and machine operators realize. It is possible to measure shock pressures using electronic test instruments. These devices are fitted with appropriate accessories to convert the pressures into electrical pulses that can be displayed as a waveform on a screen or printed out for future analysis.

The pressure rating of all conductors depends on the tensile strength of the construction materials, conductor wall thickness, and inside diameter of the conductor. Formulas exist for calculating wall thickness for conductors when the required inside diameter and material tensile strength are known. However, in most practical system design situations, manufacturer data sheets are used to select the conductors that have adequate strength to safely function.

Low flow resistance

It is desirable to have the fluid in the lines of a system move with as little resistance to flow as possible. This requires conductors that are properly sized with an inside diameter large enough to allow the needed volume of fluid to move through the lines within recommended velocity rates. Excessive velocity produces *fluid turbulence,* which increases flow resistance. See **Figure 8-8.** Smooth interior

Restriction to Flow	Basic Flow Characteristic	Contributing Factors
Minimal	Laminar flow	• Adequate conductor diameter • Recommended flow velocity • Smooth conductor surfaces
Excessive	Turbulent flow	• Small conductor diameter • Excessive fluid velocity • Inappropriate conductor fittings

Figure 8-8. The inside diameter of a conductor must be large enough to move the fluid at a velocity no greater than the maximum recommended for the line. If this velocity is exceeded, fluid turbulence and power loss can result.

surfaces and gradual bends in conductors reduce the resistance to fluid flow in a system. This helps promote *laminar,* or straight-line, flow. The selection of a fluid with the proper viscosity also helps ensure flow resistance is not too great in a system.

Resistance to flow in lines and fittings results in energy loses in a system. The energy being input into a system can be calculated using the fluid pressure and flow at the pump. As the fluid is forced through the system, the flow rate remains the same, but the pressure drops as resistance is encountered. This pressure drop is the result of the resistance to fluid flow caused by:

- Each square inch of conductor surface.
- Each bend or fitting encountered.
- Each orifice passed through.
- Level of turbulence in the fluid stream.
- Viscosity of the fluid.
- Resistance of the external load.

The pressure that is lost in the system due to all of these factors, except the external load, indicates the energy that has been lost in the system. This lost energy directly lowers the work output of the system. It also produces heat, which, in excessive amounts, can cause many system operating problems.

Installation and maintenance

To be effective, conductors must not only carry the fluid from component to component, but must be constructed in such a way as to allow economical installation and maintenance. Connectors and fittings must accommodate attachment of system components to the conductors. In actual use, the number of such units must be minimized. Each additional fitting placed in a system increases original cost and adds a point that may contribute to fluid flow resistance and turbulence.

Technicians who maintain fluid power systems often feel that planning for maintenance and service is not considered during new system design and construction. Components may be assembled in such a way as to make service difficult. However, once major maintenance is required, design variations can often be made that will allow a system to be more easily repaired during future service. An example of this is retrofitting components with flare fittings in place of the original threaded pipe connections.

Types of Conductors

A fluid power system can incorporate a number of different conductors to transmit fluid. The most common general classifications are pipe, tubing, and hose. A number of design variations, connectors, and fittings are available under each of these classifications. Also, some manufacturers classify manifolds as conductors, although these are often considered separate components. The conductor used depends on a number of factors, ranging from system operating pressure to initial cost.

Pipe

Pipe is a high-tensile-strength, rigid fluid conductor normally made from mild steel. It is normally installed by cutting and fitting various lengths into the desired configuration using threads cut directly on the pipe and threaded fittings. It is not intended to be bent.

Pipe used in hydraulic systems should be seamless, black pipe. It may be either hot or cold drawn. The internal surfaces of the pipe must be free from scale and any other contaminants. Galvanized pipe should *not* be used in a hydraulic system. The chemical ingredients in oil and its additives react with the zinc coating.

A *nominal sizing* system is used for pipes. With this system, the stated size, or nominal size, is neither the actual outside nor inside diameter of the pipe. For example, the outside diameter of a 1" pipe is 1.315", while the inside diameter varies depending on wall thickness. **Figure 8-9** includes all of the common nominal pipe sizes. It indicates the actual outside dimensions and the inside diameters of a number of the pipes rated by wall thickness.

Pipes are available in a variety of wall thicknesses, **Figure 8-10**. This is an important consideration as it is a key factor in establishing the safe maximum operating pressure of the conductor. The pipe with the thickest wall will withstand the greatest system pressure, if all other conditions are the same. Wall thickness of pipes is identified by a *schedule number* established by a standard of the American National Standards Institute (ANSI). This standard uses schedule numbers ranging from 10 to 160 to rate pipe. Hydraulic systems typically use schedule 40, 80, or 160 pipe. Schedule 40 pipe is generally used in lower-pressure parts of

Nominal Size	Outside Diameter	Inside Diameter			
		Schedule 40	Schedule 80	Schedule 160	Double Extra Heavy
1/4	.540	.364	.302	—	—
3/8	.675	.493	.423	—	—
1/2	.840	.622	.546	.466	.252
3/4	1.050	.824	.742	.618	.434
1	1.315	1.049	.957	.815	.599
1 1/4	1.660	1.380	1.278	1.160	.896
1 1/2	1.900	1.610	1.500	1.338	1.100
2	2.375	2.067	1.939	1.689	1.503

Figure 8-9. Pipe size is stated as a nominal dimension, which is neither the actual inside nor outside dimension of the conductor.

Schedule 40
(Standard)

Schedule 80
(Extra Heavy)

Schedule 160

(Double Extra Heavy)

Figure 8-10. The wall thickness of a pipe is indicated by schedule numbers that are part of an American National Standards Institute (ANSI) specification.

the systems, such as pump inlet lines and system return lines. Schedule 80 or 160 is used in the working lines, depending on the maximum system pressures expected. **Figure 8-11** shows typical burst pressure ratings for the three commonly used pipe schedule ratings. *Burst pressure* is the highest internal pressure a component can withstand without failure. Typically, the maximum recommended operating pressure is 25% of the burst pressure.

Tubing

Tubing is a relatively thin-walled, semirigid fluid conductor. It is available in a number of different materials, ranging from plastic to stainless steel. Tubing can be easily bent, which allows relatively easy conductor installation with a minimum of related fittings. The end result is a system with

fluid conductors that provide proper flow characteristics and involve a minimum amount of space while minimizing visual clutter. Care needs to be taken when selecting tubing to assure the material is compatible with system fluid and the tubing is adequately strong to withstand system pressures. For example, copper tubing should *not* be used in a hydraulic system because of the chemical reaction between copper and the oil and its additives.

The size of tubing is indicated by measurement of the actual outside diameter. The inside diameter will vary depending on wall thickness. Most tubing is manufactured to the specifications of various standardizing organizations. The American National Standards Institute (ANSI), Society of Automotive Engineers (SAE), and American Iron and Steel Institute (AISI) all have standards covering various types of tubing. A generalization that can be made for any one nominal diameter of

Nominal Size	Burst Pressure (psi)			
	Schedule 40	Schedule 80	Schedule 160	Double Extra Heavy
1/4	16,000	22,000	—	—
3/8	13,500	19,000	—	—
1/2	13,200	17,500	21,000	35,000
3/4	11,000	15,000	21,000	30,000
1	10,000	13,600	19,000	27,000
1 1/4	8,400	11,500	15,000	23,000
1 1/2	7,600	10,500	14,800	21,000
2	6,500	9,100	14,500	19,000

Figure 8-11. The maximum pressure at which a pipe can operate depends on the nominal diameter and schedule number classification. Burst pressures of the various pipe schedules vary between manufacturers.

tubing is that the smaller the inside diameter, the greater the relative thickness of the wall. A thicker wall allows a higher system operating pressure. However, care must be taken to check the specifications of each type of tubing. Construction material and construction technique can produce widely varying specifications for any given size.

Lengths of tubing are bent to form a conductor between two components in a hydraulic system. The formed tubing is then attached to the components on either end by appropriate connectors and fittings. Care must be taken in designing these tubing sections. System vibration and thermal expansion and contraction must be considered. Both vibration and expansion/contraction can be absorbed by placing a bend in the tube between the attachment points, **Figure 8-12.** Tools are available to form these bends. The tools can be used to form most angles, although angles over 90° are difficult. Generally, 90° is the maximum bend angle. The tube tends to collapse at the bend, resulting in a reduced cross section that is not round. Tube inserts made from rubber or coiled wire are sometimes used to help prevent tube collapse during bending.

Hose

Hose is a flexible conductor made from a combination of different materials that are compatible with the system fluid and sufficiently strong to withstand system operating pressures. It is commonly used in systems to make connections where

Figure 8-12. Tube can be safely bent to form angles up to 90°. This allows installations that contain fewer fittings than systems using pipe. (Used with permission of CNH America LLC)

equipment parts tilt or swivel during operation or where severe vibration is encountered. See **Figure 8-13.** To achieve the combination of flexibility and strength, a hose has a flexible inner tube to conduct the fluid, a middle layer of reinforcing materials for strength, and an outer protective layer of tough material to withstand external abrasion and abuse.

The inner tube that conducts the fluid must be smooth and sufficiently flexible to freely bend, but strong enough to resist collapse, which would cause a reduction in cross-sectional area. This tube

Figure 8-13. Flexible hydraulic hose is extensively used when fluid must be transmitted to actuators on movable machine members.
(Used with permission of CNH America LLC)

is usually made from synthetic-rubber compounds or thermoplastics. The middle reinforcement section is formed from *braided wire, spiral-wound wire,* or textile yarns that are either woven or braided. See **Figure 8-14.** Several layers may be used in this section, depending on the pressure rating of the hose. The outer section is a protective layer made from synthetic rubber, thermoplastic, or woven textile materials impregnated with rubber or thermoplastic. This outer material is formulated to resist the abuse of weather, abrasive dirt, and the chemical action of the system fluid or other materials from the environment. Additional plies of material are often used between layers to bind them together.

A wide variety of hoses are available. This makes selection difficult in many situations. Standardizing organizations, such as the Society of Automotive Engineers (SAE), provide guidelines

Figure 8-14. Flexible hydraulic hose is made up of a minimum of three layers of materials. These include an inner layer for sealing, middle layer for strength, and outer layer for protection. A—Braided wire. B—Spiral-wound wire. (Gates Corporation)

that help when comparing the products. For example, SAE standard J517 includes 16 different hydraulic hose styles and gives dimensional and performance specifications covering a full range of operating situations. Although these specifications are available, standardizing groups typically do not certify specific manufacturers or products. The standards are voluntary, but are very helpful when comparing products available from the many different hose manufacturers.

Manifolds

Manifolds are machined or fabricated machine elements used to distribute hydraulic system fluid. A manifold may be as simple as a larger-diameter pipe that serves as a common oil supply to several valves or actuators in a system. However, manifolds usually serve more-complex, multiple-valve situations and involve machined passages in blocks or multilayer subplates. These units eliminate pipe, tube, or hose between components. This allows a lighter, more-compact system.

Conductor Fittings for System Assembly

Constructing a hydraulic system requires a number of fittings to connect the conductors to the reservoir, pump, valves, actuators, and other component parts. The variety of fittings available from manufacturers for each of the various conductor types often makes selection confusing. However, selecting appropriate fittings is necessary to assure reasonable initial construction cost, efficient system operation, and minimal ongoing maintenance.

Fittings serve as the means by which the ends of individual conductor sections are attached to other conductors or system components. Adaptor fittings are often used between the conductor and the component to allow appropriate installation angles and clearances. These additional fittings also facilitate disassembly and assembly during system service.

Applications that require frequent or rapid connecting and disconnecting of lines use *quick-disconnect couplings.* These fittings are commonly found on mobile equipment that uses a variety of interchangeable hydraulic implements. The couplings allow implements to be quickly changed with little or no fluid loss, **Figure 8-15.**

Figure 8-15. Quick-disconnect couplings allow hydraulic lines to be quickly connected or disconnected. This allows, for example, implements to be quickly transferred on mobile farm and construction equipment.

Quick-disconnect couplings contain an easily operated locking device and check valves. The locking device can usually be operated without tools. The check valves prevent fluid loss when the coupling parts are disconnected. See **Figure 8-16.**

Pipe

The sections of pipe used as fluid conductors in hydraulic systems are typically assembled using threads. Thread sizes are standard, even though the outside diameter of the pipe varies slightly. A pipe thread connection includes external threads cut on the pipe ends and internal threads cut into the openings of system components and pipe fittings. The threads are tapered 3/4″ per foot to assure a positive seal at the connection. A *Dryseal standard pipe thread* should be used, rather than standard pipe threads, when assembling pipe for use as a hydraulic system conductor. This thread design provides a tight seal that prevents the leakage encountered when standard pipe threads are placed under high system pressure. Standard threads tend to leak because of a continuous internal spiral clearance that exists in the threads. See **Figure 8-17.** *Spiral clearance* is the result of space between the crests and roots of the threads when assembled. This clearance exists in a spiral along the entire length of the thread, even when the threaded parts are properly fitted and tightened.

Figure 8-16. Easily operated locking devices and effective check valves allow quick-disconnect couplings to be quickly operated without spilling fluid.

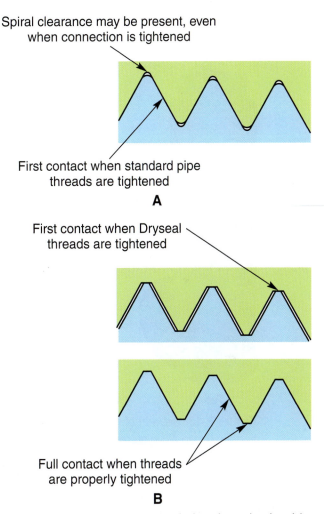

Figure 8-17. Dryseal standard pipe threads should be used when constructing hydraulic system conductors from pipe. A—Standard pipe threads. B—Dryseal pipe threads.

A wide variety of pipe fittings are available to assist in constructing the conductors needed to carry the hydraulic fluid to system components. These include tees, elbows, couplings, unions, nipples, and several other shapes, **Figure 8-18.** These fittings should be made from steel or other material that will assure a strength comparable to the schedule of the connecting pipe. All connections should be installed to permit easy removal and reassembly by common hand tools.

Union fittings allow easier connection of complex system lines. They provide a means of connecting lines using flanges that provide a positive seal. However, every attempt should be made to design pipe runs between components with a minimum of unions. It is critical to plan the system so that pipe, fittings, and system valves can be removed without dismantling large sections of the system or bending or otherwise damaging system parts.

Tubing

A wide variety of fittings are available for making connections to tubing. Manufacturers produce many different styles that are designed to provide the best means of attaching tubing to system components, **Figure 8-19.** Each must provide a proper fit and seal between the component and fitting and also between the tube and fitting.

Typically, the seal between the fitting and the system component is provided by either a tapered pipe thread or a straight thread with a separate seal. The separate seal may be an O-ring or a compression washer.

The seal between the fitting and the tube is provided by either a flared or compression connection. *Flared fittings* actually require physically flaring the end of the tube. A positive seal is provided by using a jamb nut and sleeve to force the flared tube end against a cone portion of the connection. In the United States, flared tube

Plug   Bushing   Reducer

Tee   90° Elbow   Union

Figure 8-18. Several of the common threaded pipe fittings available for assembly of fluid conductor lines in hydraulic systems.

Figure 8-19. Pipe threads or fittings with straight threads are used to attach hose fittings to hydraulic components. The seal on fittings with straight threads is obtained using an O-ring or a compression washer.

fittings are standardized at 37° and 45° angles. See **Figure 8-20.** Hydraulic systems typically use the 37° design.

A number of different designs can be found in *compression fittings* for tubing. Several

manufacturers have patented designs. Typically these include a fitting body; compression element such as an O-ring, *ferrule,* or proprietary compression sleeve; and compression nut, **Figure 8-21.** The end of the tube to be attached must be square, concentric, and free from burrs to assure proper positioning and sealing. The tube end is slipped into the fitting until it rests against an internal shoulder that provides proper positioning. Tightening the compression nut squeezes the parts to secure the tube and provide a fluid-tight seal.

The tube may be attached to the fitting by swaging, brazing, or hard soldering. Swaging joins the tube and fitting by squeezing the fitting onto the tubing. Brazing and hard soldering join the tube and fitting by using a low–melting point, nonferrous metal to secure the joint without melting the base metals. The designs usually require split flanges or other devices to secure the fitting to the component. See **Figure 8-22.**

Hose

Flexible hose is attached to a *hose-end fitting* that includes an attachment mechanism designed

Figure 8-20. A 37° or 45° flare is used to provide a seal between components and tubing. A—Hydraulic systems typically have a 37° flare. B—A 45° flare may be found in other applications. (Gates Corporation)

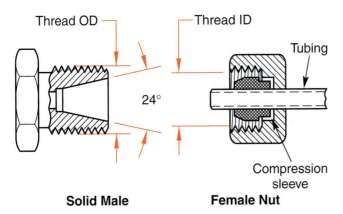

Solid Male   **Female Nut**

Figure 8-21. A number of different fittings depend on compression between the fitting and the tubing for a seal. (Gates Corporation)

to connect to other hydraulic lines, control valves, actuators, or other system components. This attachment mechanism or fitting may use pipe threads, 37° or 45° flare, flange, or other configuration. In some cases, the hose-end fitting is directly connected to system components. However, in most situations, an adapter fitting is used between the hose-end fitting and the component.

Hose-end fittings are available in permanently attached or reusable styles. A permanently attached unit consists of a nipple and socket on one end of the connector body and a threaded or other style of connection on the opposite end. The hose is forced over the nipple, filling the space between the nipple and socket. The metal of the socket is then swaged or crimped. This locks the connector over the hose and forms a permanent connection. See **Figure 8-23.** These fittings are used on

Figure 8-22. Split flange fittings that have been brazed or hard soldered to the tubing may be used to attach tubing to hydraulic components.

Figure 8-23. Permanently attached hose-end fittings are crimped or swaged onto the hose and cannot be removed. A—External view showing compression of the fitting socket. B—Cutaway showing hose materials held by the fitting nipple and compressed socket. (Gates Corporation)

off-the-shelf hose assembles. If proper equipment is available, an end user may apply these connections in the field.

Reusable hose-end fittings allow more flexibility in servicing hose assemblies. This is possible because these fittings can be attached to the hose using basic hand tools. They can also be removed from a conductor assembly that has failed and reused with a length of new hose. Reusable hose ends are generally classified as either screw-together or clamp types. See **Figure 8-24.**

The screw-together, reusable hose-end fitting consists of two parts. The first part is an outer shell fitted with internal threads that allow it to be

Figure 8-24. Reusable hose-end fittings are either screw-together or clamp type. A—Screw together. B—Clamp.

threaded over the outer diameter of the hose. One end of this outer shell contains a smaller threaded opening through which a nipple is turned into the inner tube of the hose. This forces the hose to slightly expand, producing a leakproof seal between the hose and the fitting. The end of the insert opposite of the nipple contains the connection to be used for attaching the hose to system components.

Clamp type, reusable hose-end fittings typically consist of three parts. One part is an insert that is forced into the inner tube of the hose. This insert is constructed with a nipple on one end to aid in insertion. The other end of the insert contains the connections needed for attachment to system components. Two or four bolts are used to tighten two clamp pieces around the hose. This clamping action forces the hose onto the insert and produces a positive seal.

Adapter fittings

Adapter fittings are connectors used to eliminate twisting or excessively bending a conductor during installation. They also assist in disassembly and reassembly of the system during service. Adapter fittings are often used between pipe, tube fittings, or hose-end fittings and the component. A variety of connection styles are available for the ends of these adapters based on both American and international standards. See **Figure 8-25**. The connection styles can be generally grouped into one of four categories:

- Category 1. Straight threads and a shoulder sealed with an O-ring or a metal compression washer.

- Category 2. Tapered pipe threads for both attachment and seal.

- Category 3. Flanged connection with the seal provided by an O-ring held in a groove.

- Category 4. Flared connections for the seal.

Care must be taken when working with adapter fittings as several thread sizes and flare angles exist. Attempting to mate different threads or flare angles will cause problems.

Attachment methods from the first, second, or third categories are used with adapter fittings to connect major components. These components include pumps, valves, and actuators. Attachment methods from the first, second, or fourth categories

Category 1

Category 2

Category 3

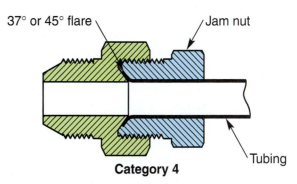

Category 4

Figure 8-25. A variety of connection methods are used to attach fluid lines and components to provide sealing against fluid leaks while promoting easy assembly and disassembly.

are used when connecting an adapter to hose and tube-end fittings. In addition, adapters are available that form 45° or 90° elbows. Swivel fittings are also available to minimize hose twisting as a component moves or rotates during system operation.

Selecting Appropriate Conductors

There is no question that fluid conductors and their associated fittings are an essential part of any hydraulic system. Many factors must be carefully considered to obtain the most appropriate type and size of conductors and fittings. However, selection is often not as carefully considered as it should be. The result can be a system with conductors that cost more than necessary, contribute to inefficient system operation, fail prematurely, or are difficult to disconnect when system service is required. Many of these problems can be reduced or eliminated by selecting conductors only after a careful analysis of system operation, the environment in which the system operates, and manufacturer data sheets for conductors.

Analysis of Circuit and System Operation

Selecting a conductor requires an examination of the hydraulic circuit involved. It also involves examining the total hydraulic system and the mechanisms operated by the system. This complete analysis should not only take into consideration the obvious elements, such as system pressure and flow, but also include machine-related factors. Machine-related factors include such things as vibration and required machine member movement that directly affect conductor performance. The following factors need to be carefully considered when selecting and installing conductors of any type:

■ Pressure rating.
■ Flow characteristics.
■ Temperature.

Pressure Rating

The recommended maximum operating pressure rating of a conductor is usually readily available in catalog materials distributed by the conductor manufacturer. These published ratings show the recommended operating pressure limits for conductors used in a continuously operating system. These recommended maximum pressures are typically no more than 25% of the burst pressure rating of the conductor. Care should be taken to avoid selecting a conductor with a recommended maximum operating pressure that is below anticipated system pressures.

Although every attempt should be made to eliminate operating situations that produce pressure surges, they need to be considered a possibility. Therefore, shock pressures also need to be considered when selecting a conductor for a system. If system pressure surges are frequent, the service life of system components will be considerably shortened. Conductor-related problems caused by pressure surges can be reduced by selecting a conductor that has a higher recommended maximum pressure rating than the normal working pressures in the system.

Flow Characteristics

A conductor must be large enough to accommodate the fluid flow required to operate the system. The inside diameter of the conductor establishes flow capacity. Flow capacity is also directly related to fluid velocity. The higher the fluid velocity, the higher the fluid flow rate for a given conductor. However, if the fluid velocity becomes too high, it can produce system operating problems. Different fluid velocities are recommended for different types of system lines and conductors. For example, the average fluid velocity in a pump inlet line should not exceed four feet per second (4 ft/sec). Average velocity in working lines should be held below 20 feet per second (20 ft/sec).

The velocity of the inlet line fluid is restricted to prevent the formation of a vacuum that is too far below normal atmospheric pressure. Excessively low pressure in an inlet line can result in pump cavitation. Cavitation can cause major pump damage. This problem does not relate to working lines, as they normally operate above atmospheric pressure. A typical working line pressure range is from slightly above atmospheric pressure to the maximum relief valve settings.

The problem with the movement of fluid through working lines is not pressure, but turbulence. Excessive fluid velocity produces turbulent flow, which causes resistance to fluid flow. This results in pressure losses in the system, which are converted into heat. The heat, in turn, results in increased system operating temperatures. The average fluid velocity in lines can be calculated using the formula:

$$V = Q \div A$$

where:

V = Flow velocity in inches per second

Q = Flow rate in cubic inches per second

A = Conductor cross-sectional area in square inches

Care needs to be taken to keep each quantity in the indicated units. The formula, for example, requires converting flow rate in gallons per minute to cubic inches per second. Cubic inches per second can be easily calculated by using the formula:

$$\begin{array}{c} \text{cubic inches} \\ \text{per second} \\ \text{(CIS)} \end{array} = \dfrac{\text{gallons per minute} \times 231 \text{ cubic inches per gallon}}{60 \text{ seconds per minute}}$$

For example, calculate the velocity of the fluid in a conductor with a 1/2″ inside diameter that is moving five gallons per minute.

1. Calculate the inside cross-sectional area of the conductor:

   area = $\pi\, r^2$

   = 3.1416 × .25″ × .25″

   = .196 in²

2. Calculate the flow rate in cubic inches per second:

   CIS = 5 gpm × 231 cubic inches ÷ 60 seconds

   = 19.25 CIS

3. Calculate the average fluid velocity in the conductor:

   velocity = 19.25 CIS ÷ .196 in²

   = 98.21 inches per second

   or

   8.19 feet per second

Temperature

Although pipe and tubing are generally not affected by temperature, both ambient and system operating temperatures must be carefully considered when selecting hose, **Figure 8-26.** Very high or very low temperatures can adversely affect a hose.

Extreme ambient temperatures can reduce hose service life by causing deterioration of the hose cover and reinforcement materials. Continuous system operating temperatures that are at or exceed the rated temperature of a hose will cause deterioration of both the hose inner tube and cover. This also results in a shortened service life. Hose specifications should be carefully examined to assure the hose selected for the system can tolerate the temperatures that will be encountered during system operation.

Interpreting Data Sheets from Manufacturers

A sizable number of companies produce products related to hydraulic fluid conductors. The *Fluid Power Handbook and Directory* lists over 20 product categories for pipe, tubing, and hose. For

Figure 8-26. Temperature-sensing devices range from simple thermometers to elaborate temperature recording devices. System operating temperatures must be carefully considered when selecting hose. (Used with permission of CNH America LLC)

example, the Hydraulic Hose category identified in the directory lists approximately 40 companies that produce hose. In addition to these 20 product categories are a considerable number of other categories related to conductors such as fittings and seals. These companies publish a wide variety of catalog materials and technical data sheets that describe their products in detail.

The published materials usually provide a good deal of information on the products of a company. It can be difficult to obtain comparative information on similar products because no standard format is required by the industry. However, the materials do provide information that can be readily used to select items for a system. Common information includes:

- Construction details and typical applications.
- Dimensions.
- Acceptable operating temperatures and pressures.
- Flow rates for different conductor sizes and pressure drops.

Typically, the information provided for pipe and tubing relates to the material the item is made from and the finished dimensions. The information also specifies the rating of the product in relation to an SAE specification or one from another standardizing group.

Many companies produce hose with a wide variety of features not included in the specification of a rating organization. As such, the information provided on data sheets for hydraulic hose usually includes additional information not found on pipe and tubing data sheets. Hydraulic hose catalogs and data sheets often include construction details, typical applications, and acceptable operating temperature ranges. They also include materials and information related to specification numbers of standardizing organizations. Detailed information concerning the fittings required to connect conductors and specific system components are also available in company catalogs.

Care must be taken to carefully examine catalog data. The exact type of information supplied will vary from one manufacturer to another manufacturer. Hydraulic service personnel need to be reasonably proficient in interpreting catalog information. This proficiency will increase as you work with a variety of catalogs on the job.

Conductor Installation

When selecting conductors for hydraulic-powered equipment, installation factors must be considered in addition to pressure and flow requirements and temperature conditions. When examining the equipment, consider required component movement, conductor flexibility, and potential operating conditions involving vibration and abrasion. These factors influence the types of conductors selected and the methods used to install them in the equipment.

Pipe

When installing pipe as the conductor in a hydraulic system, the primary considerations are:

- Matching the centerline of the pipe to the centerline of the component connections.
- Determining the proper length of conductor.

Pipe sections should fit without being distorted or placed under tension. Both of these conditions will cause stress on the pipe and other components. This stress could result in material fatigue and lead to part failure. Careful planning must also be done to allow easy removal of components during system service. This may require the use of unions to provide an easy means of removing pipe, fittings, and components.

Tubing

Tubing is more expensive than pipe. However, it is more easily installed, lighter in weight, produces less flow resistance, and generally provides a neater installation. The number of fittings can be reduced in a tube installation by bending the tubing whenever possible. Manufacturer specifications indicate the minimum radius of bends for each type, diameter, and wall thickness of tube. Kinks, flattened spots, and wrinkles can be avoided when fabricating tube sections by using the proper bending equipment. Both hand-operated and power benders are available for making up tube sections.

Straight-line connections between components should be avoided in tube installations. These connections place excessive stress on the tube, tube fittings, and system components. The use of strategically placed bends provides connections that will perform very well in most installations.

Long lengths of tube should be supported by brackets or clamps to secure the conductor to the equipment. This is done to prevent fatigue caused by vibration or the weight of the tube and the hydraulic fluid it contains.

Care must also be taken when fabricating sections of tubing to assure the completed tube sections are the correct length and the angles of the bends are appropriate. Tubing sections that are the incorrect length or do not contain the proper bend angles will cause stress on all parts involved. This occurs when they are sprung into position or forced into alignment when fittings are tightened.

Hose

Hose allows the transmission of hydraulic fluid between stationary and moveable machine components. Although hoses are flexible to provide this connection, they must be installed according to manufacturer specifications to obtain maximum service life.

A wide variety of materials and manufacturing methods are used by hydraulic hose producers. This makes selection of a specific hose for an application more complicated. Basic factors to consider when selecting a hose for an application are:

■ Pressure and flow ratings.

■ Chemical resistance to the fluid used in the system.

■ Resistance to external abrasion.

■ Flexibility.

Once these factors have been established, specific installation factors must be considered.

Changes in system pressure must be considered when working with hose. As the pressure increases, the length of the hose tends to decrease and its diameter tends to increase. Always provide some slack in a hose during assembly to allow for length change. See **Figure 8-27.** When the hose passes through openings in machine members, be sure adequate clearance is provided for increases in hose diameter. Also, be careful to avoid any twists in hoses when the installation is complete. See **Figure 8-28.** Pressure tends to straighten the hose, which can actually loosen fittings. In addition, the constant twisting caused by this tendency to straighten under pressure can result in separation of the hose reinforcement and lead to hose failure.

Hose fittings, elbows, and angle adapters should be used to relieve the strain on hose assemblies and maintain as neat and uncluttered of an appearance as possible. See **Figure 8-29.** Also, use standoff brackets wherever hoses must pass close to hot machine parts. These brackets hold the hoses away from the hot parts to allow air circulation. Direct contact with hot parts can shorten the service life of a hose.

Clamps should be used to support long hose runs. This is especially true when there is a possibility of abrasive damage from the hose being rubbed by moving machine members. Do *not*

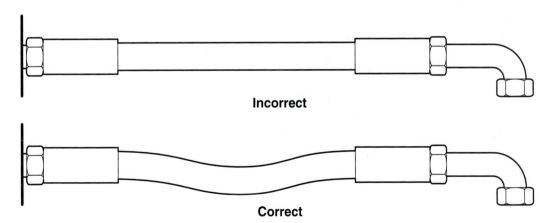

Incorrect

Correct

Figure 8-27. When installing hydraulic hose, always provide slack in lines. This counters the effect of the hose expanding in diameter and decreasing in length as system pressure increases. (Gates Corporation)

Hydraulic systems are capable of easily transmitting and applying power directly to the required work location. In addition, the systems have an advantage over other power-transmission methods because of the ability to readily provide both linear and rotary motion. Major factors relating to these capabilities are the actuators or work devices. Actuators range from large units that can generate huge force or torque outputs to miniature components that produce extremely low power. This chapter deals with basic cylinder and motor designs as well as specialized actuators, such as limited-rotation devices.

Objectives

After completing this chapter, you will be able to:

- Describe the construction and operation of basic hydraulic cylinders, limited-rotation actuators, and motors.
- Compare the design and operation of various types of hydraulic cylinders.
- Select appropriate cylinder design options available for mounting hydraulic cylinders and reducing hydraulic shock.
- Compare the design and operation of various types of hydraulic motors.
- Contrast the operation of fixed- and variable-speed hydraulic motors.
- Describe the construction and operation of a basic hydrostatic transmission.
- Size hydraulic cylinders and motors to correctly meet system force and speed requirements.
- Interpret manufacturer specifications for hydraulic cylinders.

Selected Key Terms

The following terms will be used in this chapter. As you read the text, record the meaning and importance of each. Additionally, you may use other sources, such as manufacturer literature, an encyclopedia, or the Internet, to obtain more information.

barrel
cap
clevis mount
closed-loop circuit
cushioning
double-acting cylinder
effective piston area
fixed-centerline mount
head
hydrostatic drive
limited-rotation actuator
mill cylinders
open-loop circuit
orbiting gerotor motor
parallel circuit
pivoting-centerline mount
rack gear
replenishment circuits
series circuit
single-acting cylinder
telescoping cylinders
threaded-end cylinders
tie-rod cylinder
trunnion mount
wear plates

Internet Resources

www.eaton.com/Eaton/ProductsServices/ProductsbyCategory/Hydraulics/Transmissions/Index.htm
Eaton Corporation
Product literature library on light-duty hydrostatic transmissions designed for applications such as lawn tractors. Includes general descriptions, features, benefits, and performance data.

www.helac.com/products-services/rotary-actuators
Helac Corporation
Provides details, specifications, and applications for hydraulic rotary actuators manufactured by the company.

www.hydraulicspneumatics.com/200/TechZone/HydraulicPumpsM/Article/False/6429/TechZone-HydraulicPumpsM
Hydraulics & Pneumatics Magazine
Examines several designs of low-speed/high-torque hydraulic motors.

www.hydraulicspneumatics.com/200/TechZone/HydraulicPumpsM/Article/False/6472/TechZone-HydraulicPumpsM
Hydraulics & Pneumatics Magazine
Provides information on a number of circuits basic to effective system operation. Actuator braking, replenishment, and series/parallel connections are covered in detail.

www.iqsdirectory.com/hydraulic-motors/hydraulics-motors
Industrial Quick Search Manufacturer Directory
Provides a variety of information on hydraulic motors. Includes links to other manufacturer sites that feature small motors; large, industrial motors; and mobile units.

www.parker.com/literature/Literature%20Files/pneumatic/Literature/Actuator-Cylinder/HY03-1800.pdf
Parker Hannifin Corporation
Catalog of hydraulic rotary actuators manufactured by the company. Provides product information, applications, and a selection guide.

www.peninsularcylinders.com
Peninsular Cylinder Company
Homepage of a fluid power cylinder manufacturer. Explore the hydraulic cylinder section of the site, which provides details of cylinder features.

Hydraulic Cylinders

A hydraulic *cylinder* is the hydraulic system component that converts fluid pressure and flow into linear mechanical force and movement. See **Figure 9-1.** Often called a *linear actuator,* in a sense it extracts energy from the hydraulic system fluid and converts it into force and mechanical movement. The *force* generated by the cylinder is determined by the area of the piston operating within the cylinder and the system operating pressure. The *speed* of movement is controlled by the effective volume of the cylinder per unit of travel and the rate of fluid flow entering the component. Effective volume is the volume of the cylinder minus the volume of the rod (if applicable). The maximum *distance* of movement is controlled by the physical length of the cylinder. Controlling system pressure and flow rates with even the most simple designs allows a wide range of cylinder forces and speeds.

Figure 9-1. The primary purpose of a cylinder is to convert fluid pressure and flow into linear mechanical movement.

Basic Cylinder

A basic cylinder consists of a movable element made up of a piston and piston rod assembly. The piston and piston rod assembly operate within a close-fitting cylindrical tube. One end of the tube is closed, while an opening in the opposite end allows the piston rod to extend outside of the component for attachment to machine members. Each cylinder end also contains a port connected to the system allowing fluid to be forced into and out of the cylinder. Seals are incorporated in the cylinder design to prevent fluid leakage around the piston. This increases cylinder efficiency and prevents the leakage of fluid to the exterior of the cylindrical tube.

This design forms two sealed chambers inside of the cylinder that are separated by the piston. See **Figure 9-2.** Forcing system fluid into the chamber between the closed end of the cylinder and the piston causes the piston and piston rod assembly to move, enlarging the volume of that chamber. The piston movement reduces the volume of the second chamber, forcing fluid out while extending the piston rod. Forcing fluid into the chamber on the opposite side of the piston reverses the process, causing the piston rod to retract. The piston rod is usually attached to the movable machine element. The body of the cylinder is typically anchored to a stationary machine member.

Each of the parts of a basic cylinder are identified by specific terms:

- Barrel.
- Head.

Figure 9-2. Sealed chambers are formed on either side of the piston of a cylinder. Fluid pressure acting on the piston surfaces moves the piston and rod assembly.

- Cap.
- Piston.
- Rod.
- Seals.
- Rod wipers.

These names are typically used in the fluid power industry. The next sections describe the materials and processes commonly used for the construction of these parts.

Barrel

Figure 9-3 shows a parts list for a typical cylinder. The tube portion of a cylinder is called the *barrel*. It is made from metal tubing that has been machined to provide the desired bore diameter. The surface finish on the walls of the bore is obtained by honing. A proper finish produces a minimum of piston seal wear as the cylinder extends and retracts during operation. Some cylinder manufacturers chrome plate the barrel bore surface to increase the service life of the cylinder.

Wall thickness and the type of material used for construction of the barrel are two important factors that establish the maximum safe operating pressure of a cylinder. Manufacturers must consider the weight, strength, and bulk of materials in order to provide a cylinder that is as light and compact as possible, while providing adequate strength for safe operation at specified system operating pressures.

REPLACEMENT PARTS			
ITEM	**DESCRIPTION**	**ITEM**	**DESCRIPTION**
1, 1A, & 1B	ROD SEAL KIT **	10	REAR CAP
1, 1A, 1B, & 2	ROD GLAND KIT **	11	TIE RODS AND NUTS
3	RETAINER RING/PLATE	13	CUSHION NEEDLE ASSEMBLY
5	BALL CHECK ASSEMBLY	14	FRONT HEAD
7	PISTON ROD	15	CYLINDER TUBE (BARREL)
8	CUSHION SPUD W/ SEAL	16	PISTON
9 & 9A	PISTON SEAL KIT	17	CUSHION STAR ASSEMBLY

Figure 9-3. The component parts of a cylinder must withstand the expected maximum system operating pressure and provide a positive seal to prevent fluid leakage. (Yates Industries, Inc.)

Head and cap

The ends of the barrel are closed by component parts called the head and the cap. The *head* closes the end through which the cylinder rod passes. The *cap* closes the opposite end, the end without the rod. The head is often called the *rod end*, while the cap is called the *blind end*. Seals are provided at both ends of the cylinder to prevent fluid leakage between the barrel and the head and cap.

The hole that allows the cylinder rod to pass through the head contains a bearing surface to support the rod during extension and retraction. It also contains a seal to prevent leakage around the rod. The head and cap also contain the ports that allow system fluid to enter and exit the cylinder during extension and retraction. The cap is often used to attach the cylinder to a stationary machine member.

Piston and rod

The piston and rod are usually thought of as the working parts of the cylinder. The *piston* separates the cylinder into two, variable-size chambers. A positive seal is provided between the chambers by piston rings or other seals located between the piston and the bore surface of the barrel. Applying pressure on one side of the piston produces the force needed to either extend or retract the cylinder. Applying force on the opposite side of the piston reverses the cylinder.

The piston is machined to a specific diameter to accommodate the type of seal used by the manufacturer. The width of the piston is also carefully considered to provide an adequate bearing surface to withstand side loading as the cylinder extends. *Side loading* is force applied to the rod/piston assembly that is not parallel to the axis of the rod.

The *rod* is connected to the piston and transfers the force generated to act against the load. Rods are usually made from high-tensile-strength steel to withstand heavy operating loads. The surface of a rod is often chrome plated to extend service life under dirty operating conditions.

Seals and rod wipers

Leakage in any part of a hydraulic system results in reduced operating efficiency, reduced cleanliness, and increased safety hazards. Seals are incorporated in several locations in a basic cylinder to prevent internal and external leakage of system fluid, **Figure 9-4**. These locations include:

- Contact points between the barrel and the head and cap.

- Clearance between the piston and barrel.

- Clearance between the rod and the opening in the head.

Seals can be classified as static or dynamic. *Static seals* are materials compressed between two rigid, nonmoving parts. They may be flat, paper gaskets; molded synthetic O-rings; or a variety of other materials or forms. *Dynamic seals* function between parts that are moving relative to each other. These seals may be formed from natural material (such as leather), molded synthetic materials, or machined metal compression rings. These seals are subject to wear because at least one of the moving parts must move across the seal material.

The seal between the rod and the opening in the cylinder head is often made up of two parts. The first seal prevents external leaks. The second is a *wiper seal* or *scraper* that prevents water, dirt, or other contaminants from entering the system as the rod is drawn back into the cylinder during the retraction stroke. See **Figure 9-5**.

Classification of Cylinders

A number of variations from the basic cylinder design are available for use in fluid power systems. This variety is necessary to provide the type of linear movement and force application required by the many machines using fluid power. Two common methods to classify these cylinder variations are by basic operating principles and construction type. A number of other cylinder forms that do not fit into these classifications are also discussed in this section.

Figure 9-4. A wide variety of seal designs are used to assure leak-free cylinder operation.

Operating principles

Cylinder operation may be classified as single acting or double acting. A *single-acting cylinder* exerts force only on extension or retraction. It depends on some outside force to complete the second movement. A *double-acting cylinder* can exert force on both extension and retraction. See **Figure 9-6.**

Figure 9-5. When a cylinder operates under dirty conditions, a wiper seal or scraper can be used to prevent contaminants from entering the system. A—A wiper seal on a cylinder. B—A detailed cross section of a wiper seal. (Used with permission of CNH America LLC; IMI Norgren, Inc.)

Figure 9-6. Cylinders may be classified as either double-acting or single-acting. A—A single-acting cylinder must use external force to return the piston. B—The piston in a double-acting cylinder can be returned by pressurizing the opposite side of the piston.

Mounts are not on the
cylinder centerline

Figure 9-21. Fixed-noncenterline cylinder mounting can cause stress to machine members because the force generated attempts to rotate the actuator around the mounting point.

Clevis

Figure 9-22. Pivoting-centerline cylinder mounting allows the cylinder to follow a curved path as it applies force to a moving machine member. This is a clevis mount.

off-center attachment position causes the forces generated during system operation to attempt to rotate the cylinder around the mounting points. This rotating force can apply undesirable stress to machine and actuator parts, causing service problems, unless cylinders are carefully sized for an application.

The feet can be angle brackets bolted to the cylinder head and cap, lugs cast as part of or welded to the head and caps, or holes drilled and tapped into the head and cap. Each of these designs allows direct mounting of the cylinder using bolts or machine screws.

Pivoting centerline

Equipment often includes arms or other machine members that move in an arc as they perform their task. These machine elements are often powered by a cylinder with a *pivoting-centerline mount,* which allows the cylinder to pivot to follow the movement. Two mounting methods are used to allow the cylinder to follow this curved path. The lines of force caused by cylinder loading are concentric with the centerline of the cylinder in both of these mounts.

The first mounting method is the clevis mount. A clevis is a U-shaped shackle. A *clevis mount* includes a clevis on the cylinder cap. The cap is normally attached to the body of the machine. Another clevis is included on the end of the cylinder rod. This is attached to the movable machine member. A clevis mount is shown in **Figure 9-22.**

The second mounting method is the *trunnion mount.* This mount has pivot pins, or trunnions, mounted on the sides of the cylinder. See **Figure 9-23.** The trunnions may be positioned on head, cap, or at any point in between in order

to obtain a desired pivoting action. The trunnions are attached to the body of the machine. A clevis or similar device is used to attach the cylinder rod to the movable machine member.

Cylinder Calculations

The selection of cylinders for a particular application requires that a series of calculations be completed. These calculations establish that the cylinders can perform at the levels indicated by system specifications using the system power unit and recommended pressure levels and flow rates. Calculations include:

■ General cylinder extension/retraction force and speed rates.

■ Speed, pressure, and flow rates needed to generate force and speeds for specific cylinders.

Force

The force a cylinder can generate depends on system pressure and the effective area of the piston. *Effective area* is defined as the surface area of the piston exposed to system pressure. See **Figure 9-24.** During the extension of a double-acting cylinder, the effective area is equal to the area of the piston on the side without the rod. During retraction, the effective area is the area of the piston *minus* the cross-sectional area of the piston rod.

This formula is used to calculate the force of a cylinder on the extension stroke:

extension force (Ef) = system pressure (Sp)
× piston area (Pa)

Trunnions

Fixed mount

Load resistance

A

B

Figure 9-23. A—A trunnion mount allows a cylinder to pivot if a machine member is required to move through an arc in its operation. B—This is a large cylinder with a trunnion mount. (Yates Industries, Inc.)

End	Effective Area	Calculation
Blind end of cylinder		Effective area = Total area of piston face
Rod end of cylinder		Effective area = Total area of piston face – Area of rod cross section

Figure 9-24. An understanding of the effective area of the cylinder piston is critical when calculating cylinder force, speed, and required flow rates.

This formula is used to calculate the force of the cylinder on the retraction stroke:

retraction force (Rf) = system pressure (Sp)
× [piston area (Pa) – rod area (Ra)]

where:

Rf = total retraction force in pounds

Sp = system pressure in pounds per square inch

Pa = piston area in square inches

Ra = piston rod area in square inches

A closely associated, additional calculation that is often desired is the system pressure needed to produce a required force while using a cylinder of a specific size. This calculation is accomplished by modifying the above formulas:

$$\text{system pressure (Sp)} = \frac{\text{cylinder force (Cf)}}{\text{effective piston area (Ea)}}$$

where:

Sp = system pressure in pounds per square inch

Cf = force in pounds needed to be generated by the cylinder

Ea = effective area of the piston in square inches

where:

Ef = total extension force in pounds

Sp = system pressure in pounds per square inch

Pa = piston area in square inches (area = π × radius²)

Speed

The speed at which a cylinder rod extends or retracts depends on the volume displaced per inch of piston travel and the amount of fluid entering the cylinder. The effective area of the piston, as defined in the cylinder force calculations above, must be used to determine cylinder speed.

The extension speed of a cylinder is calculated using this formula:

$$\text{extension speed (Es)} = \frac{\text{flow delivery rate (Fr)} \times \text{cubic inches in one gallon (Cg)}}{\text{piston area (Pa)}}$$

where:

Es = extension speed in inches per minute

Fr = flow rate in gallons per minute

Cg = 231 cubic inches per gallon

Pa = piston area in square inches (area = $\pi \times \text{radius}^2$)

The following formula is used to calculate the speed of the cylinder on the retraction stroke:

$$\text{reaction speed (Rs)} = \frac{\text{flow delivery rate (Fr)} \times \text{cubic inches in one gallon (Cg)}}{[\text{piston area (Pa)} - \text{rod area (Ra)}]}$$

where:

Rs = retraction speed in inches per minute

Fr = flow rate in gallons per minute

Cg = 231 cubic inches per gallon

Pa = piston area in square inches

Ra = piston rod area in square inches

Another closely associated calculation is the system flow rate needed to produce a desired extension or retraction speed for a cylinder of a specific size. This calculation is accomplished by modifying the above formulas:

$$\text{system flow rate (Fr)} = \frac{\text{effective cylinder area (Ea)} \times \text{cylinder speed (Cs)}}{\text{cubic inches in one gallon (Cg)}}$$

where:

Fr = flow rate in gallons per minute

Ea = effective area of the piston in square inches

Cs = cylinder speed in inches per minute

Cg = 231 cubic inches per gallon

Interpreting Cylinder Manufacturer Specifications

A large number of different companies produce hydraulic cylinders. The online *Fluid Power Manufacturers Directory* (http://manuf.hydraulics-pneumatics.com) lists approximately 60 companies that manufacture various types of hydraulic cylinders. The products are grouped into 14 categories to help in searching. To see the list, select cylinders subcategories, then hydraulic types subcategories. With the amount of information available about this wide selection of cylinders, it is often difficult to interpret and compare the data. This section discusses cylinder specifications often included in manufacturer literature or data sheets.

Typical specifications

Manufacturers generally provide detailed specifications for each of their cylinder lines. The format for presenting the specifications varies considerably between manufacturers. However, the following factors are often readily available from company literature.

- **Duty service.** Light-, medium-, or heavy-duty based on the specifications of national standards groups.

- **Construction type.** Tie-rod, mill, threaded-end, or one-piece designs.

- **Pressure rating.** Depends on the bore size of the cylinder barrel.

- **Fluid rating.** Hydraulic oil or other specialized fluid that may require special materials for seals.

- **Acceptable operating temperature.** May range from below 0°F to over 165°F.

- **Bore size.** The diameter of the cylinder barrel bore may range from 1″ to well over 8″, depending on the cylinder series.

- **Piston rod diameter.** The rod diameter may range from 1/2″ to over 6″, depending on the cylinder series.

- **Mounting styles.** Most cylinder series provide a variety of mounting configurations to facilitate attachment to machines, including flange, lugs, clevis, trunnion, and so on.

- **Stroke.** Most manufacturers indicate that their cylinders are available in any practical stroke length.
- **Cushions.** Usually optional at either end of the cylinder.
- **Rod end configuration.** Several thread styles and sizes are considered standard, with several others available as a custom option.
- **Ports.** NPTF pipe threads are normally used, but other styles, such as O-rings, are acceptable.

Display of cylinder specifications

Manufacturer literature typically includes a series of descriptions, illustrations, and tables that contain the information required to select a cylinder to meet the specifications of an installation. **Figures 9-25** through **9-27** are materials taken from the cylinder section of a cylinder manufacturer's catalog. These materials are representative of the approaches used by companies to indicate the characteristics of their products. This informa-

tion is critical when a machine designer is developing equipment. It is also important when service personnel are selecting cylinders to replace failed components or upgrading components to improve equipment performance.

The cylinder section of a typical catalog or manufacturer website includes dimension drawings of each basic cylinder design. Rather than showing specific sizes on the drawings, the dimensions are labeled with code letters that can be cross-referenced with dimensions shown on the tables associated with the drawings. These tables include information on all of the critical dimensions of a cylinder, including the exterior envelope, fittings for mounting, rod diameter and length, and rod end fittings. **Figure 9-25** illustrates a typical basic cylinder type and identifies critical dimensions. The table below the drawing contains the dimensions for each bore size available for this specific cylinder model. The drawings and tables provide the machine designer or technician with all of the dimensional information required for selecting a cylinder.

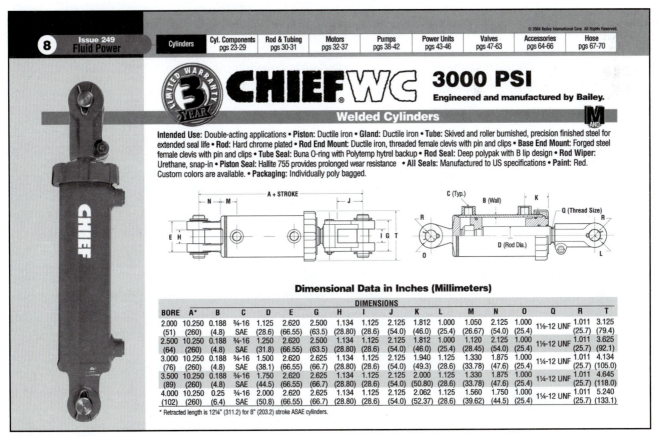

Figure 9-25. Manufacturer catalogs provide a wealth of information about the construction and size of their cylinder models. (Bailey International Corporation)

BAILEY NO.	STROKE	ROD DIAMETER	RETRACT	COLUMN LOAD	PORT SIZE	PIN DIAMETER	SHIP WT.
2" Bore • 3000 PSI							
286-001	4"	1⅛"	14¼"	Full PSI	SAE 8	1"	12
286-002	6"	1⅛"	16¼"	Full PSI	SAE 8	1"	15
286-003	8" ASAE	1⅛"	20¼"	Full PSI	SAE 8	1"	17
286-004	8"	1⅛"	18¼"	Full PSI	SAE 8	1"	16
286-005	10"	1⅛"	20¼"	Full PSI	SAE 8	1"	17
286-006	12"	1⅛"	22¼"	Full PSI	SAE 8	1"	19
286-007	14"	1⅛"	24¼"	Full PSI	SAE 8	1"	20
286-008	16"	1⅛"	26¼"	Full PSI	SAE 8	1"	21
286-009	18"	1⅛"	28¼"	8100 lbs.	SAE 8	1"	23
286-010	20"	1⅛"	30¼"	6880 lbs.	SAE 8	1"	26
286-011	24"	1⅛"	34¼"	5140 lbs.	SAE 8	1"	27
286-014	30"	1⅛"	40¼"	3550 lbs.	SAE 8	1"	31
286-015	36"	1⅛"	46¼"	2600 lbs.	SAE 8	1"	39
286-100		1⅛"		WC Packing Kit			
2½" Bore • 3000 PSI							
286-016	4"	1¼"	14¼"	Full PSI	SAE 8	1"	17
286-017	6"	1¼"	16¼"	Full PSI	SAE 8	1"	18
286-018	8" ASAE	1¼"	20¼"	Full PSI	SAE 8	1"	17
286-019	8"	1¼"	18¼"	Full PSI	SAE 8	1"	20
286-020	10"	1¼"	20¼"	Full PSI	SAE 8	1"	21
286-021	12"	1¼"	22¼"	Full PSI	SAE 8	1"	23
286-022	14"	1¼"	24¼"	Full PSI	SAE 8	1"	25
286-023	16"	1¼"	26¼"	Full PSI	SAE 8	1"	27
286-024	18"	1¼"	28¼"	12500 lbs.	SAE 8	1"	28
286-025	20"	1¼"	30¼"	10610 lbs.	SAE 8	1"	29
286-026	24"	1¼"	34¼"	7920 lbs.	SAE 8	1"	33
286-027	30"	1¼"	40¼"	5470 lbs.	SAE 8	1"	38
286-028	36"	1¼"	46¼"	4000 lbs.	SAE 8	1"	46
286-101		1¼"		WC Packing Kit			

Figure 9-26. Manufacturer catalogs provide details covering cylinder bore, stroke, and other basic envelope dimensions. (Bailey International Corporation)

Figure 9-27. Many cylinder catalogs provide illustrations covering the configurations available for a cylinder rod end. (Bailey International Corporation)

Limited-Rotation Hydraulic Actuators

Machine shafts and other components often require torque in situations where rotation is limited to one or two revolutions or less. If the demanded rotation is less than 180°, a hydraulic system may use a standard linear cylinder attached to a lever to produce the required motion and torque. If this arrangement cannot produce the desired rotation and torque or the components are too bulky for the space allowed in the machine design, a limited-rotation actuator may provide a desirable solution. A *limited-rotation actuator* provides restricted rotary or reciprocating rotary motion for lifting/lowering, opening/closing, or indexing operations. These devices are sometimes called *torque motors*.

Design and Construction

A variety of designs are used in limited-rotation actuators. The three most common designs are:

- Rack-and-pinion.
- Vane.
- Helical shaft.

Other designs that may be found are units consisting of a cylinder, piston, connecting rod, and crankshaft or a cylinder, piston, chain, and sprocket. The three most common design types are discussed in the next sections.

Rack-and-pinion

A rack-and-pinion design consists of a rack gear and a pinion gear powered by two concentric cylinders. The *rack gear* is machined into the cylinder rod with the teeth along its length. The *pinion gear* is a round gear with teeth that mate with the rack. The pistons of the two cylinders are attached to the ends of the cylinder rod containing the rack gear. See **Figure 9-28**. The cylinder chambers are located on either end of the rack gear/piston assembly.

The housing also contains the pinion gear attached to a power-output shaft. The axis of the output shaft is perpendicular to the axis of the rack gear located on the cylinder rod. The housing holds the pinion gear in constant mesh with the rack gear.

Pressurizing the cylinder chamber on one end forces the rack gear/piston assembly to move. This movement turns the pinion gear, thus rotating the output shaft. Pressurizing the opposite cylinder chamber forces the rack gear and piston assembly in the opposite direction. This reverses the direction of rotation of the output shaft.

Figure 9-28. A common type of limited-rotation actuator uses a rack-and-pinion to produce rotary motion. (IMI Norgren, Inc.)

external-tooth drive gear attached to the motor output shaft. It also contains a meshed follower gear that turns freely on a fixed bearing shaft attached to the housing, **Figure 9-36.** The output shaft is supported on either end by bearings located in the housing. The drive and follower gears are closely meshed and tightly fit into the housing. There are minimal clearances between the sides of the gears and the housing and the tips of the gear teeth and the housing.

The primary housing cavity is oval, **Figure 9-37.** The end portions of the cavity match the diameter of the drive and follower gears. This primary cavity is further divided into a number of small, generally triangular cavities formed by the space between the gear teeth and the cavity wall. The middle portion of the primary cavity does not follow the curve of the gear diameters, rather it forms two larger, triangular, open cavities. The inlet and outlet ports of the motor are located in these cavities. Because of the precision fit between the gear teeth, gear sides, and walls of the housing, a minimal amount of fluid leakage occurs between all of these cavities. Separate pressure- or spring-loaded wear plates are used in some gear motor designs. These *wear plates* control gear end play and leakage around the sides of the gears.

Figure 9-37. The primary housing cavity for an external gear motor is oval.

When system fluid is directed into the inlet port of the motor, the fluid pressure begins to increase. This is an attempt to overcome the resistance to motor rotation caused by friction of the internal parts of the motor and the load on the output shaft. When the pressure difference between the inlet side and the outlet side is high enough to generate adequate force on the gear teeth, the motor turns. As a result, torque is produced at the output shaft. The pressurized fluid entering the motor inlet cavity is carried in the small cavities formed by the space between the gear teeth to the outlet cavity. It is then returned to the system reservoir at near atmospheric pressure. The motor continues to turn as long as the force generated by system pressure can overcome the internal motor resistance and the shaft load. Reversing fluid flow though the motor reverses shaft rotation.

Internal gear motors

Internal gear hydraulic motors use a basic gerotor configuration with several component variations in their construction. These gear motors are fixed displacement and pressure unbalanced.

The construction of the gerotor involves a housing that contains a cavity fitted with three rotating elements. See **Figure 9-38.** The first element rotates, while using the walls of the housing cavity as a bearing surface. This element is a disk-like component

Figure 9-36. A basic external gear motor design.

Figure 9-38. The internal gear motor uses a gerotor design similar to the gerotor pump. (Courtesy of Eaton Fluid Power Training)

containing specially shaped internal gear teeth. The second element is a smaller-diameter disk constructed with external gear teeth of a matching design. The third element is the power output shaft. This is placed off center in the opening of the disk with the internal teeth. The gear with the external teeth is securely mounted on the output shaft. This gear has one less tooth than the gear with the internal teeth. As a result, the gear teeth can mesh and the shaft and gears can freely turn. The key to the motor's operation is the fact that the tips of the external teeth are always in contact with the surface of the internal teeth. This produces a series of individual chambers that increase and decrease in volume as the unit rotates.

The final design feature of the gerotor motor is the placement of an inlet and an outlet port. These crescent-shaped ports are located in an area of the housing that contacts the side of the rotating gear elements. The leading edge of the inlet port is located at the point where the volume of each chamber begins to enlarge. The trailing edge of the inlet port is located where the chamber has reached its maximum volume. The leading edge of the outlet port is located where the chamber volume begins to decrease. The trailing edge of the outlet port is located near the point where the chamber volume is the smallest.

When system fluid is directed into the inlet port, the fluid pressure begins to increase to overcome

internal motor friction and the external load. Several chambers formed by the clearance between the internal and external gear teeth are exposed to this pressure. Once the pressure increase has generated adequate force to rotate the motor, the system pump supplies the fluid needed to fill the motor chambers as they enlarge during motor rotation. As the motor rotates, the chambers connected to the outlet port decrease in volume. This pushes the spent fluid, which is near atmospheric pressure, into the system lines leading to the reservoir. The motor continues to rotate as long as the system continues to supply fluid and the pressure is sufficiently high to overcome the internal motor resistance and system load.

A variation of the gerotor motor is the *orbiting gerotor motor.* In this design, the gear with the internal teeth is fixed and does not rotate. The gear with the external teeth rotates. The center point of this gear follows an elliptical path around the center point of the fixed gear element. Instead of crescent-shaped ports, this design has a rotating valve plate that turns at the same speed as the rotating inner gear. The openings in the valve plate allow the appropriate chambers to be pressurized as they are increasing in volume. It also connects the chambers to the reservoir when they are decreasing in size.

A design variation of orbiting gerotor motors incorporates modifications to both gear elements. See **Figure 9-39.** The gear that normally has

Figure 9-39. A variation of the gerotor design is the orbiting gerotor. This design provides reduced wear by using rollers. (Courtesy of Eaton Fluid Power Training)

Figure 9-45. A circuit with two hydraulic motors in series.

system operation results in relative speed changes. **Figure 9-46B** shows a parallel circuit where pressure-compensated flow control valves are used to divide pump fluid output. This assures uniform flow to the motors. A variety of design variations can be used with parallel connections to provide a range of individual flow and pressure settings needed to obtain the performance required from individual motors in a circuit.

Braking circuits

When hydraulic-motor-driven equipment involves a heavy rotating load, the inertia of that load must be considered in designing the system. *Inertia* is the tendency of an object to remain either in motion or at rest. A heavy, rotating object has a lot of inertia. Under these conditions, stopping the motor involves more than placing a directional control valve in the inlet or outlet lines of a hydraulic motor. This is because the inertia of the rotating load will tend to maintain the rotation of the hydraulic motor even after the directional control valve has been shifted to stop fluid flow from the pump. This continued application of rotational force causes the motor to act as a pump. It attempts to draw fluid in through the inlet port and move fluid through the outlet port to other sections of the circuit. The result can be cavitation in the inlet port area, if the pressure drops too low, or very high shock pressures in the outlet area, if the discharge fluid has no escape route. To prevent these

conditions, a *motor braking circuit* can be used to allow a motor to be driven under normal load situations, then coast to a stop when desired or quickly stopped using hydraulic braking when a shorter deceleration time is required.

Figure 9-47 shows a basic motor braking circuit that uses a three-position directional control valve to provide the three operating conditions. Position 1 for the directional control valve is the normal drive position for the motor. Pump flow passes through the motor and produces rotation until the load causes system pressure to increase to the point where the system relief valve functions. The relief valve then returns fluid to the reservoir and the motor stalls. When the directional control valve is shifted to the center position (position 2), both sides of the motor and the pump are connected to the reservoir. This allows the motor and its load to coast to a stop.

However, when the directional control valve is in position 3, the pump outlet is directly connected to the reservoir. This allows the pump output to be dumped without going through the motor. The outlet of the motor is also connected to the reservoir, but through the brake relief valve. The inlet of the motor remains connected to the pump outlet line. If there is adequate inertia from the load, the motor continues to rotate and begins to act as a pump. This draws fluid in from the pump line and forces it through the brake relief valve to the reservoir. When the directional control valve is in position 3, motor deceleration is provided by using the

Figure 9-46. A circuit with two hydraulic motors in parallel. A—A basic parallel circuit. The motors and loads must be balanced. B—This parallel circuit provides adjustment for each motor branch in the circuit.

pressure setting of the brake relief valve. Also, a ready supply of fluid is ensured in the motor inlet port to eliminate cavitation.

Replenishment

Replenishment circuits replace the fluid lost through system leakage. They are used in hydraulic systems that continuously circulate fluid through the motor and pump without returning it to a large-volume system reservoir. Typically, the reservoir of these systems is very small. A replenishment circuit provides makeup fluid to replace any fluid that is lost during system operation. These circuits are also used to make up the fluid dumped over the relief valve in motor-braking circuits.

Figure 9-48 illustrates a replenishment circuit. This circuit works in conjunction with a braking system for a bidirectional motor. A variable-displacement pump powers the circuit. The system maintains the volume of the hydraulic fluid in the line functioning as the connection between the low-pressure motor outlet and pump inlet. The relief valves are set to maintain the desired operating pressure in that line. The replenishing check valves allow fluid to be supplied to either side of the system, depending on the position of the variable-displacement system pump. The position of the system pump controls the direction of motor rotation.

Hydrostatic Drives

Designing a compact drive unit that allows easy and accurate control of output shaft power is a challenge for any machine designer. A relatively simple solution to this challenge has emerged in

Valve Position
1 = Power
2 = Freewheeling
3 = Braking

Figure 9-47. A basic braking circuit using a three-position directional control valve.

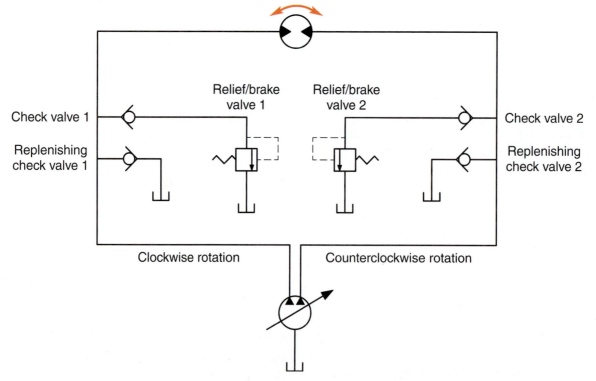

Figure 9-48. A replenishment circuit may be used to make up fluid that is lost during the operation of closed-loop hydraulic motor circuits.

the form of hydrostatic drive devices, **Figure 9-49.** A *hydrostatic drive* is simply a drive unit where a pump provides fluid to drive a motor. Like other drive units, these devices are not prime movers. They depend on an electric motor, internal combustion engine, or some other device for power. However, hydrostatic drives efficiently transmit power. They also have configurations that allow them to easily adjust and accurately control output

A

B

Figure 9-49. Hydrostatic drives provide an easy means of transmitting and controlling power in consumer products as well as heavy industrial applications. A—A common lawn tractor with a hydrostatic drive. B—A skid steer with a hydrostatic drive. (MDMA Equipment—Menomonie; Used with permission of CNH America LLC)

shaft speed, torque, horsepower, and direction of rotation. In addition, when compared to conventional transmissions, these units:

- Have a high power output–to–size ratio.
- May be stalled under full load with no internal damage.
- Accurately maintain speed under varying load conditions.
- Provide an almost infinite number of input/output speed ratios.

Conceptual Circuit Construction and Operation

Hydrostatic drives consist of basic components typical of other hydraulic motor circuits. These include a hydraulic pump that generates fluid flow, a hydraulic motor that produces the rotary motion, lines to transmit fluid between components, valves to control direction and pressure, and a reservoir to store fluid. These components may be arranged using either open-loop or closed-loop circuit configurations.

The *open-loop circuit* configuration has a layout typical of a basic hydraulic motor circuit. See **Figure 9-50.** The pump inlet is connected to the reservoir and the outlet to a directional control valve. The directional control valve is connected to the motor with the fluid returning from the motor directed to the reservoir. A relief valve controlling maximum system pressure is located between the pump and directional control valve. This design allows the motor to be stopped or reversed by shifting the directional control valve. Motor speed depends on the relationship between the displacement of the pump and motor. The reservoir serves to compensate for any fluid lost from system leakage.

The *closed-loop circuit* configuration uses a layout where the outlet of the pump is directly connected to the inlet of the motor. The outlet of the motor is plumbed to the inlet of the pump. The only control valve used in this configuration is a relief valve that sets the maximum pressure that may be developed in the system. A small replenishment pump and reservoir are used to make up any fluid lost during system operation. As a result, these systems are much more compact than the open circuit design. See **Figure 9-51.**

Figure 9-50. A circuit configuration typical of a basic, open-circuit hydrostatic drive.

When the pump rotates in the closed circuit design, the system fluid is moved directly to the inlet of the motor. As the motor rotates, the fluid passing through the unit is returned directly to the inlet of the pump. The speed of the motor is established by the relationship between the displacement of the pump and motor. The replenishment pump replaces any fluid that leaks from the lines or other components.

Pump-Motor Arrangements

All hydrostatic drives involve a hydraulic pump and hydraulic motor that work together to transmit and control primer mover power. The pumps and motors may be either fixed or variable displacement. There are four possible pump/motor, fixed/variable combinations. The following discussion is based on controlling maximum circuit operating pressure using a relief valve with fixed-displacement pumps and pressure compensation with variable-displacement pumps. The discussion also assumes the prime mover is running at a constant speed.

In the first pump/motor arrangement, both components have fixed displacements. This combination results in a drive unit where the maximum horsepower, torque, and output shaft speed are fixed. These cannot be adjusted.

In the second pump/motor arrangement, the pump has a fixed displacement and the displacement of the motor is variable. A drive unit using this combination has a fixed maximum horsepower output, but torque and speed are variable. Output shaft rotation may be reversed if the

variable-displacement motor is reversible. The need for a relief valve to control maximum system pressure lowers the efficiency of any drive unit using a fixed-displacement pump.

The third pump/motor arrangement has a drive unit with a variable-displacement pump and a fixed-displacement motor. This arrangement produces a hydrostatic drive having variable horsepower and speed capability. However, the torque output is constant. Output shaft rotation may be reversed if the design of the pump allows the flow to be reversed.

In the fourth pump/motor arrangement, the pump and motor have variable displacements. In this drive, the horsepower, torque, and output shaft speed and direction can be changed. This arrangement is the most versatile of the pump/motor combinations. It allows a wide selection of torque- and speed-to-power ratios.

Hydrostatic Transmissions

A hydrostatic drive is typically considered a *hydrostatic transmission* when the pump, motor, or both have a variable displacement. Without the variable feature, the drives can be compared to a basic gear train that simply transmits power from the prime mover to the load. With the variable feature, hydrostatic transmissions can be manually or automatically adjusted to control torque, speed, and power.

Two general types of construction are used with hydrostatic transmissions: integral and nonintegral. *Integral construction* combines all of the transmission parts into a single

Figure 9-51. A circuit configuration typical of a basic, closed-circuit hydrostatic drive. A—Schematic with symbols. B—Pictorial schematic. (Sauer-Danfoss, Ames, IA; Used with permission of CNH America LLC)

housing. This includes the pump, motor, fluid lines, valving, and accessories. In addition, the housing may contain axles and mountings for the prime mover and machine parts. This type construction is generally used in transmissions in lower-horsepower devices. In *nonintegral construction,* the pump is attached to the prime mover and the motor is attached to the workload. Hoses or tubing assemblies connect the pump, motor, and accessories.

which the leakage passes is not properly drained. If the fluid cannot escape from the chamber, there is the possibility of fluid buildup. This will cause an increase in pressure, which can result in the valve elements not properly functioning. Valve failure does not usually occur. When it does, the valve will not properly operate. It may not respond when an operator attempts to adjust a pressure or flow setting.

Valves may be internally or externally drained to remove the fluid that results from internal leakage. See **Figure 10-10.** Internal drains connect the valve chambers that need to be drained to the normal outlet of the valve. Valves with outlet or return lines that are *never* subjected to system operating pressure can be internally drained. External drains connect the valve chambers that need to be drained to an external line connected to a low-back-pressure return line leading to the system reservoir. If a valve does not have a separate return line and its outlet line is subjected to system operating pressure, it must be externally drained.

Identifying Normal Valve Position

Control valves are often classified by their *normal operating position.* This refers to the position the internal elements assume when the system is shut down. Although it is difficult to classify all control valves in relation to their normal operating position, the classification can be useful when analyzing the operation of a hydraulic component or circuit.

Typically, pressure control valves are either fully open or fully closed when a hydraulic system is shut down. A *normally open* valve shifts to a fully open position when the system is shut down. A *normally closed* valve shifts to a fully closed position when the system is shut down. See **Figure 10-11.** When the system is operating, the internal pressure-control elements shift to provide the desired system performance. Depending on the valve setting and the load on the system, the position of the internal elements may range from fully open to fully closed during system operation. A valve's normal position classification relates only to the position when the valve is not subjected to system load.

It may be more difficult to classify the normal position of directional control valves. A primary

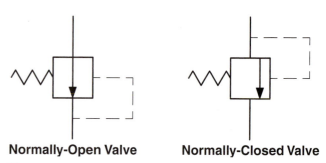

Normally-Open Valve   **Normally-Closed Valve**

Figure 10-11. Pressure control valves are typically classified by the position the internal parts of the valve assume when the system is not operating. These positions are identified as normally open or normally closed.

System return line
to reservoir

Internally Drained

Pressurized   Direct line
system line   to reservoir

Externally Drained

Figure 10-10. Hydraulic fluid leakage around the internal moving parts of valves serves to lubricate valve parts. This fluid must be returned to the reservoir to eliminate back pressure that will cause a valve to malfunction.

reason for this is the fact that directional control valves may be simple on/off valves or may be more complex valves having two, three, or more positions. See **Figure 10-12.** Another complicating element is the valves may hold a position until they are physically shifted, even though the system has been shut down. Being aware of the variations of directional control valves and how they react when shifted or when the system is shut down can be helpful when analyzing system operation.

The normal position of flow control valves is also not as easily identified as pressure control valves. Flow control valves range from manually operated needle valves to complex pressure-compensated valves. However, in general terms, these valves may be considered normally open even though the size of the opening varies in some designs. Manually operated valves are set by a machine operator. The opening remains fixed when cycled though a variety of loads during operation or when the system is turned off. Pressure-compensated valves also remain open, although the opening varies in size to provide a uniform flow rate as the load on the system changes.

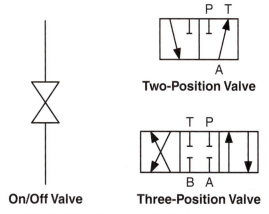

Two-Position Valve

On/Off Valve

Three-Position Valve

Figure 10-12. Directional control valves are typically classified by the number of internal operating positions in which the valve may function. Normally, these are identified by graphic symbols as on/off, two position, and three position. More than three positions is possible.

Valve Operation and Springs, Fluid Pressure, and Fluid Flow

Springs, pressure, and flow are fundamental to the operation of hydraulic system control valves. Pressure acting against the surface of internal valve components produces the force needed to compress a spring to open, close, or adjust many of these valves. Likewise, basic fluid-flow concepts are used to produce conditions that allow a valve to function within design specifications. Awareness of spring, pressure, and flow concepts is necessary to fully understand the operation of valves and, in some cases, their accessories.

Springs and Their Use in Valves

Springs can take many different forms in hydraulic valves. They can be constructed using leaf, coil, or spiral forms and operate through expansion,

compression, or torsion. They are used in hydraulic control valves to perform a number of functions, **Figure 10-13.** The three basic functions are:

- Moving the spool of a directional control valve back to its normal position after an operator has released the shifting force.
- Establishing the maximum pressure at which a pressure control valve operates.
- Serving as a biasing device in pressure-compensated flow control valves to maintain a constant pressure drop across a metering orifice.

One basic spring characteristic that influences the operation of hydraulic valves is the change in tension as a spring is activated. As a spring is compressed, stretched, or twisted, the tension increase per unit of movement is not equal. The most uniform tension change occurs in the middle movement range of the spring. Manufacturers of hydraulic components must take this concept into consideration to produce valves that maintain the most consistent pressure, flow, or movement control.

Fluid Pressure Used as a Valve Operator

Fluid pressure acts on internal surfaces in actuators to generate a force. This is generally understood to be the function of hydraulic cylinders and earlier will be related to specific valve types and other, more-specialized characteristics will be discussed.

System Maximum Pressure Control

Pressure control valves and devices that control maximum system pressure are referred to as relief valves, safety valves, and hydraulic fuses. *Relief valves* are used to prevent system pressure from exceeding the desired maximum, normal operating pressure. This operating pressure is established by the pressure needed to perform required machine operations. Normal operating pressure is usually well below the level where circuit, prime mover, machine, or product damage will occur. *Safety valves* and *hydraulic pressure fuses* are incorporated into hydraulic systems to prevent damage resulting from excessive system pressure caused by relief valve malfunctions. They are set at a pressure higher than the system relief valve and their design assures a minimal failure rate.

Relief valves and safety valves remain closed until the pressure setting is reached. Then they open, dumping fluid to the reservoir to prevent further pressure increases. When system pressure drops below the pressure setting, the valves close. This allows continuing fluid flow to operate the system actuators.

Hydraulic pressure fuses, on the other hand, rupture when the pressure setting is reached. The fluid is dumped to the reservoir to prevent further pressure increases. The pressure-sensitive fuse material is often called a *rupture disk.* The rupture disk must be replaced after system pressure exceeds the pressure setting and the disk ruptures.

Relief valves

Relief valves are normally closed valves that open as system pressure approaches the maximum, normal operating pressure. They are included in systems using fixed-displacement pumps to prevent the system pressure from exceeding a preselected maximum operating level. Various manufacturers produce a variety of designs to meet the needs of the many hydraulic systems in operation today. However, relief valves can generally be identified based on one of the two internal modes of operation.

Direct-Operated Relief Valve. The most basic of the operating modes is the direct-operated type.

It uses the pressure generated by the resistance to fluid movement in the actuators to directly compress a spring and open the valve. This action maintains a selected maximum system pressure by allowing fluid to return to the reservoir.

Figure 10-16 shows the internal structure of a typical direct-operated relief valve. These valves consist of:

- Body containing passageways with pressure and reservoir ports.
- Ball or poppet and a seat.
- Heavy spring.
- Spring-adjustment mechanism.

The body may be a casting specially designed for the valve or it may be made from bar stock cut to length with the internal passageways and ports machined into the block.

When the assembled valve is operating in a system, the pressure port is connected to the main operating portion of the circuit. The reservoir port is connected to a line directly leading to the system reservoir. The operating pressure of the valve is set by adjusting spring tension using the spring-adjustment mechanism. The spring holds the ball or poppet against the valve seat.

During system operation, the pressure in the system increases as the load on the system actuators increases. System pressure directly acts on the end surface of the ball or poppet, generating a force that attempts to compress the spring. When

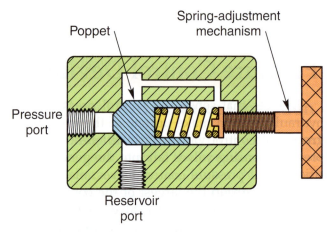

Figure 10-16. The internal structure of a typical direct-operated relief valve. These valves depend on spring tension to establish maximum system pressure.

the force is high enough, the spring begins to compress and the ball or poppet is forced off of the seat. A portion of the system fluid is then returned to the reservoir through the valve.

As actuator resistance increases, the fluid pressure increases. This generates more force on the ball or poppet, which moves further off of the seat as the spring is compressed more. The result is an increase in flow through the relief valve until the actuators stall and all pump flow is returned to the reservoir through the valve.

Balancing-Piston Relief Valve. The relief valves in the second design group use a balancing-piston mode of operation. These valves are commonly called *compound relief valves.* They have internal designs that are more complex. Compound relief valves commonly use a pilot stage and a balancing piston for operation. Pressure differences are used to operate internal components to dump the fluid required to prevent the system from exceeding the selected maximum operating pressure.

Figure 10-17 shows the internal structure of a typical compound relief valve. These valves consist of a pilot section and a balancing-piston section combined in a single valve. The pilot section is a small, direct-operated relief valve that controls the position of a balancing piston. The balancing piston is a primary part of the second section. The two sections work together to produce a valve that is generally more efficient and quieter than direct-operated designs.

The pilot section is at the top of the valve. The balancing-piston section is located between the pilot section and the chamber connected to the hydraulic system. The balancing piston is a critical operating part of this section. It typically includes:

- Poppet valve attached to the system side of the balancing piston.
- Cavity on the pilot side of the piston that contains a small balancing spring.
- Metering orifice that connects the pilot section and hydraulic system chamber.

The exact design of these features varies between valve manufacturers.

The face of the poppet valve attached to the balancing piston rests against a machined seat in the valve body. The balancing spring located on the opposite side of the balancing piston keeps the valve face in contact with the seat. This contact prevents fluid flow back through the relief valve until a selected system operating pressure is reached.

The metering orifice plays an especially critical part in the operation of this type of valve. It is located in a passageway that connects the pilot-valve side and system side of the balancing piston. The primary function of the orifice is to create a pressure difference between the pilot and system sides of the balancing piston.

Figure 10-17. The internal structure of a typical compound relief valve. These valves depend on a pilot section to produce an internal pressure differential that moves the balancing piston, which establishes maximum system pressure.

Actuator Sequence Control

The need to sequence the operation of multiple actuators or sections of a circuit is a fairly common requirement. This need can be met using direct-operated pressure control valves to limit fluid flow to an initial portion of a circuit until a pre-determined operation is completed. Once this first operation is completed, system pressure increases until the control valve opens. This allows fluid flow to activate a second portion of the circuit, which then performs the next operation in the sequence. The fact that these valves are pressure activated requires that each step in the sequence increase the system pressure to a higher level to initiate the next step in the process. Valves that control this type of movement are called *sequence valves.*

The simple clamp-and-drill hydraulic circuit shown in **Figure 10-21** illustrates the concept of sequencing. The clamp (cylinder 1) must secure the workpiece before the drilling operation (cylinder 2) can begin. This is achieved by using a sequence valve to block flow to the actuator that advances the drill. When the machine operator activates the circuit, hydraulic fluid flows only to the clamp actuator. This is because the sequence valve is closed, blocking flow to the drill actuator. When the clamp closes, pressure increases in the initial portion of the circuit to the level of the sequence valve setting. This causes the sequence valve to open, which allows flow to the drill actuator. The drill actuator then advances and completes the operation. Maximum system pressure is determined by the setting of the relief valve. This setting depends on the pressure required to provide effective operation of the sequence valve as well as the clamping and drilling force requirements.

Figure 10-22 shows a cutaway of a typical sequence valve. The operating concepts of these valves are similar, even though design variations exist between manufacturers. These valves consist of a body containing passageways and ports, which are attached to the initial- and secondary-movement portions of the circuit; a spool precision fit in a bore; a spring and spring-adjustment mechanism; and passageways allowing pilot and drain functions. The body may be a casting especially designed for the valve or it may be machined from bar stock.

Sequence valves are basically normally closed valves that block the movement of fluid between the initial and secondary ports until a preset pressure is reached. The pressure needed to open the

Figure 10-21. A clamp-and-drill hydraulic circuit is an example of sequence control. The clamp portion of the circuit functions before the drill is activated. The drill withdraws before the workpiece is unclamped.

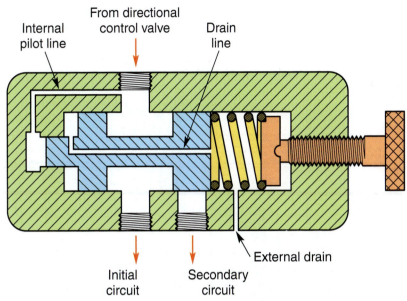

Figure 10-22. Sequence valves are normally closed valves that block the movement of fluid between the initial and secondary portions of a hydraulic circuit.

valve is set by adjusting the tension of the spring that holds the spool in the closed position. In the basic clamp-and-drill circuit shown in **Figure 10-21,** the sequence valve remains closed until the clamp actuator has closed the clamping device. Once the clamp has closed, increased resistance to fluid flow causes system pressure to increase. The valve uses a pilot line and chamber to sense this pressure and apply force to the end of the spool opposite of the spring-adjustment mechanism. When the force is high enough, the spring begins to compress and the spool slides within the bore. This allows system fluid to move into the second portion of the circuit, starting the drilling operation. As the drill actuator encounters greater resistance, system pressure increases and generates more force on the pilot end of the spool. The increased force causes the spool to continue to slide in the bore until the passageway between the two sides of the valve is fully open. This allows maximum flow through the valve.

In sequence valves, the chamber containing the spring and adjustment mechanism must be externally drained. This is required as both the initial and secondary sides of sequence valves are subject to full system pressure at the same time during some phases of circuit operation. If an internal drain is used, the pilot and drain chambers will be exposed to equal pressure. This will generate equal forces on the ends of the spool, allowing the spring to move the spool into the closed position. A circuit malfunction results.

Sequence valves are normally closed valves that do not allow reverse fluid flow without the addition of a check valve. **Figure 10-23** illustrates a sequence valve containing an integral check valve. This is a common method used to allow reverse flow to bypass the valve. Sequence valves are also available with balancing-piston operation.

Restrained Movement Control

Restrained movement control refers to the control of actuator movement in a hydraulic circuit by placing a pressure control valve in the outlet line leading from an actuator. These normally closed, direct-operated valves are used to restrain the movement of both cylinders and hydraulic motors. The valves provide both increased hydraulic system control and operator safety, **Figure 10-24.** Typically, the valves are called *counterbalance valves* when used with cylinders and *brake valves* when the actuator is a motor.

These valves are very similar in construction to the sequence valves discussed in the previous section. The primary differences are found in the

Flow returns to directional
control valve and reservoir

External
drain

Check
valve

Initial
circuit

Secondary
circuit

Figure 10-23. Check valves are commonly incorporated in sequence valves to allow reverse flow through the system to bypass the valve.

Figure 10-24. Counterbalance valves are often used in the actuator portion of load-lifting equipment. They restrain the uncontrolled drop of a load due to equipment failure or improper operation.

pilot and drain connections and their applications in a hydraulic system. Internal, external, or multiple pilots may be used in either counterbalance or braking valves, depending on the system demands. Internal drain connections are appropriate for most circuits. Like sequence valves, these valves are normally closed. This requires a check valve to allow reverse flow to freely move around the valve. The check valve may be an integral part of the valve or a separate component. See **Figure 10-25.**

Counterbalance valves

The simple circuit shown in **Figure 10-26** illustrates the location of a counterbalance valve in a hydraulic press. Variations of this circuit can be found in mobile equipment, such as a cherry picker, or other equipment used in manufacturing operations. In this example, when the press is raised, fluid in the rod end of the cylinder and the line leading to the counterbalance valve is trapped. Even if the system pump is not operating, this trapped fluid is pressurized because of gravity acting on the weight of the press platen and any load. This pressure is equal to the weight of the platen and the load divided by the surface area of the blind side of the piston.

If the cracking pressure of the counterbalance valve is set at a pressure higher than that generated by this weight, the load will not drop even if the directional control valve is shifted. In order to lower the load, the system pump needs to be operating with the directional control valve shifted to extend the cylinder. The pressure increase caused by the fluid coming from the pump added to that generated by the weight of the platen and load is sufficient to open the counterbalance valve, extend the cylinder, and lower the press platen.

However, without the counterbalance valve in place, the platen and load will drop if the directional control valve is accidentally shifted, even without the pump operating. Gravity acting on the platen and load causes them to move downward, extending the cylinder. This movement creates a negative pressure in the blind end of the cylinder. The negative pressure draws fluid out of the reservoir to fill the blind end, while fluid in the rod end is returned to the tank. The platen and load may fall uncontrolled, causing damage or operator injury.

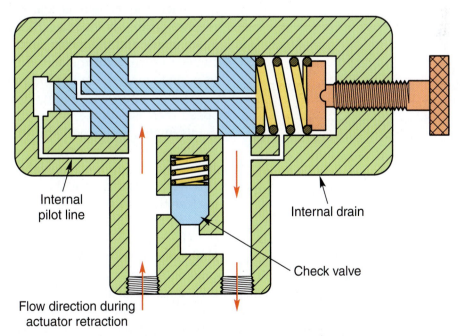

Figure 10-25. The internal structure of a counterbalance pressure control valve. These valves include either an internal or external pilot and, in most cases, an integral check valve. In this example, the check valve allows the free movement of fluid around the valve during actuator extension.

Figure 10-26. Placing a counterbalance valve in the outlet line of a cylinder prevents the uncontrolled drop of a load, even if the directional control valve is improperly shifted or a line fails between the directional control and counterbalance valves.

Caution: Service personnel must be aware of circuit designs that may allow unexpected actuator movement, even without system pump operation.

Brake valves

The hydraulic motor circuit shown in **Figure 10-27** illustrates the location of a brake valve in a circuit. Brake valves are variations of normally closed counterbalance valves. They are designed for use in hydraulic motor circuits. Their specific purpose is to prevent an *overrunning load* from turning the motor at a higher speed than desired.

Brake valves use two pilot lines to sense pressure at the inlet, or remote, side and the outlet, or load, side of the motor. The pilot lines are connected to two chambers that generate force on the end of the valve spool that is opposite of the control spring. **Figure 10-28** shows the construction of the valve. Notice the location and difference in size of the pistons in the two pilot chambers. The valve spring is used to set the pressure of the braking load.

In the following example, the braking load for the valve is set at 500 psi. The pistons in the inlet and load pilot chambers have an area ratio of 5:1 (inlet to load).

When the system is started and 100 psi is reached in the motor inlet pilot line, the braking valve opens and the motor turns. The motor continues to turn, with the actual line inlet pressure varying to meet the motor-load demand.

If the motor load begins to run away, motor speed increases to the point where fluid requirements match or exceed the pump output. The result is inlet line pressure drops below the 100 psi needed to keep the valve open. This pressure drop in the inlet pilot allows the valve spring to close the brake valve, restricting flow out of the motor.

The rotational momentum of the motor and load momentarily turns the motor into a pump, causing a pressure increase in the motor outlet line. Because of the difference in areas between the two pilot pistons, the outlet line pressure must reach 500 psi before the load pilot opens the valve. Once open, fluid is allowed to again pass to the reservoir from the motor outlet.

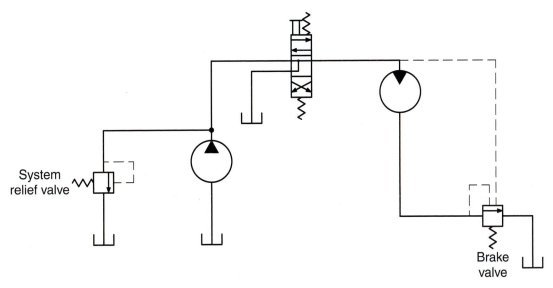

Figure 10-27. Placing a brake valve in the outlet line of a hydraulic motor prevents overrunning loads from turning the actuator at a higher-than-desired speed.

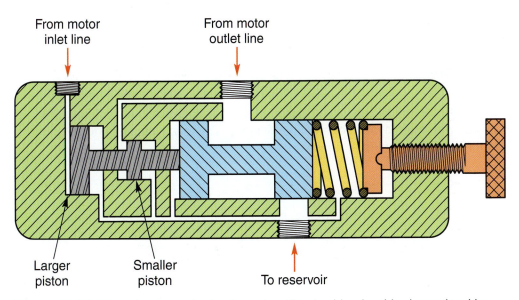

Figure 10-28. The structure of a brake valve. The braking load is determined by the tension of the spring and the ratio of the piston areas in the inlet and load pilot chambers.

The 500 psi back-pressure load has a braking effect on the overrunning speed of the motor. This continues until the motor speed has dropped to a level where pressure in the motor inlet line again reaches 100 psi. At that point, the motor has been slowed to the desired operating speed and continues normal operation until another overrunning-load condition occurs.

Pump Unloading Control

Operating a hydraulic system pump at full flow and pressure levels results in maximum energy and maintenance costs. This is wasteful of resources, especially when many systems do not require the pump to supply full-flow output under maximum pressure on a continuous basis. Stopping the system prime mover and pump when fluid flow is unneeded is not usually a practical solution. This is also expensive and shortens the service life of equipment. One technique that is often used is a design called a *pump unloading control.* This design uses a pressure control valve to monitor a direct, low-pressure flow path between the pump and the reservoir. See **Figure 10-29.** Unloading valves are commonly used in circuit designs using dual pumps or accumulators.

Unloading valves are classified as normally closed valves as they prevent the return of fluid to the reservoir until a desired operating pressure is achieved. They must also have an external pilot as the operating mechanism must sense pressure at a remote location in the system. See **Figure 10-30.** These valves are available using a direct-operated design, similar to the sequence and counterbalance valves discussed earlier. They may also be more complex, unloading relief valves following a variation of the design found in compound relief valves.

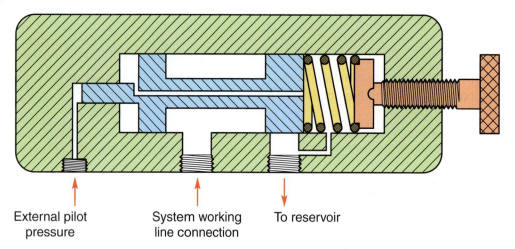

External pilot pressure   System working line connection   To reservoir

Figure 10-29. Unloading valves are often used in dual-pump applications that provide maximum system operating efficiency. These valves use external pilots as they sense pressure at remote locations.

Large forming press with heavy platen

Press platen

Unloading relief valve

Figure 10-30. Unloading relief valves are often used in circuits where accumulators supply additional fluid volume or hold an actuator under pressure for extended periods of time.

Direct-operated unloading valve

The high-low pump circuit shown in **Figure 10-31** illustrates one application of pump unloading control that uses a direct-operated unloading valve. The circuit has two pumps, a check valve, a relief valve, and the direct-operated unloading valve. It provides high-volume, low-pressure fluid for the initial operation of a circuit and then automatically switches to low-volume, high-pressure output in the second stage of operation. During the second stage, the unloading valve dumps the output of the largest pump to the reservoir at near-zero pressure to reduce energy consumption and heat buildup.

When the system starts, the low-volume pump moves fluid directly to the system. The flow from the high-volume pump passes through the check valve before blending with the low-volume pump output. When the actuator in the system encounters resistance, system pressure begins to increase. The external pilot line of the unloading valve senses system line pressure near the low-volume pump. When that pressure reaches the cracking pressure of the unloading valve, flow from the high-volume pump begins to flow to the reservoir. As the system actuator resistance increases, the pressure continues to increase until the unloading valve is fully open. At that point, all of the output of the high-volume pump is passing though the unloading valve to the reservoir.

As flow from the low-volume pump attempts to reach the low-pressure area of the unloading valve, it closes the check valve. With the check valve closed, the low-volume pump continues to supply fluid to the system actuator until the pressure setting of the system relief valve is reached. The unloading valve remains open because of the high pressure in the pilot line. The high-volume pump continues to return all fluid output to the reservoir, as long as system pressure is above the pressure setting of the unloading valve.

Unloading relief valve

When greater sensitivity is needed to provide pump unloading control, a valve with the expanded design features of a balancing piston is used. **Figure 10-32** shows the construction of an

Figure 10-31. A high-low pump circuit can use a direct-operated unloading valve to deliver two different flow rates under two different pressures. Maximum system pressure is controlled by the relief valve located in the low-volume-pump section of the circuit.

unloading relief valve used for such situations. This valve is very similar in basic construction to a compound relief valve. However, additional features are incorporated into the valve:

■ Integral, spring-loaded check valve in the outlet line of the valve stops reverse fluid flow.

■ Plunger mechanism in the pilot section of the valve assists in operating the balancing piston.

The addition of these two elements provides a valve that is compact and has the unloading feature desirable in accumulator and other circuits.

Figure 10-33 shows a basic accumulator circuit using an unloading relief valve described above. This diagram and the internal structure shown in

Figure 10-32 are used to explain the operation of the valve.

The integral, spring-loaded check valve is located in the passageway between the inlet and outlet ports of the valve. It is in close proximity to the pilot section of the unloading valve. This check valve is used to stop reverse fluid flow from the actuator portion of the system. A plunger mechanism is located in the pilot section of the valve. It is designed to assist in unseating and holding open the poppet of the small, direct-operated relief valve in the pilot section. The plunger is activated by pressure from an internal pilot sensing pressure in the passageway between the check valve and outlet port.

Figure 10-32. An unloading relief valve provides effective control of a single-pump unloading system. This is achieved by using a plunger mechanism to assist in the operation of the direct-operated relief valve located in the pilot section of the valve.

In the circuit shown in **Figure 10-33,** the directional control valve is centered, blocking flow to the system. When this occurs, the pump supplies fluid to the accumulator through the built-in check valve. As the accumulator approaches its fluid-storage capacity, pressure in the system begins to rise. The small, direct-operated relief valve in the pilot section of the valve senses this pressure increase through the orifice in the balancing piston of the unloading valve. At the same time, the pilot line leading from the plunger mechanism is sensing pressure in the passageway between the check valve and the outlet port of the valve. As system pressure increases, the force generated on the poppet of the small, direct-operated relief valve begins to compress the spring of that valve. Pilot pressure in the plunger mechanism also generates force that is transmitted to the end of the poppet by a rod extending from the end of the plunger. These two forces combine to unseat the poppet. This begins the process of shifting the balancing piston to unload pump output directly to the reservoir.

As the balancing piston shifts, allowing pump output to pass to the reservoir, the accumulator begins to discharge. This reverses flow as the fluid attempts to reach the low-pressure area of the reservoir. The spring-loaded check valve immediately closes to block the reverse flow. This traps pressurized fluid in the accumulator and the lines leading to the check valve and directional control valve. The plunger mechanism continues to hold the poppet of the small, direct-operated relief open because of high pilot pressure sensed in the passageway between the check valve and the outlet of the valve. Under these conditions, the unloading valve fully opens and dumps pump output to the reservoir at minimal back pressure. This produces considerable energy savings.

Once the directional control valve is shifted to activate system actuators, the accumulator discharges fluid to operate the system. This results in a system pressure drop that eventually allows the poppet in the small, direct-operated valve to close. The unloading valve then closes and the process is repeated. The amount of pressure drop required to reset the valve depends on the difference in the effective area of the plunger in the plunger mechanism and the area of the poppet in the small, direct-operated relief valve. For example, one manufacturer designs valves of this type to reseat at 85% of the maximum pressure setting.

Reduced Pressure Control

Some circuits require that a branch of the circuit operate at a maximum pressure lower than the setting of the system relief valve. This lower-pressure branch may involve a single actuator or a more extensive portion of the circuit. The reduced

Figure 10-33. This is an accumulator circuit using an unloading relief valve. This combination of components allows the pump to operate at near-zero pressure while the accumulator maintains a high holding pressure in the system.

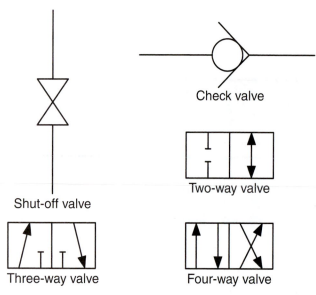

Figure 10-36. Directional control valves are classified under one of four general groups: shut-off or two-way valves, check valves, three-way valves, and four-way valves. These graphic symbols are used to indicate each valve type.

Shut-Off Valves

The primary purpose for shut-off valves is to block fluid flow in a system line. They are commonly used to isolate a component or a circuit branch during maintenance. These valves have a single internal passageway connecting two ports. Designs include globe, gate, ball, spool, and needle valves. All of these valves are simple in design. Although these valve designs can be used to throttle (regulate) fluid flow, they are generally considered inefficient as flow-control devices. They should be treated as valves that function best in the fully open or fully closed position, except needle valves.

Globe valves

Globe valves are very commonly used in water, air, and hydraulic systems. **Figure 10-37** shows the internal structure of these valves. The passageway through the valve is slightly offset with a seat located at the offset. This offset produces fluid turbulence, which results in increased pressure drops at higher flow rates. A *globe* or disk with a machined surface that matches the seat is mounted on a threaded shaft attached to an external handwheel. Turning the handwheel allows an operator to adjust the valve opening from fully closed to fully open.

Gate valves

The internal structure of a *gate valve* provides a straight line between the inlet and outlet ports. See **Figure 10-38.** The device that opens and closes the valve is a rectangular piece, called the gate, mounted on a threaded shaft attached to an external handwheel. The gate is fitted into a slot that is perpendicular to the valve passageway. Turning

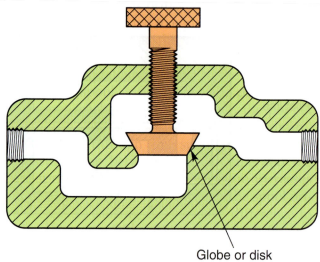

Figure 10-37. The internal structure and flow path through a basic globe valve. This valve design produces high resistance to fluid flow.

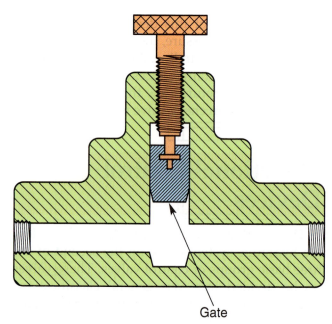

Figure 10-38. The internal structure and flow path through a basic gate valve. When fully open, this valve produces little flow resistance.

the handwheel allows the operator to slide the gate to a point where it fully blocks the passageway to stop fluid movement. Turning the handwheel in the opposite direction can completely retract the gate from the passageway. This allows unrestricted movement of fluid through the valve.

Ball valves

Ball valves consist of a valve body with a straight-line passageway connecting the inlet and outlet ports. The valve mechanism is a spherical piece attached to a shaft. See **Figure 10-39.** The ball contains a precision-machined hole that is the same size as the passageway through the valve body. When the valve parts are assembled, the shaft holds the hole through the ball in alignment with the passageway through the valve body. A handwheel or other handle on the shaft allows an operator to rotate the ball. Turning the ball 1/4 turn (90°) moves the valve from the fully open to the fully closed position. When the valve is in the fully open position, there is unrestricted movement of fluid through the valve.

Spool valves

A *spool valve* consists of a valve body with a precision-machined bore and passageways connecting inlet and outlet ports. A precision-fit spool is placed in the bore to provide a means by which the inlet and outlet ports can be connected and disconnected. **Figure 10-40** shows the internal structure of a spool valve. Different spool configurations and shifting methods may be used to provide different operating characteristics. These valves can be designed to function with minimal flow resistance, resulting in a low pressure drop across the valve even at maximum flow capacity.

Needle valves

Needle valves are very commonly used in fluid power systems as both shut-off and metering

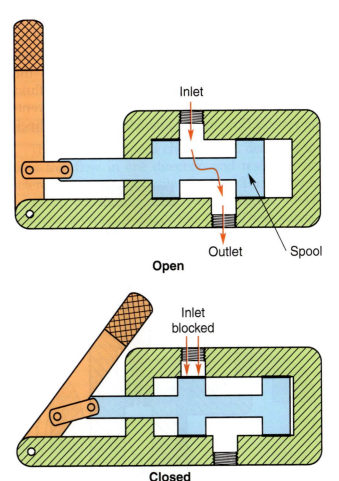

Figure 10-40. The internal structure and flow path through a basic spool valve. Linear movement of the valve handle provides rapid adjustment of this valve from fully open to closed.

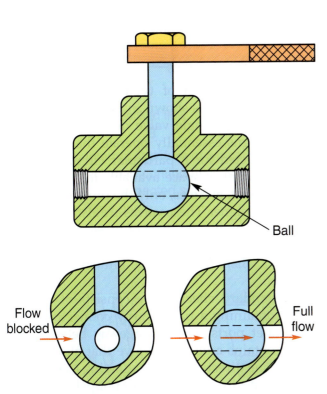

Figure 10-39. The internal structure and flow path through a basic ball valve. Only a 1/4 turn of the handle is required to adjust this valve from fully closed to fully open.

The pressure increases until the check valve opens to allow free flow of fluid to the system. When fluid flow attempts to reverse direction, the poppet seats to block flow. However, the orifice in the poppet allows a measured amount of fluid to return through the valve. The exact amount of return flow is determined by the size of the opening. This is specified when selecting the valve for an application.

Restriction check valves may be found in large forming presses where they are used to control the lowering speed of the large cylinders that move the platen. They are also used to decompress large hydraulic systems where the sudden release of high system pressure can cause undesirable shock to components.

Pilot-operated check valves

Pilot-operated check valves have all of the features of standard check valves. However, they may be opened to allow reverse flow through the valve. In some designs, the pilot feature may also be used to prevent flow through the valve in either direction.

Figure 10-45A illustrates a check valve with a pilot that can be used to open the poppet to allow reverse flow through the valve. When pilot pressure is not applied, the valve operates as a standard check valve, limiting fluid flow to only one direction. When pilot pressure is applied, the pilot

Figure 10-44. A restriction check valve has an orifice machined into the movable sealing element. This orifice allows a measured reverse flow through the valve.

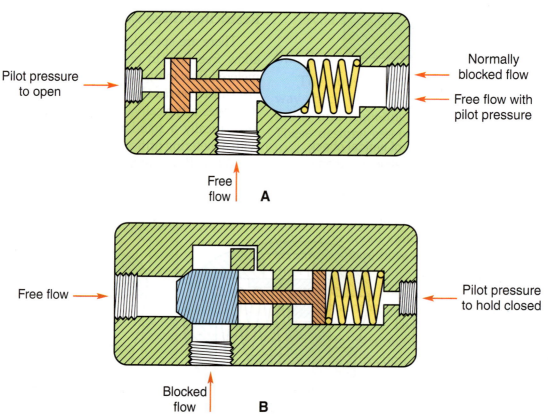

Figure 10-45. Pilot-operated check valves. A—Pilot pressure to open the valve against system pressure. B—Pilot pressure to hold the valve shut against normal reverse flow.

piston unseats the ball against the force generated by the valve spring and the system pressure on the ball. As long as sufficient pilot pressure is applied, the valve remains open to allow free flow in either direction.

Figure 10-45B illustrates another version of a pilot-operated check valve. It is used to prevent the movement of fluid through the valve in either direction. In this way, it acts as a shut-off valve. When the pilot section of this valve is activated, the poppet is forced against the seat to prevent fluid flow in either direction.

The pressure required to energize the pilot in pilot-operated check valves depends on the force needed to unseat the movable part to allow reverse flow or to close it and hold it seated. When opening the valve to allow reverse flow, the pilot must overcome the force generated by the strength of the spring and system pressure acting against the area of the movable part. When acting as a shut-off valve, the pilot pressure acting on the movable part must produce a force greater than the force generated by system pressure attempting to unseat it.

Three-Way Valves

Three-way valves are directional control valves used to operate rams and single-acting cylinders. They provide a means to extend the actuator using hydraulic system pressure, while using the weight of the external system load, a spring return in the cylinder, or the weight of a machine member to power the return stroke.

Three-way valves have three external ports with connections to the system pump, one port of the cylinder, and a system line leading to the reservoir. Shifting the valve in one direction connects the pump to one side of the cylinder. This allows system fluid to extend the cylinder. In this shifted position, the port leading to the reservoir is blocked. Shifting the valve in the opposite direction blocks the pump port and connects the cylinder port to the reservoir. External forces acting on the cylinder causes it to retract. This forces the fluid contained in the cylinder to flow to the reservoir. See **Figure 10-46.**

Figure 10-47 shows the structure of a basic three-way valve. The valve consists of a valve body containing a precision-machined bore. The bore is connected to pump, actuator, and reservoir ports. Fitted in the bore is a sliding spool designed to direct the flow of fluid through the valve. A clearance machined into the spool allows ports to be connected to produce the desired fluid passageway through the valve.

The system actuator is activated by moving the spool so the pump and actuator ports are connected through the clearance on the spool. At the same time, the reservoir port is blocked by one of the full-diameter sections of the spool. These sections are called *lands.*

Shifting the spool in the opposite direction connects the actuator and reservoir ports. The pump port is blocked by a spool land. Under this condition, the fluid from the actuator flows thorough the valve to the reservoir. This allows the cylinder

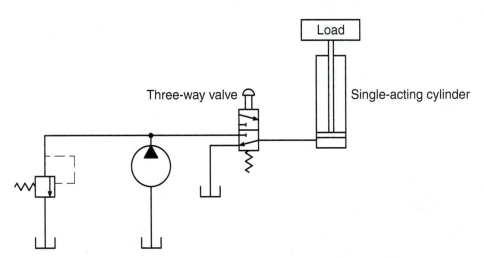

Figure 10-46. Three-way valves provide pressurized fluid to power one side of a single-acting cylinder. External force returns the actuator to its original position when the valve is shifted to the starting position.

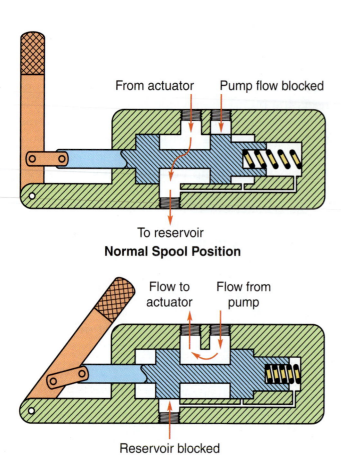

Normal Spool Position

From actuator

Pump flow blocked

To reservoir

Flow to actuator

Flow from pump

Reservoir blocked

Shifted Spool Position

Figure 10-47. Three-way directional control valves have three ports, providing connections to the pump, actuator, and reservoir.

to retract while the pump output flow is returned to the reservoir through the system relief valve. The retraction speed of the actuator depends on the force produced by the external factors acting on the single-acting cylinder.

Located on either end of the spool is a spring and an actuating mechanism, such as a solenoid or lever. The actuating mechanism is selected to best fit the valve application. The spring and mechanism allow easy and accurate control of the spool position in relation to the valve ports.

Three-way valves may be constructed as either normally open or normally closed. A valve is considered normally open if it connects the pump to the actuator when the actuating mechanism is not activated. If the valve connects the pump to the reservoir in the normal position, it is normally closed.

Four-Way Valves

Four-way valves are used in circuits with hydraulic motors or double-acting cylinders. They provide a means of using system pressure to both extend and retract the cylinder. See **Figure 10-48.** Likewise, the valve design allows hydraulic motors to be powered in either rotation direction.

Four-way valves have four external ports with connections to the system pump, inlet and outlet

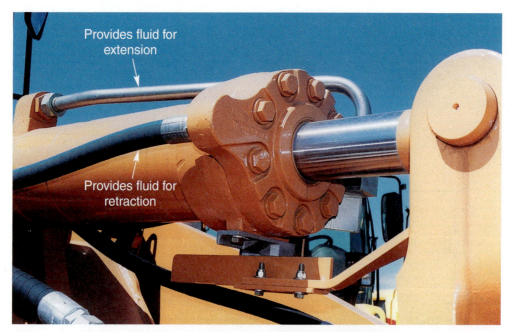

Provides fluid for extension

Provides fluid for retraction

Figure 10-48. Four-way directional control valves provide pressurized fluid to power double-acting cylinders during both extension and retraction.

lines of the actuator, and a system line leading to the reservoir. Shifting the valve in one direction connects the pump to the cylinder, allowing system fluid to extend the cylinder. In this shifted position, the valve also connects the opposite end of the cylinder to the reservoir. This allows fluid from that chamber to return to storage with minimal flow resistance. Shifting the valve in the opposite direction connects the pump to the rod end of the cylinder and the blind end of the cylinder to the reservoir. This forces the cylinder to retract. A comparable combination of connections allow a hydraulic motor to be powered in a clockwise or counterclockwise direction.

Two-position, four-way valve

A basic two-position, four-way valve causes the actuator to move in one direction while in the first position and to reverse the operating direction when shifted into the second position. This valve consists of a valve body containing a precision-machined bore connected to internal passageways leading to the pump, reservoir, and two actuator ports. The ports are often identified as P for the pump port, T for the reservoir port, and A and B for the actuator ports. Fitted in the bore is a sliding spool designed to direct the flow of fluid through the valve. Clearances machined into the spool allow the ports to be connected to produce the desired fluid movement through the valve.

Shifting the valve to the first position connects the pump port (P) to the actuator port (A) leading to the blind-end of the system cylinder. See **Figure 10-49A.** Fluid from the pump enters the valve and passes through the internal passageways and clearance on the spool. The fluid is directed through the actuator port (A) and enters the blind end of the cylinder. As the cylinder extends, fluid from the rod end is forced back to the directional control valve through the actuator port (B). This internally directs fluid to port (T), which returns the fluid to the system reservoir.

Shifting the valve spool to the second position connects the pump to the rod end of the cylinder. See **Figure 10-49B.** This supplies the fluid needed to retract the actuator, while routing the fluid returning from the blind end of the cylinder to the reservoir.

Four-way valve center position

Three-position valve design variations provide four-way directional valves with a third control that goes beyond the simple extension and retraction of cylinders and rotational direction of motors. This third control position is located between the two extreme positions described in the previous section. See **Figure 10-50.** It is often called the *center position*. This position is sometimes referred to as the *intermediate position*. In some specialized industries, such as the mobile equipment field, two or more of these positions are added to provide desired operating characteristics.

The spool configuration and internal passageways in the valve body provide the operating characteristics desired for the valve center position. The location of the valve body passageways, the spacing of the grooves and lands on the spool, or passageways machined into the center of the spool are used to produce the desired operating characteristics.

Several center position configurations are available from valve manufacturers. However, this discussion is limited to five commonly available center design options. These include closed, open, tandem, floating, and regenerative centers. The symbols of these center positions are shown in **Figure 10-51.**

The *closed-center* configuration provides a center position where the pump (P), reservoir (T), and actuator ports (A and B) are all closed. With the pump port closed, the pump output is returned to the reservoir over the system relief valve at full system pressure. With both actuator ports blocked, the actuator is held in a locked position. Because the system is operating at relief-valve pressure, the full capacity of the pump may be used to operate another actuator in the system. If the system is using multiple actuators with closed-center, four-way valves, simultaneous operation is possible, but system response time is reduced. Another characteristic of this center position is the performance of the system as the valve is moved from either of the activated positions. Shifting the valve from an activated position to the center position abruptly blocks fluid flow. This may produce high shock pressures in the circuit. Under some operating and load conditions, this may lead to component damage.

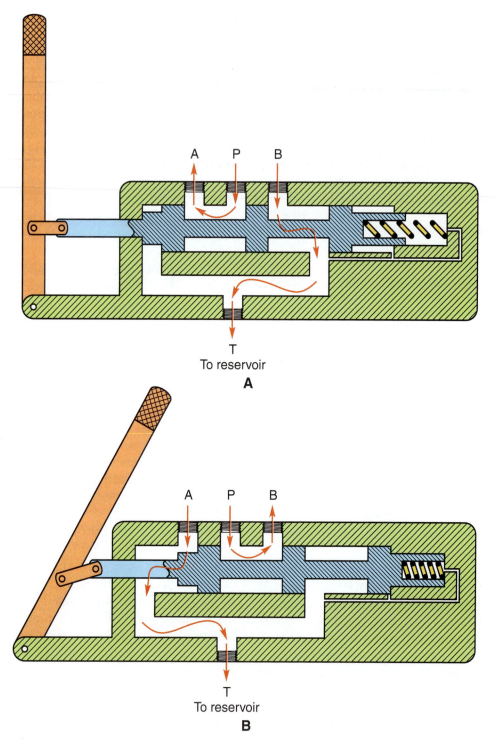

Figure 10-49. A two-position, four-way directional control valve showing the relationship of the ports. A—Fluid is directed to port A. Port B is open to the reservoir. B—Fluid is directed to port B. Port A is open to the reservoir.

The *open-center* configuration connects the pump (P), reservoir (T), and actuator ports (A and B). This interconnection allows the pump to return fluid to the reservoir at minimal pressure while permitting fluid to freely move into and out of either side of the actuator. This configuration minimizes energy consumption by the prime mover when the system is idle. However, it eliminates the possibility of the pump output operating other system actuators when the directional

Figure 10-50. A three-position, four-way directional control valve. The two extreme positions of the valve direct fluid the same as the two-position valve shown in Figure 10-48. The center position shown here provides a third alternative, which blocks all ports of the valve.

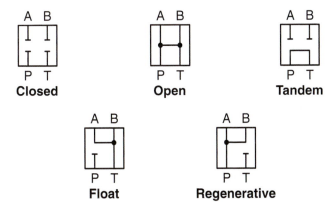

Figure 10-51. Many center design options are available for three-position directional control valves. However, the five center-design options shown here are widely available.

control valve is centered. Also, with both sides of the actuator connected to the pump and reservoir, the actuator is free to float in response to external forces applied by the load.

The *tandem-center* configuration connects the pump (P) and reservoir (T) ports and blocks both actuator ports (A and B). This results in the pump

output being transferred to the reservoir at minimal pressure when the system is idle, while the actuator is locked in place. This allows minimal energy consumption by the prime mover while the system is idle. In addition, two or more of these valves may be used in series by connecting the reservoir and pump ports of alternate valves. In a series configuration, a single valve may be operated without diminished performance. However, simultaneously operating valves modifies system performance.

In the *floating-center* configuration, the pump port (P) is blocked while the actuator ports (A and B) are connected to the reservoir port (T). The blocked pump port allows the full capacity of the pump to be used in other system segments. Since both actuator ports are connected to the reservoir, the actuator is free to float in response to external loads.

The *regenerative-center* design connects the pump port (P) to the actuator ports (A and B), while the reservoir port (T) is blocked. This configuration is used in situations where a rapid advance is required for a double-acting cylinder. When the directional control valve is shifted into this center position, both actuator ports are pressurized.

is especially important to have a means to compensate for changes in fluid temperature. Without *temperature compensation*, accurate flow control is not possible. Controlling changes in flow rate due to changes in fluid viscosity caused by fluid temperature variations is accomplished using either a:

- Specially designed, sharp-edged orifice.
- Heat-sensitive metal rod for the automatic adjustment of the opening in an orifice.

The sharp-edged orifice is considered viscosity insensitive (unaffected by viscosity). It does not physically adjust valve components to obtain a uniform flow, **Figure 10-65**. The theory of the design is that the orifice minimizes changes in flow rates even when temperature and viscosity changes occur in system fluid. These orifices are difficult to design and manufacture. However, some technical data sheets claim accuracies within 2%, even under substantial viscosity changes.

The heat-sensitive metal rod changes dimensions as the fluid temperature changes. The expansion rate and the geometry of the other mechanism parts must correlate to the viscosity changes that occur in the fluid. The design must be accurate so the device can modify the orifice size as fluid viscosity varies with temperature to maintain a constant flow rate. The orifice used with this design may be a traditional donut orifice with the heat-sensitive metal positioning a needlelike device for the desired control. See **Figure 10-66.** Other manufacturers have designs with specially shaped flow control devices to provide uniform flow.

Valve performance

Figure 10-67 illustrates a basic hydraulic circuit using a restrictor-type, compensated flow control valve. The flow control valve is placed between the

Spool shifts back and forth to maintain a constant pressure drop across the flow-control device

Pilot pressure

Pilot pressure

Varying Load Conditions

Spool stabilizes to maintain a constant pressure drop across the flow-control-device

Forces equal on each end of spool

Constant Load Conditions

Figure 10-64. The operation of a compensator maintains a constant pressure drop across the flow-control device in a pressure-compensated flow control valve.

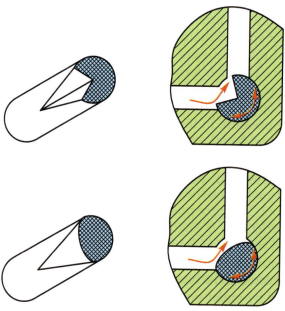

Figure 10-65. Manufacturers often use specially designed, sharp-edged orifices in flow control valves to greatly reduce flow variations resulting from viscosity changes produced by system temperature fluctuations. Two styles of hardware using this general concept are shown here.

blind end of the cylinder and the directional control valve. The pump has an output of 5 gpm, the relief valve is set at 500 psi, the flow control setting is 3 gpm, and the biasing spring tension requires 50 psi to overcome. This example circuit uses the same component layout as the noncompensated valve circuit shown in **Figure 10-62.** The difference is a pressure-compensated flow control valve is substituted for the noncompensated valve.

Figure 10-66. A heat-sensitive metal rod in conjunction with a needle can be used as a temperature-compensating device.

In this circuit, when the cylinder is extending under a no-load condition, the flow control valve delivers 3 gpm to the cylinder. The extra 2 gpm of pump output is forced through the relief valve. This is the same as in the noncompensated circuit. The potential pressure drop across the flow control device is a consistent 50 psi. This is established by the tension of the biasing spring. When the cylinder is subjected to varying loads, the pressure drop across the flow control device continues to remain constant. The flow rate, therefore, remains constant as long as the valve is operated within specifications. The result is a constant actuator speed. This is possible because the pressure compensator built into the valve maintains a constant pressure drop between the inlet and outlet sides of the internal flow control device.

Bypass-Type Flow Control Valves

Bypass-type flow control valves provide accurate flow to actuators by maintaining a consistent pressure drop across the flow control device in the valve while directing any excess flow from the pump to the reservoir. A relief valve is incorporated in these circuits, but it only functions when the resistance from an actuator load is sufficient to increase system pressure to the valve cracking pressure.

Figure 10-67. Pressure and flow conditions in a hydraulic circuit using a basic restrictor-type, compensated flow control valve.

Bypass valves are often referred to as *three-port flow control valves* as they have inlet, outlet, and bypass ports. In some applications, the bypass port is connected to a line leading to an additional function. More commonly, however, it is directly connected to the reservoir. In most models, the bypass connection cannot:

■ Be pressurized.

■ Effectively deliver fluid to another system branch.

■ Accommodate reverse fluid flow.

Valve construction and operation

A bypass flow control valve consists of a valve body; flow-control device; pressure-compensating device; and inlet, outlet, and bypass ports. See **Figure 10-68.** The flow control device in this valve is typically the same type of fixed orifice or adjustable needle valve used in restrictor-type flow control valves. The pressure compensator in this valve is similar to the compensator used in restrictor-type valves. Both types of valves are designed to provide a uniform pressure drop across the opening of the flow control device. However, in a bypass valve, this device dumps fluid to the bypass port rather than restricting flow through a compensator passageway.

The compensator consists of a spool fitted into a precision-machined bore in the valve body and a biasing spring that establishes the preselected pressure drop across the flow control device. A pilot line connects the end of the spool with the biasing spring to the outlet port of the valve. The inlet and bypass ports of the valve are connected via a passageway in the bore in which the compensator spool operates. The flow control device is located in a passageway between the inlet and outlet ports of the valve.

As in the restrictor-type valve, compensator operation is based on the biasing spring tension and forces generated by fluid pressure acting on the ends of the spool. In this compensating device, the force sliding the spool to restrict flow to the bypass port is established by the area of the spool end multiplied by the pressure of the fluid in the outlet port plus the force generated by the biasing spring. This force is counteracted by the force generated by inlet port pressure of the valve acting on the end of the spool facing the valve inlet.

When the hydraulic system is not operating, the force generated by the bias spring slides the spool of the compensating device to block the passageway between the bypass port and inlet port of the valve. Once the system is operating, the pump delivers fluid to the actuator through the flow control device.

As an example, assume the pump has a fixed displacement of 10 gpm, the flow control device is set to 5 gpm, the tension of the biasing spring requires a system pressure of 50 psi to function, and there is no load on the actuator connected to the valve output port. Refer to **Figure 10-69** as you read the following description.

When the system is started, the pump immediately directs 10 gpm of fluid to the valve. Part of the fluid moves through the flow control device and outlet port on the way to the actuator. The remaining fluid enters the passageway leading to the compensator where it encounters the end of the spool. With no outlet to this passageway, the pressure of the fluid quickly increases to the 50 psi required to shift the compensator spool. When the 50 psi pressure is reached, the spool begins to shift. This exposes the passageway to the bypass port. At this point, the flow control device is passing

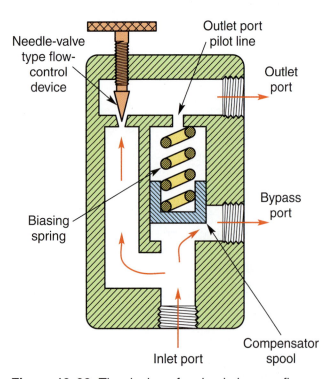

Figure 10-68. The design of a simple bypass flow control valve. Note the internal components and their relationship to the inlet, outlet, and bypass ports of the valve.

Spool position fluctuates
to produce correct flow

Pressure drop across
metering device varies

System pressure established
by actuator load plus
biasing spring tension

Outlet pressure established
by actuator load

Excess flow
varies

Outlet flow
varies

A

Spool position steady

Pressure drop across
metering device steady

Forces equal on
opposite sides of spool

Pressure drop across
metering device established
by tension of biasing spring

Excess flow
steady

Outlet flow
steady

B

Biasing spring slides spool
closed, blocking bypass port

No flow across
metering device

Pump output returned
to reservoir through
system relief valve

Pressure maintained
on actuator by
system relief valve

No flow through
bypass port

No flow through
outlet port

C

Figure 10-69. The operation of a bypass pressure-compensated flow control valve during three stages of system operation. A—Increasing or decreasing actuator load. B—Steady actuator load. C—Stalled actuator.

5 gpm to the actuator, while the other 5 gpm of pump output is returned to the system reservoir through the bypass port. The maximum inlet port pressure is approximately 50 psi, reflecting the activation pressure of the compensator. The pressure at the valve outlet port is near zero, as there is no load on the actuator.

When the actuator load increases, the pressure in the outlet port also increases. The pilot line connecting the end of the spool with the biasing spring exposes the diameter of the spool to the increased pressure. This increased pressure generates an increased force on the spool, moving it to decrease the size of the opening to the bypass port. This reduced opening increases the resistance to fluid flow through the opening, which, in turn, increases the fluid pressure in the valve inlet port.

The inlet port pressure may momentarily fluctuate, but stabilizes at a pressure required to move the load plus overcome the tension of the biasing spring. Using the example pressures given above, the inlet pressure stabilizes at 50 psi above whatever outlet pressure is required to move the load. The flow through the flow control device stabilizes at 5 gpm because of the consistent 50 psi pressure drop between the inlet and outlet of the valve.

Valve performance

Figure 10-70 illustrates a basic hydraulic circuit using a bypass-type flow control valve. The flow control valve is placed between the blind end of the cylinder and the directional control valve. In this example, the pump has an output of 5 gpm, the relief valve is set at 500 psi, the flow control valve is set at 3 gpm, and the biasing spring requires 50 psi to overcome the tension. This example circuit uses basically the same component layout as the noncompensated and compensated restrictor valve circuits shown in **Figures 10-62** and **10-67**. The primary difference is the bypass port connection that leads directly to the system reservoir.

In this circuit, when the cylinder is extending under a no-load condition, the flow control valve delivers 3 gpm to the cylinder. The extra 2 gpm of pump output is returned to the system reservoir through the bypass port of the valve. Under a no-load condition, the total pressure drop across the flow control valve is approximately the 50 psi needed to overcome the biasing spring tension, rather than the 500 psi setting of the relief valve. When the cylinder is subjected to an increased load, the pressure in the inlet and outlet port areas increases to produce the force needed to move the load.

Figure 10-70. Pressure and flow conditions in a hydraulic circuit using a bypass-type, pressure-compensated flow control valve.

However, the pressure compensator maintains a 50 psi pressure drop between the ports. This allows the flow control device of the valve to maintain the selected fluid flow rate. Therefore, a constant actuator speed is maintained, even under varying actuator load conditions.

The bypass-type of flow control is more efficient than the restrictor type. The surplus pump fluid returned to the reservoir is at a pressure below the relief valve setting. This reduces heat generation, which prolongs fluid life and lowers energy consumption. However, care must be taken when designing and operating a circuit of this type. The pressure required to operate the actuator must be accurately determined. In order for a bypass flow control valve to properly function, the cracking pressure of the system relief valve must be higher than the required actuator operating pressure plus the pressure needed to operate the compensator. If this is not true, the system will not operate properly.

Flow Divider Valves

Flow divider valves are used to divide one fluid supply between two circuit branches or subsystems. The valves are often used in mobile equipment where pump output flow varies with vehicle engine speed. Many of these valves can also be used to combine the flow from two different circuits. Both *priority* and *proportional* designs are available.

Priority divider valves

In hydraulic equipment with circuit subsystems, it is sometimes necessary to be certain a preset flow level is always available to a vital actuator. This may be related to the function of an important step in the machine operation or a safety issue in case of unexpected flow demands or a sudden pressure drop in the remainder of the circuit. A priority divider valve provides flow to one port before providing flow to a second port, **Figure 10-71.** These valves are often used in mobile equipment where pump output is controlled by engine speed. Because of safety issues and the absolute need for easy maneuverability, the power steering unit of mobile equipment is often operated by priority flow coming from such divider valves.

Figure 10-72 shows the basic structure of a priority-type flow divider valve. The valve consists of a body containing a machined bore, pump inlet port, priority and secondary outlet ports, spool, metering orifice, and biasing spring. The priority and secondary ports are positioned so that sliding the spool slightly in either direction changes the size of the openings between the bore and the ports. Moving the spool in one direction reduces the size of the opening to the priority port and enlarges the size of the opening leading to the secondary port. Movement in the opposite direction causes the opposite size changes. The metering orifice located in the center of the spool establishes the priority flow rate of the valve. The biasing

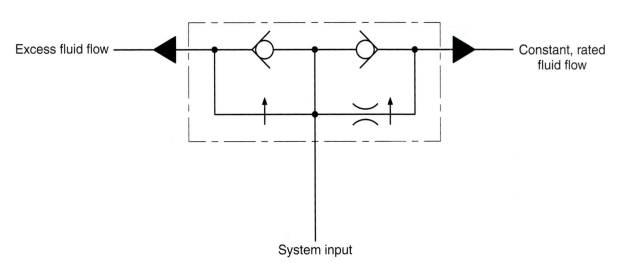

Figure 10-71. Symbol for a nonadjustable, priority flow divider valve. The check valves allow uncontrolled reverse flow through the valve. (Brand Hydraulics)

spring maintains a constant pressure drop across the orifice. This maintains a consistent flow rate to the priority port.

When the system is not operating, the biasing spring slides the spool toward the secondary port. This fully exposes the priority port and fully closes the secondary port. When the system pump begins to operate, fluid is supplied to the inlet port. Fluid flows through the metering orifice and out the priority port to the subsystem requiring preferential flow. As the flow increases, the pressure drop across the metering orifice increases. This generates a force on the spool that causes it to compress the biasing spring and move toward the priority port. As the spool moves, the size of the opening to the priority port is reduced. Also, the opening to the secondary port begins to be exposed.

The spool rapidly shifts back and forth until the forces generated by system pressure on the ends of the spool and the tension of the biasing spring are in equilibrium. The pressure drop across the metering orifice then produces the priority flow rate through the orifice. Any additional fluid coming from the pump flows out the secondary outlet port to operate other elements of the system.

Proportional divider valves

A proportional divider valve splits input port flow into two proportional output flows. These flows are directed to individual circuit subsystems or actuators. The ratio between the output flows may be fixed or variable. Proportions ranging from a 50-50 ratio to a 95-5 ratio are available from various valve manufacturers. A ratio of 50-50 is most common. See **Figure 10-73.**

A basic proportional-type flow divider valve consists of a body containing a machined bore, pump inlet port, and two outlet ports. Internally, these valves typically contain a spool fitted with control orifices leading to control chambers. During operation, valve inlet flow passes through the control orifices, control chambers, variable metering orifices, and on to the system through the outlet ports of the valve. The spool

Figure 10-72. The internal structure of a priority-type flow divider valve.

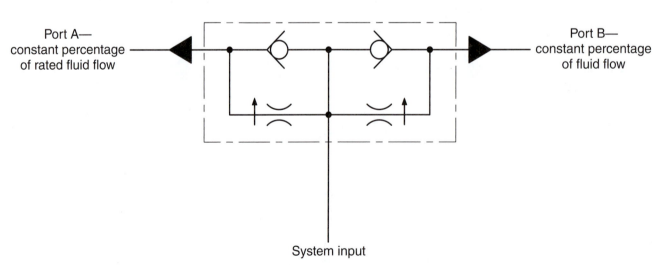

Figure 10-73. Symbol for a nonadjustable, proportional flow divider valve. Ports A and B receive a predetermined percentage of the total system input based on the characteristics of the orifice and spring. (Brand Hydraulics)

stays centered in the bore as long as the pressure generated by the load on each circuit subsystem is equal.

Increasing pressure in one of the control chambers produces an imbalance in the forces holding the spool in equilibrium. The spool rapidly moves back and forth to adjust the variable metering orifices until the pressures in the control chambers are again equalized. The result is an equal pressure drop across the control orifices. This produces the proportional flow.

Valve performance

Figure 10-74 illustrates a basic hydraulic circuit using a priority flow divider valve. The divider valve is placed between the pump and the circuit subsystems. In this example, the pump has a maximum output of 10 gpm and the relief valve is set at 500 psi. The priority flow setting of the valve is 2 gpm.

The priority divider valve delivers a maximum of 2 gpm to the subsystem connected to the priority port (the motor). Any excess flow, up to 8 gpm, is diverted to the second subsystem (the cylinder). The actuator in the cylinder subsystem operates on any excess flow until system pressure increases to the system relief valve setting. At that point, any flow over 2 gpm is returned to the reservoir through the system relief valve and the actuator in the second subsystem stops. The subsystem with the motor continues to receive the 2 gpm priority flow, as long as pump speed is sufficient to produce that

Figure 10-74. In this circuit, the priority outlet of the divider valve discharges 2 gpm. The remaining pump output goes to the second circuit. If pump delivery drops below 2 gpm, all fluid output is delivered through the priority port.

volume. The spring-loaded valve in the motor subsystem serves as a relief valve to set the maximum operating pressure of the motor.

The subsystem with the motor represents a steering mechanism on mobile equipment. As you can see, the steering system has priority over the second subsystem. That subsystem may be used to extend a load, for example.

Chapter Test

1. Control valves allow a hydraulic system to produce the type of _____ and level of _____ needed for machine operation.

2. What is the primary purpose of pressure control valves in a hydraulic circuit?

3. The _____ serves as the holder of all the working elements of the valve.

4. Define *fixed orifice*.

5. What are the two sources of internal force typically used to position control valve elements?

6. Why is it important to remove fluid that leaks into internal clearances and control chambers?

7. When a valve is classified as normally open or normally closed, it is only in relation to the position of the internal elements when the system is _____.

8. List the four basic tasks that fluid pressure performs in the operation and control of a hydraulic system.

9. Define *pressure override*.

10. What are the two basic internal modes of operation for pressure control valves?

11. Name the two types of pressure control valves that are used to prevent damage to hydraulic systems in cases where the system relief valve fails, causing excessive system pressure.

12. Describe the primary function of the metering orifice in the pilot section of a compound relief valve.

13. Safety valves are typically set at a pressure _____% above the setting of the primary system relief valve.

14. Multiple operating steps actuated in a preselected order can be obtained in a hydraulic circuit by the use of _____ valves.

15. Describe what is needed to lower a load that is held in position by a counterbalance valve.

16. Unloading pressure control valves are commonly found in hydraulic circuits that use accumulators or _____.

17. A pressure-reducing valve is normally open and must be _____ drained.

18. A(n) _____ valve is used primarily to block fluid flow through a hydraulic system line.

19. _____ valves may be adjusted from the fully open position to the fully closed position by turning the control handle only 90° (1/4 turn).

20. What is the most common application of a check valve in a hydraulic system?

21. A three-way directional control valve has three ports. Describe what each port is connected to in the system.

22. A four-way directional control valve has four ports. Describe what each port is connected to in the system.

23. The middle position in a three-position, four-way directional control valve is often called the _____ position. It provides a control feature beyond cylinder extension and retraction.

24. Which type of a three-position, four-way directional control valve has a middle position with the pump, reservoir, and actuator ports connected?

25. The _____ method of directional control activation uses system pressure, which may be remotely sensed, to shift the valve.

26. In some directional control valves, the spool is held in a selected position by a locking device called a(n) _____.

27. What are the two general operating designs of flow control valves?

28. In an operating fluid power system, the _____ of the fluid will change as the operating temperature of the system changes.

29. The simplest of the adjustable, noncompensated flow control valves is a _____.

30. What is a *pressure compensator*?

31. Identify the two ways to accomplish temperature compensation in a flow control valve.

32. During system operation, when actuator pressure requirements are below the cracking pressure of the relief valve, a(n) _____-type flow control valve directs any excess pump output flow directly to the reservoir via a port on the valve.

33. List three limitations that must be observed when incorporating most bypass-type flow control valves into hydraulic systems.

34. Proportional flow divider valves are available with a ratio of 50-50 to as high as _____.

35. A system pump is delivering 5 gpm to the inlet port of a priority-type flow divider valve. What amount of fluid is delivered to the priority and secondary ports when the priority outlet is rated at 2 gpm? What is the delivery if the pump output drops to 3 gpm?

Pump Output	Priority Port	Secondary Port
5 gpm	_____	_____
3 gpm	_____	_____

Equipment shaping the final contour of tunnels in mining and construction applications. (Atlas Copco)

Basic Functions of an Accumulator

A simple hydraulic system that is made up of a prime mover, pump, lines, control valves, actuators, and a liquid hydraulic fluid cannot *store* energy. The prime mover must continually operate to keep the system energized. When the prime mover is operating, fluid flow is created. This, in turn, powers the actuator to move a load. Stopping energy input from the prime mover stops system operation and system pressure drops to zero. The noncompressible nature of liquid causes this type of system response. The noncompressible aspect of a liquid is a positive feature desirable for hydraulic systems. However, there may be times when it is desirable to store energy for short periods during some phases of system operation.

Accumulators are the fluid power components that absorb and store energy in a system, **Figure 11-1.** The stored energy is then used to:

- Maintain system pressure.
- Produce limited flow to expand pump output.
- Supply fluid to operate system subcircuits when the prime mover and pump are not functioning.

Accumulators are also used to prevent damage to the system by absorbing the energy of sudden system pressure increases caused by pressure surges. *Pressure surges* result from sudden changes to actuator loads or the rapid closing of a system valve.

Accumulators store energy using one of three methods:

- Lifting a weight.
- Compressing a spring.
- Compressing a gas.

Gravity acting on the lifted weight produces pressure on the system fluid. Likewise, the tension of the compressed spring maintains fluid pressure in spring-loaded units. Gas-type accumulators store energy using the compressible nature of gases. In each design, the stored energy maintains pressure on the noncompressible hydraulic fluid for later use in the system.

Figure 11-1. These accumulators allow the system to apply the noncompressible characteristics of liquids while having the ability to store energy and smooth flow and pressure fluctuations during critical system operations. (MTS Systems Corporation)

Accumulator Design and Operation

Accumulators use weights, springs or compressed gas to provide the energy storage needed to perform their function. New materials and applications of hydraulic systems have somewhat changed the appearance of accumulators over the years and may have expanded their tasks, but the basis concepts of operation remain the same. This section groups accumulators under the three energy storage methods of weight loaded, spring loaded, and gas charged.

Weight-Loaded Accumulators

A *weight-loaded accumulator* can be described as a large, vertically mounted cylinder with a heavy weight attached to the end of the cylinder ram. See **Figure 11-2.** The blind end of the cylinder is plumbed to the hydraulic system. The weight is lifted when force generated by system pressure acting on the blind end of the ram is greater than the force generated by gravity acting on the weight. This stores energy in the accumulator. Likewise, the ram lowers when the system pressure drops below that needed to support the weight.

Weight-loaded accumulators are generally very large units. This is due to the weight needed to provide the pressure. They are typically used in central hydraulic systems powering applications such as heavy, stationary presses. In these applications, constant pressure and large fluid volumes are necessary. Weight-loaded accumulators may be found in large, forming-press operations. However, they are seldom, if ever, used today because of their massive size and weight.

A weight-loaded accumulator is unique as it is the only type of accumulator that can deliver its entire fluid volume at a near-constant pressure. However, an undesirable operating characteristic sometimes encountered with this design is high shock pressures. *Shock pressure* is the result of the inertia of the heavy weight when a control valve suddenly closes during the time the accumulator is delivering fluid to the system.

Spring-Loaded Accumulators

A *spring-loaded accumulator* uses one or more springs to store energy. The unit can be described as a cylinder with a sliding piston that divides the cylinder into two chambers. See **Figure 11-3.**

Figure 11-2. Weight-loaded accumulators use gravity acting on heavy weights to store energy and smooth pressure surges.

Figure 11-3. Spring-loaded accumulators typically have a coil spring that stores energy, which can be used to enhance system flow and pressure.

Figure 11-6. A—When a bladder-type accumulator is fully discharged, a poppet valve is often used to prevent the bladder from being extruded into the connecting line. B—The bladder is charged with an inert gas through a valve that also positions the bladder in the accumulator shell.

slide toward the gas-chamber end of the unit. This movement reduces the volume of the isolated gas chamber. As a result, the gas pressure increases to match system pressure. As gas-chamber volume decreases, the volume of the system-port chamber increases. Hydraulic fluid enters the chamber and is stored under pressure. This filling process continues until the maximum system pressure established by the relief valve is reached.

Precharging a bladder-type, gas-charged accumulator expands the bladder to fill the accumulator shell. The poppet valve located at the system port closes to protect the bladder from damage that may occur if the bladder is forced into the port. During system operation, the pressurized system fluid enters the accumulator shell through the system port. There, it begins applying pressure to the bladder. When system pressure exceeds the precharge pressure of the gas inside of the bladder, the bladder begins to compress. Reducing the volume of the bladder increases the gas pressure to match system pressure.

As in a piston-type accumulator, fluid and energy are stored in a bladder accumulator whenever system operating pressure exceeds the precharge pressure. Whenever pressure is decreasing in the system, the higher gas pressure inside of the bladder expands the bladder. The expanding bladder fills the accumulator shell and forces fluid back into the system. This condition continues until the expanding bladder closes the poppet valve, blocking flow out of the unit.

The operation of the diaphragm accumulator is very similar to the piston and bladder designs. The flexible diaphragm is used to separate the system-port chamber and isolated gas chamber. The diaphragm moves to change the volume of the chambers. This provides basically the same operation as the piston and bladder designs. A metal disk is often attached to the center of the diaphragm to prevent extrusion of the diaphragm into the port as the unit discharges.

subjected to actual system operating pressure. This pressure acts on the surface area of the piston. The generated force attempts to slide the piston toward the gas-charged end of the unit. However, the gas in the isolated gas chamber generates an opposing force. When the force generated by system pressure becomes higher than the opposing force generated by the gas precharge pressure, the piston begins to

Note: A general rule to remember is that an accumulator begins to store fluid and energy when system operating pressure exceeds the gas precharge pressure. After that point is reached, an accumulator is storing fluid and energy as system pressure is increasing and dispensing fluid and energy as system pressure decreases.

Nature of compressed gases

The gas used in an accumulator responds to pressure, temperature, and volume changes in the same way as other gases. You may wish to review the discussion of the *ideal gas laws* in Chapter 3. Understanding these laws will help in understanding the way an accumulator functions in a hydraulic system.

The terms isothermal and adiabatic are often used in explaining the operation of accumulators. These terms relate to temperature and heat. An *isothermal* process is completed without a temperature change, although heat may be added or removed during the process. An *adiabatic* process is completed without heat gained from or lost to surrounding materials. These principles apply to the compression and expansion of a gas. Examining the operation of an accumulator should illustrate the importance of these factors.

An isothermal condition exists when an accumulator is filled over an extended period of time. In this situation, the temperature of the accumulator gas remains constant. Any heat generated from the compression of the gas is transferred to the surrounding materials. In this case, the volumes of the system-port chamber and isolated gas chamber can produce the volume of hydraulic fluid predicted by applying the ideal gas laws.

However, rapid filling of the accumulator can produce an adiabatic situation. Heat transfer cannot happen quickly enough to maintain a constant gas temperature. In turn, the pressure of the gas is increased beyond that predicted by the ideal gas laws. This increase in gas pressure increases gas volume beyond that produced by the isothermal process. The increased gas volume in the isolated gas chamber is reflected in a similar reduction in the volume of the system-port chamber. The end result is a reduction in the amount of pressurized hydraulic fluid available from the accumulator.

Figure 11-7 compares the differences in volume and pressure during isothermal and adiabatic compression of a gas. The area between the resulting curves is caused by the retention of heat in the gas. In actual practice, most accumulators function somewhere between the two curves. However, having a good understanding of these characteristics can be helpful in understanding manufacturer specifications and troubleshooting accumulator circuit performance.

Functions of Accumulator Circuits

An accumulator is the component that allows hydraulic systems to maintain the advantages of a noncompressible liquid while using weights, springs, or a gas to enhance system operation. The selective use of these elements can increase the efficiency of a hydraulic system by:

■ Providing an easy means to store energy.

Figure 11-7. The final volume of identical volumes of a gas compressed under isothermal and adiabatic conditions. The heat retained during adiabatic compression results in a larger volume of gas when an identical pressure is maintained in each test condition.

- Controlling pressure and flow variations.
- Maintaining system pressure.
- Supplementing pump output flow.

This section examines each of these functions and provides basic examples.

Storing Energy

In a number of situations, a hydraulic system must be able to store energy for a period of time in order to increase system performance or efficiency. Even though a liquid can be slightly compressed, it is generally treated as incompressible. However, a lifted weight, compressed spring, or compressed gas can provide a means to maintain a significant amount of potential energy. Using one of these devices, an accumulator is able to effectively act as an energy-storage unit in a hydraulic system.

In one sense, all accumulator applications involve the storage of energy. This is simply because any accumulator is storing energy as it fills. It does not matter if that potential energy is in the form of a raised weight, compressed spring, or compressed gas. The energy is there and may be called on to perform a task.

The simple hydraulic motor circuit shown in **Figure 11-8** illustrates how an accumulator can be used to store energy for extended periods. The hydraulic motor may be a starting motor for an internal combustion engine that powers the pump for hydraulic equipment on mobile equipment. The engine drives the pump to supply hydraulic fluid to all aspects of the system, including filling the accumulator. When the engine is stopped, flow from the accumulator closes the check valve. This traps pressurized fluid in the accumulator section of the circuit. The energy needed to restart the engine is held in the accumulator as a pressurized fluid. When engine operation is again required, the directional control valve is shifted to direct fluid flow to the hydraulic motor. The motor then spins the engine crankshaft to start the engine.

Dampening Pressure and Flow Variations

Under varying load conditions, certain hydraulic components and circuits can produce substantial

Figure 11-8. A simple hydraulic circuit containing an accumulator that stores energy to be used later to operate a motor when the hydraulic power unit is not operating.

pressure and flow variations. These variations can be great enough to:

- Reduce the efficiency of a system.
- Cause damage to system components.
- Be a safety issue that could harm machine operators or other workers in the area.

Extreme pressure increases can be caused when an actuator encounters a heavy load or when a control valve suddenly shifts to limit or totally block fluid flow. These situations may briefly produce pressures that are several times higher than the system relief valve setting. In addition, some pump designs inherently produce a pulsating flow. Under normal operating conditions, these flow variations are not critical. However, as system pressures increase, they may reduce system performance and shorten the service life of the pump and other components.

Accumulators can help reduce the above system problems. The effectiveness of an accumulator in controlling these problems involves careful sizing, precharging, and placement of the accumulator in the circuit. **Figure 11-9** shows the placement of an accumulator in a circuit to reduce shock

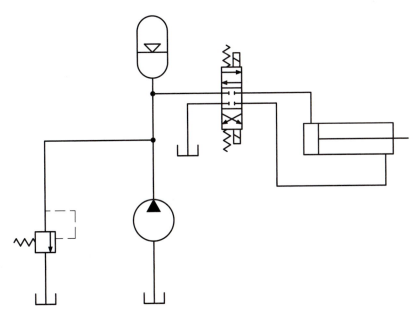

Figure 11-9. A hydraulic circuit designed to reduce shock pressures in the system when a directional control valve suddenly closes or an actuator suddenly encounters a heavy load.

pressures when a closed-center directional control valve is shifted to the center position. This placement also reduces shock when the cylinder reaches the end of its stroke in either direction. High pump flow rates even in simple circuits such as this may require a carefully placed accumulator to prevent high shock pressures.

Maintaining System Pressure

The operation of many hydraulic systems involves a holding phase. During this phase, pressure must be maintained for an extended period of time with little or no actuator movement. Continuously operating the pump under full system pressure in this situation consumes an unnecessary amount of energy. However, without a small amount of fluid being added to the system, internal leakage in components may quickly cause the pressure to drop below the desired level.

Closely associated with this type of circuit design are changes in system pressure resulting from temperature variations. Isolating the fluid in the actuator area of a system for long periods of time causes the fluid to be affected by ambient temperature. As the fluid temperature changes, the fluid volume changes. This, in turn, alters the fluid pressure. Raising the temperature increases the pressure, while decreasing it reduces the pressure.

A properly sized and precharged accumulator can resupply the actuator in a holding circuit with the volume of fluid needed to maintain the desired system pressure. **Figure 11-10** shows a simple holding circuit used with a press. The accumulator is filled as the press closes and system pressure increases to the setting of the system unloading relief valve. After the unloading valve opens, the accumulator maintains the pressure in the actuator and resupplies any fluid lost around the piston. This condition exists until sufficient fluid has leaked around the piston to drop the pressure to the point where the unloading valve closes. Then, the pump again supplies fluid to the system.

The same circuit can be used as an example of maintaining pressure in a system where ambient temperature is a factor. When the fluid in the press cylinder and the isolated lines cools, the accumulator supplies any fluid needed to maintain the holding pressure. Likewise, the accumulator absorbs any increase in fluid volume caused by an increase in fluid temperature due to a high ambient temperature.

Figure 11-10. A simplified circuit for a hydraulic press that must maintain a high pressure for an extended period of time. Unloading the system power unit significantly reduces energy consumption.

Supplementing Pump Output

Another common problem with hydraulic systems is the intermittent need of high fluid-flow rates. A straightforward approach to satisfying this need is to size the pump so it adequately delivers the required flow throughout the entire operation of the system. However, the result is higher-than-needed initial hardware cost and operating cost related to both equipment service and energy consumption.

A solution that is often used is placing an accumulator in the circuit. The accumulator is filled during the low-flow-need portion of the operating cycle of the machine. The stored fluid and the fluid output of the pump are then combined during the high-flow-need portion of the operating cycle. The combined fluid output provides the high flow rate needed to efficiently operate the machine.

This type of accumulator application is often found in die casting and plastic injection molding machines. These machines require high fluid-flow rates to rapidly inject the metal or plastic into the mold. However, the high-flow-rate demand is for only a small portion of the total machine cycle. **Figure 11-11** is a simple example of a circuit that uses an accumulator to provide a higher-than-pump output flow for both cylinder extension and retraction.

Selecting and Sizing Accumulators

The variety of accumulator uses and the performance variations of each design make a single selection and sizing procedure impractical, **Figure 11-12**. Most manufacturers provide detailed data concerning the accumulator models they produce. Most companies also provide suggested procedures to follow for the selection and installation of their products in various types of applications. The discussion in this section is limited to topics such as storage capacity, pressure ratings, and the usable volume of accumulators.

Rated Capacity and Pressure Ratings

The *rated capacity* of gas-charged accumulators is based on the gas volume when all liquid

Figure 11-11. A portion of a circuit showing an accumulator used to supplement pump output. Adequate time to fill the accumulator must be provided when planning the cycle time of these circuits. (Used with permission of CNH America LLC)

has been expelled. The actual value of liquid that can be stored is well below this rated capacity. The actual volume will vary somewhat between manufacturers. Rated capacities range from under 1 pint to over 200 gallons. However, some companies produce models that have even higher capacities.

Accumulators using the diaphragm design generally have the most limited capacity. Typically, the capacity of these designs ranges from about 5 cubic inches to approximately 1 gallon. Accumulators using the bladder design typically have capacities ranging from 10 cubic inches to over 40 gallons. Piston-type accumulators are available in an even wider range of capacities. Capacities range from under 10 cubic inches up to 150 gallons. This size range means piston accumulators vary in size from a small unit that is

approximately 2 1/2″ in diameter and 8″ long to a large unit that is 16″ in diameter, 20′ in length, and weighing 3700 pounds dry.

Accumulators are also available in a number of pressure ratings. The *pressure rating* is based on the maximum safe operating pressure for the unit, as determined by the manufacturer. This varies from a maximum operating pressure of 2000 psi to over 10,000 psi. One major manufacturer rates their units at either 3000 psi or 5000 psi. Other companies manufacture units rated at 2000 psi, 2500 psi, 3000 psi, 5000 psi, and 10,000 psi. These rated pressures are maximum system operating pressures, *not* accumulator precharge pressures. Generally, the precharge pressure should not exceed 50% of the rated pressure. For example, an accumulator rated at 3000 psi should be precharged to no more than 1500 psi.

Figure 11-12. Accumulators are available in a wide range of capacities. Stated capacity indicates the maximum volume of fluid that can be held in the unit, rather than usable operating volume. (HYDAC Technology Corporation)

Usable Operating Volume

The rated capacity of an accumulator does not indicate the usable volume of hydraulic fluid the unit can supply in a specific system application. The *usable volume* is the volume of hydraulic fluid the unit can supply in a specific system application. It is determined for a specific accumulator installation by three pressure levels:

■ Maximum operating pressure, which is usually determined by the system relief valve setting.

■ Minimum operating pressure established by the lowest level of pressure needed to perform the task involved.

■ Precharge pressure, applied weight, or spring characteristic.

These pressures effect the amount of fluid that is stored in the unit and, therefore, the amount that can be discharged during the system task supported by the accumulator.

Most manufacturers recommend a precharge pressure that is 100 psi below the minimum operating pressure required to perform the task. This pressure difference allows the compressed gas of the accumulator to maintain at least a slight pressure on the fluid throughout the task.

Assume that the task to be supported requires 5 gallons of usable, pressurized hydraulic fluid from an accumulator. The pressure range is from a maximum system pressure of 2000 psi to a minimum operating pressure of 1600 psi. The first step is to select an accumulator model that can store 5 gallons of fluid between the two pressure levels. This selection can be easily done by using tables available from manufacturer catalogs. Next, after installation in the system, the selected accumulator needs to be precharged to 1500 psi. This follows the general recommendations of precharge pressure 100 psi less than the minimum operating pressure. The following series of events then occurs in the system during the charging and discharging of the accumulator.

1. Once the hydraulic system pump is operating, the pressure in the system increases to the precharge level of the accumulator gas.

2. The accumulator begins to fill and system pressure begins to climb toward the setting of the relief valve.

3. As system pressure increases, the gas in the accumulator is compressed to match the pressure of the hydraulic fluid.

4. As the accumulator continues to fill, system pressure increases to the setting of the relief valve.

5. The system pump can be stopped.

6. The compressed gas of the accumulator forces fluid out of the accumulator to perform the designated task in the system. The accumulator provides the five gallons of usable volume of fluid before dropping to the minimum pressure of 1600 psi.

7. The system pump again needs to be operated to refill the accumulator and repeat the cycle.

Setup and Maintenance of Accumulators

This section discusses general factors that can increase the operating efficiency of systems containing accumulators. The safe operation of hydraulic-powered and hydraulic-controlled equipment containing accumulators is also discussed.

> **Caution:** Hydraulic accumulators are pressurized containers that need to be handled with a great deal of care. Only individuals with specific training should install or repair accumulators.

Mounting Position

The best mounting position for any style of accumulator is vertical with the hydraulic connection facing downward. See **Figure 11-13**. This reduces collection in the accumulator of contaminants from the hydraulic fluid. It also promotes the even loading of internal parts. The weight-type accumulator must *always* be vertically positioned because of the use of gravity. The two other designs can function in vertical, horizontal, or inclined positions.

Horizontal mounting tends to decrease the service life of the bladder in bladder-type accumulators. This is because of increased contact with the shell as the bladder expands and contracts. This contact increases in units used in systems with high cycle rates. It tends to be more critical in units that have a large storage capacity.

Precharging

The precharge pressure is critical to the proper operation of the accumulator. The correct precharge is necessary for the accumulator to store energy, dampen system flow and pressure variations, maintain system pressure, or supplement fluid output of the pump.

The precharge pressure depends on the function of the accumulator in the hydraulic system. The example in the previous section recommended a precharge of 100 psi below the required minimum operating pressure. However, when the accumulator is being used as a shock or surge suppressor, a precharge of 100 psi *over* normal system operating pressure is often recommended. Consult the accumulator manufacturer to establish the initial precharge pressure for the application at hand. Adjustments can then be made to that pressure as needed for maximum system efficiency.

Figure 11-13. The preferred mounting position for accumulators is vertical. This position provides the least stress on the internal components of the units. (MTS Systems Corporation)

Accumulator manufacturers recommend a regular check of the gas precharge pressure. A loss in precharge pressure reduces system efficiency. In some situations, pressure loss can also cause damage to the accumulator bladder, diaphragm, or piston seals.

A charging and gauging unit is shown in **Figure 11-14.** This unit is used to precharge the accumulator with dry nitrogen. It is also used to check the existing precharge pressure in an installed unit. The charging/gauging units typically include the gauges, valves, hoses, and

fittings to connect the accumulator gas valve to a nitrogen bottle, **Figure 11-15.** A typical gauge is shown in **Figure 11-16.**

Caution: The procedure suggested by the equipment manufacturer should be carefully followed to assure accuracy and personal safety during either gauging or charging. A gas-charged accumulator should only be charged after the hydraulic system has been reduced to 0 psi.

Figure 11-14. The components of a charging and gauging unit. (HYDAC Technology Corporation)

Figure 11-15. Precharging a gas-charged accumulator requires pressurized nitrogen and an appropriate charging and gauging unit.

Figure 11-16. A typical gauge from a charging and gauging unit. (HYDAC Technology Corporation)

Working Safely with Accumulators

Whenever a hydraulic system includes an accumulator, the system should be treated with special respect. A charged accumulator has the ability to supply pressurized hydraulic fluid even when the system may appear to be powered down. The training of machine operators must include this type of information. In addition, signs should be prominently displayed to remind operators of the proper procedure for discharging the accumulator and locking out the system when shutting down machines. This process is a critical safety issue to prevent both injury to personnel and damage to equipment.

Safety bleed valves may be recommended by the accumulator manufacturer. These valves assure an accumulator is discharged when the electrical power is shut off in the system. Electricity to the solenoid of the bleed valve is cut at the same time as the electricity to the motor of the hydraulic power unit. The spring in the bleed valve then shifts the valve to connect the accumulator directly to the reservoir. The pressurized hydraulic fluid remaining in the accumulator bleeds off to the reservoir. A flow control valve in the bleed-off line limits the rate of discharge flow from the accumulator to prevent an excessively high flow rate in the line to the reservoir.

Caution: Gas-charged accumulators are the most popular designs used today in both industrial and mobile hydraulic applications. The following is a list of safety factors that must be followed to provide a safe working environment when using gas-charged units.

- Only individuals with specific training should install accumulators or perform repair work.

- Never fill an accumulator with oxygen. An explosion may result if a mixture of oxygen and oil is pressurized.

- Never fill an accumulator with air. Three factors make the use of air in an accumulator undesirable. First, there is a risk of explosion as the oxygen in the air is compressed in the system. Second, the moisture contained in the air can condense, rusting system parts and reducing hydraulic fluid life. Third, oxygen in the air causes oxidation of the hydraulic oil.

- Always fill an accumulator with an inert gas, such as dry nitrogen. Using dry nitrogen eliminates the above problems.

- Never exceed the precharge pressure recommended by the accumulator manufacturer.

- Use soapy water to check for exterior gas leaks. Apply soapy water to all fittings and seams exposed to pressurized gas in the accumulator. If bubbles form, there is a leak.

- Internal gas leaks in the system may be identified by the appearance of bubbles in the reservoir.

- To remove all accumulator-related pressure from the hydraulic oil, first shut down the pump. Then, cycle an actuator in the accumulator portion of the circuit to relieve oil pressure. If the circuit is equipped with a bleeder valve, simply open that valve and bleed the hydraulic system pressure to 0 psi.

- Be certain to remove all gas pressure from an accumulator before removing it from the system.

Chapter Test

Answer the following questions. Write your answers on a separate sheet of paper.

1. An accumulator permits _____ to be absorbed and stored in a hydraulic system.

2. _____-loaded accumulators use the force of gravity to allow the storage of energy in a hydraulic system.

3. List the three designs of gas-charged accumulators used in hydraulic systems.

4. Air or _____ should *never* be used in gas-charged accumulators.

5. From what material is the bladder in a bladder-type accumulator constructed?

6. In bladder-type gas-charged accumulators, what is the purpose of the poppet valve?

7. In a gas-charged accumulator, the volume of the isolated gas chamber _____ as the volume of the system-port chamber increases.

8. A(n) _____ process is one that is completed without heat gained from or lost to surrounding materials.

9. The volume of fluid available from an accumulator is greatest if the unit is charged and discharged as a(n) _____ process.

10. List three results that may occur if pressure and flow variations are great enough in a system under varying load conditions.

11. Accumulators are available with rated fluid capacity ranges from under 1 pint to over _____ gallons, although some specialized accumulators may have much greater capacities.

12. What are the three pressures used to establish the usable volume of an accumulator?

13. Which accumulator design must be mounted in a vertical position?

14. Precharging a gas-charged accumulator should be done only after the hydraulic side of the system has been reduced to _____ psi.

15. List three reasons why air cannot be used as a precharge gas.

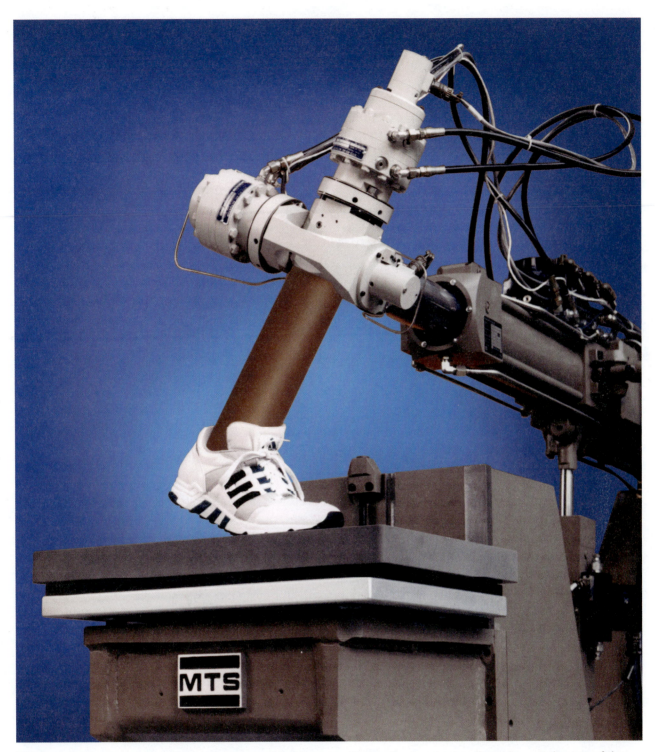

This is a testing apparatus that makes use of fluid power. Notice the two hydraulic motors at the top of the apparatus and the various cylinders. Testing equipment typically requires very precise control of operating conditions. A machine such as this may make use of accumulators to store system pressure or reduce pressure surges. (MTS Systems Corporation)

Hydraulic systems operate in a wide variety of workplaces and under various load conditions. These conditions subject the system to dust, dirt, moisture, and heat. Each of these elements can be harmful to system components. This can reduce system efficiency and lead to major component damage. Although a clean, cool surrounding is ideal for system operation and service life, that condition is not possible in most industrial situations. This chapter provides information on the various filtration devices and heat exchanger equipment used to condition system fluid so it is clean and at a desired operating temperature.

Objectives

After completing this chapter, you will be able to:

- Identify the typical contaminants found in hydraulic system fluid and describe the source of each.

- Explain the source of the energy responsible for increasing fluid temperature during hydraulic system operation.

- Describe how reservoir design can be used to reduce fluid contamination.

- List and compare the various filter media used for hydraulic fluid filtration.

- Describe the filtration rating methods used with hydraulic filters and strainers.

- Compare the characteristics of the various filter locations and circuits that may be used in hydraulic systems.

- Explain the function of heat exchangers in a hydraulic system.

- Describe the design and structure of heat exchangers commonly used in hydraulic systems.

- Identify the factors that must be considered when determining the need for a heat exchanger in a hydraulic system.

Chapter 12

Conditioning System Fluid
Filtration and Temperature Control

Selected Key Terms

The following terms will be used in this chapter. As you read the text, record the meaning and importance of each. Additionally, you may use other sources, such as manufacturer literature, an encyclopedia, or the Internet, to obtain more information.

absolute rating
absorbent filter
adsorbent filter
bonnet
brazed-plate heat exchanger
bypass valve
cleanout
contaminant
depth-type filter
filtration system
finned conductor
fluid conditioning
full-flow filtration
heat exchanger
immersion heat exchanger
micron
nominal rating
off-line filtration
proportional filtration
shell-and-tube heat exchanger
sludge
strainer
suction filter
sump strainer
surface-type filter
varnish

Internet Resources

Need for Fluid Conditioning

Unlike the other major areas of hydraulic systems, fluid-conditioning equipment and circuits do not directly contribute to the development and control of system power output. However, these elements must be considered critical to the efficient and cost-effective operation of a system, **Figure 12-1.** These elements, in a sense, provide critical services to the components of the system and the fluid that is the life blood of the system.

Fluid conditioning provides the system with fluid that is clean and at an acceptable operating temperature. Without effective methods to provide these conditions, the operating efficiency and life of any hydraulic system is greatly reduced. It is often claimed that 75% of hydraulic system failures are a direct result of fluid contamination. Any system failure results in substantial increases in operating costs. This is due to system maintenance and the loss of production during the equipment downtime.

Figure 12-1. Filters and heat exchangers prolong the operating life of complex systems and increase operating efficiency. (MTS Systems Corporation)

Contaminants and Their Sources

A hydraulic system must be protected from a wide variety of contaminants. *Contaminants* are particles or liquids that should not be in the system. They can reduce the operating efficiency or increase the rate of wear of component parts. Contaminants include those:

- Built-in into the components during manufacture.
- Entering the system during operation.
- Resulting from component wear.
- Resulting from the breakdown of the hydraulic fluid.

The processes used to produce equipment can result in a variety of manufacturing-, fabrication-, and shipping-related contaminants. These contaminants can include materials such as:

- Grains of sand from sand molds and cores used to cast the body.
- Rust chips and paint overspray related to cleaning and painting the product.
- Dust and water picked up during shipment of the assembled equipment to the final installation location.

It is the responsibility of the *filtration system* to remove contaminants. In new or reconditioned equipment, the filtration system initially needs more frequent attention to assure the above built-in contaminants are removed as quickly and efficiently as possible.

Contaminants are also introduced into a hydraulic system during normal system operation, **Figure 12-2.** The cylinder rod of linear actuators may be exposed to dust, dirt, and liquid contaminants. On cylinder retraction, these may be introduced into the system. Also, dust and moisture may enter through the reservoir breather. Foreign materials may also enter a system during the disassembly of components for service and the routine refilling of the reservoir. Service personnel should always be careful to use clean work surfaces and containers.

An operating hydraulic system also internally produces contaminants during normal operation. These include particles released due to wear on bearings. Packing and seals also contribute contaminants as they wear.

Figure 12-2. External conditions expose a hydraulic system to dirt and water that can cause rapid deterioration of system fluid and components. (Reprinted courtesy of Caterpillar, Inc.)

In systems exposed to humid conditions, condensation may introduce water into the system during periods of shutdown. This contributes to the breakdown of the fluid and the formation of sludge and acids. *Sludge* is a viscous, often gel-like material that forms when water and other contaminates interact with the hydraulic fluid. Although neither sludge nor acid is abrasive, they both contribute to system service problems.

Effects of Contamination

Hydraulic fluid contamination results in reduced system efficiency and physical harm to system components. These are caused by the reduced ability of the fluid to:

- Provide lubrication of moving parts.
- Seal the clearances between moving parts.
- Transfer heat.
- Transfer energy in the system.

Chapter 6 discusses each of these functions in detail.

Dirt particles reduce the ability of the fluid to provide lubrication. Reduced lubrication leads to excessive wear of component parts, **Figure 12-3.** In turn, bearing and spool clearances are increased. Excessive clearances cause internal component leakage and can lead to bearing failures in pumps and motors.

Sludge accelerates the build-up of residue on the surface of parts. It also leads to the gradual clogging of valve passageways. Acids attack metallic surfaces. This results in pitting and an increased rate of wear on bearing surfaces.

System Operating Temperature

System operating temperature is a major concern when designing and operating hydraulic systems. Systems function under a wide range of ambient temperatures. Systems also typically generate considerable heat during operation. This heat is a reflection of the relatively poor efficiency of these systems. Whenever the relief valve of a system functions, efficiency is reduced. The energy used to overcome the resistance of the valve is turned into heat. When all of this is considered, even well-designed systems convert 20% or more of horsepower input into heat. The heat is absorbed by the hydraulic fluid. Once started, this causes the temperature of the system to quickly increase.

System operating temperatures are usually monitored by measuring the temperature of the

Excessive wear

Figure 12-3. Dirt particles and acid in hydraulic fluids can cause rapid wear of bearing surfaces. (Schroeder Industries LLC)

hydraulic fluid in the reservoir. In most systems, reservoir-oil temperature ranging from 110°F–140°F is considered desirable. Extreme system operating temperatures, either high or low, are undesirable in any hydraulic system. One manufacturer indicates that the temperature should not exceed 145°F for even a short period of time without special considerations for fluids and seal materials.

High hydraulic oil temperatures can cause the system oil to break down. This leads to the formation of *varnish*. This material adheres to internal surfaces of components and clogs orifices, contributes to the formation of acid, and promotes the formation of sludge. Also, as temperature increases, the viscosity of the oil decreases. Excessive oil temperature may reduce viscosity to the point where the lowered lubricating qualities lead to increased bearing surface wear and decreased component service life. Excessive system operating temperature also shortens the life of system seals. This leads to increased fluid leakage and increased overall maintenance costs.

Low hydraulic oil temperatures present another set of problems that can adversely affect system operation. Oil temperature lower than the recommended level causes larger-than-desired pressure drops and sluggish system operation. The end result is an inefficient system. Fluid in equipment exposed to cold weather can have enough of a reduction in fluidity that major problems result, **Figure 12-4.** For example, highly viscous oil in the pump inlet line can cause cavitation. This leads to serious damage to internal pump parts. When ambient temperatures are low, a system should be operated under a no-load condition until the oil temperature approaches the recommended operating level.

Figure 12-4. Low system operating temperatures cause a special set of service problems, including pump cavitation during system startup. (Deere & Company)

Hydraulic systems need to be designed to balance the system heat input with the rate of heat removed by conduction, radiation, or convection. A properly designed system should provide the means to balance this input and output of heat. If a hydraulic system is operating at a temperature higher or lower than the recommended, maximum equipment performance will not be achieved.

Controlling and Removing Contaminants

The best route to follow to assure a clean operating system is preventing contaminants from entering a hydraulic system. However, because of the many extreme operating conditions in which systems must operate, this solution is not easily achieved. It is possible to reduce contamination if designers, machine operators, and maintenance personnel are aware of the sources of contaminants and how they enter a system. Using this information, each of these individuals can help reduce contamination and improve performance and service life of systems.

The designer must have adequate information concerning the conditions in which a machine must function. With this information, designs can provide adequate seals and baffles to reduce the entrance of dust, dirt, and water. Machine operators need to be aware of the sources of contaminates. Situations that exceed the ability of the system to block the entrance of contaminants must be avoided or promptly reported to supervisory personnel. Maintenance staff also needs to be alert to source of contamination. These sources should be eliminated before a replacement component is placed in service.

This section examines the components and basic circuits that deal with the control of contamination. This includes terminology, design factors, operating theory, installation, and service.

Contamination Control and the Reservoir

The reservoir is not only the fluid storage unit of a hydraulic system, it is considered a basic contamination-control component. The reservoir can make a large contribution to a clean and efficient operating system. Careful reservoir design and construction provide features that prevent the entrance of dust, dirt, and water into the system. Features can also be provided to help remove contaminants that do gain access to the system. See **Figure 12-5.** Review the reservoir portion of Chapter 8 for information on design features and additional points relating to the role of the reservoir.

Figure 12-5. The reservoir serves as the primary fluid storage unit of the system, but also acts as a basic fluid conditioning component. It helps remove dirt, water, entrained air, and heat from the system.

The traditional position for a return line filter is in the return line connecting the directional control valve return port and the reservoir. **Figure 12-17** illustrates several additional locations where these filters may be placed in a system. These locations include a pump case drain line, the relief valve return line, and the return line of a bypass flow control valve. During typical system operation, these return line filter locations are not subjected to the maximum operating pressure of the system. This reduces the need for a high-pressure filter housing, which reduces component cost.

The flow rate in return lines varies in relation to the operating phases of a system. For example, the fluid being returned to the reservoir may exceed pump output. This may happen if the rod diameter is large in relation to cylinder bore. Careful analysis of return line flow must be done to assure the selection of a filter with an adequate flow rating.

An inadequately sized filter will produce excessive backpressure in the line. This reduces system operating efficiency and the effectiveness of the filter.

Routing Fluid Flow for Filtration

Three routing methods are used to direct system fluid through system filters:

- Proportional filtration.
- Full-flow filtration.
- Off-line filtration.

Variations of these methods are also used, such as duplex filters. *Duplex filters* provide two or more filters in a single housing. A control valve provides an easy switching method allowing the fluid flow route to be alternated through the various filters. An advantage of this variation is the filter not currently in use can be changed without shutting down the system.

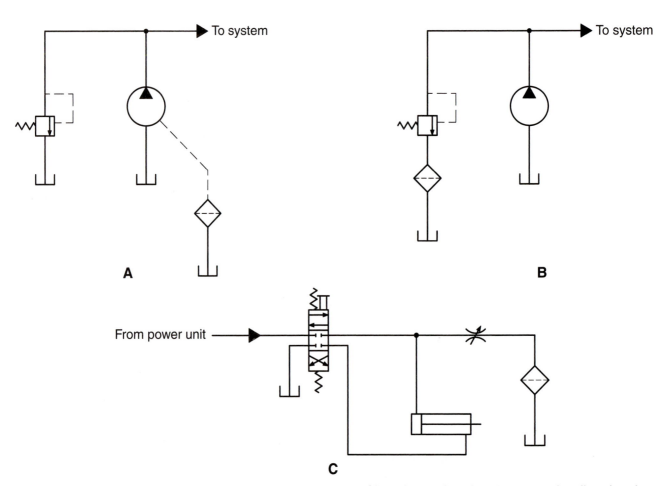

Figure 12-17. Return-line filters may be placed in a variety of locations other than between the directional control valve and reservoir. The locations shown have specific advantages and disadvantages in relation to filtration and service. A—Pump case drain. B—Relief valve return. C—Bypass flow control.

Proportional filtration directs a measured amount of system fluid through a filter. **Figure 12-18** shows a circuit diagram of a simple circuit using proportional filtering. In this circuit, a line leading from a system working line contains a metering valve and the filter. Part of the pump output passes through the filter and is returned directly to the reservoir. Filtration is continuous. The theory is that eventually all system fluid will pass through the filter and be cleaned. This method has become less popular as system tolerances have increased, which require more positive, finer filtration of system fluid.

Full-flow filtration filters all of the fluid circulated by the system pump. The location of filter in the system and the level of filtration varies by system design. However, all fluid pumped into the system is filtered before returning to the reservoir. In some systems, filters are placed in several locations. This results in portions of the fluid being filtered more than once as it is circulated. An example of this is a system using a sump strainer and a return line filter. This provides double filtration.

The use of full-flow filters has become more critical with the increased precision of hydraulic component parts. This increased filtration requires bypass valves to prevent excessive pressure drops across the filter as it becomes blocked by contaminants. Excessive pressure drops in the pump inlet line can cause pump cavitation. An overall loss of system efficiency occurs when excessive pressure drops occur at other filter locations.

Off-line filtration consists of a separate circulating pump and filtering system, **Figure 12-19.** This type of filtration is sometimes called *recirculating, kidney-loop,* or *side-stream filtration.* It should be considered a subsystem that continuously pumps fluid out of the reservoir, through a filter, and then back to the reservoir. As a general rule, the circulating pump should have an output flow rated at 10% of the reservoir volume.

Off-line filtration systems can maintain a clean system if pump flow and filter size are matched to the operating hydraulic system. It is relatively easy to retrofit an existing system with an off-line filtration system. Also, they can be easily serviced without shutting down the overall system.

Figure 12-20 shows a portable, off-line system. These can easily be moved from point to point. They can be used to provide filtration in emergency situations.

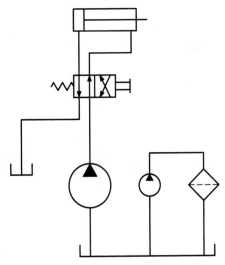

Figure 12-19. An off-line filtration system involves a separate pump that continually circulates reservoir fluid through a filter.

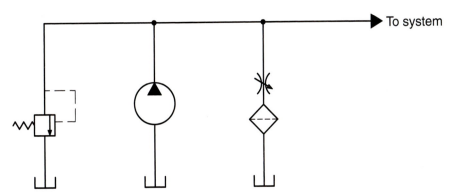

Figure 12-18. This diagram illustrates one method of proportional filtration in which a percentage of total pump output is diverted through a filter on a continuous basis.

Multiple Operating Pressures in a System

Hydraulic circuits often provide *multiple maximum system operating pressures.* In some systems, varying the pressure is the only practical way to produce the varying force needed to complete the work. In other systems, the varying pressure settings may be related to reducing energy consumption during extended *system cycle times.* **Figure 13-5** illustrates a simple circuit that can provide one maximum operating pressure for

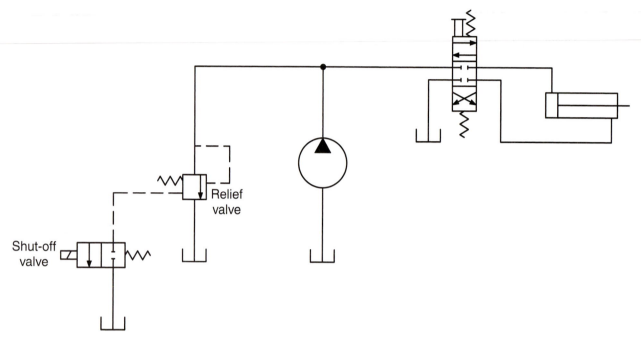

Figure 13-3. The pressure in a hydraulic circuit may be controlled from a remote location by using the vent feature of a compound relief valve. Opening the solenoid-operated shut-off valve drops system pressure to near zero.

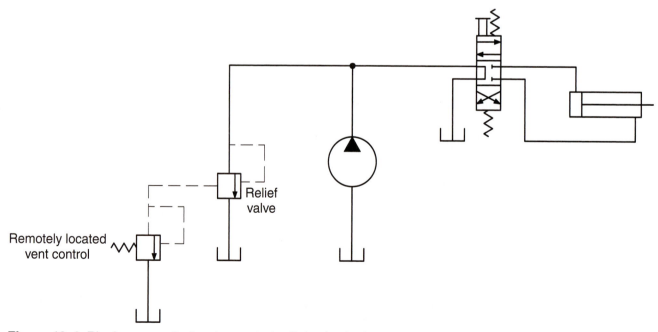

Figure 13-4. Placing a small, direct-operated relief valve in the vent line of a compound relief valve provides a remote means of adjusting the operating pressure of the system.

High-pressure relief valve

Low-pressure relief valve

Figure 13-5. Different maximum operating pressures may be obtained for cylinder extension and retraction by locating a second, lower-pressure relief valve in the working line between the directional control valve and the cylinder.

extension and a different one for retraction. This system uses two relief valves with full system flow capacity. The valve located on the pump side of the directional control valve must be adjusted for the higher pressure.

Figure 13-6 illustrates a circuit that can provide three different maximum system operating pressures. The primary relief valve, which is set for the highest maximum system pressure, is a vented, compound relief valve. This valve must have flow capacity adequate to handle the rated output flow of the pump. The two remaining relief valves can be direct operated and only large enough to accommodate the fluid discharged through the vent of the primary valve. The fourth valve shown in this circuit is a three-position, four-way directional control valve with a closed center. In this example, the directional control valve is spring centered and solenoid shifted.

When the solenoid-controlled directional control valve is not activated, the primary relief valve controls maximum system pressure. This is because the vent line of the valve is blocked by the closed center of the directional control valve. When the directional control valve is shifted in either direction, the vent line is connected to one of the small, direct-operated relief valves. Fluid from the control chamber of the primary valve then flows through the smaller direct-operated valve. This sets a reduced operating pressure for the system. Shifting the directional control valve to connect the second direct-operated relief valve produces the third pressure level.

Reduced Pressure in a System

Some hydraulic circuits contain sections requiring a maximum operating pressure that is lower than the setting of the primary relief valve. This *reduced-pressure section* requires a normally open valve that closes to restrict fluid flow into the section once the reduced pressure is reached. In order to function, these *pressure-reducing valves* must be externally drained with the pilot sensing pressure from the downstream side of the valve.

Figure 13-7 shows the placement of a pressure-reducing valve to control the maximum pressure of the clamp cylinder in a clamp-and-machining sequence. Once the pressure at the clamp cylinder reaches the setting of the pressure-reducing valve, the valve allows only enough hydraulic fluid to pass to maintain the reduced pressure.

Figure 13-16. This circuit uses control valves to sense pressure in the working lines leading to the actuators and then accurately controls the operating sequence of the circuit actuators.

When cylinder 2 encounters resistance during retraction, the pressure in the line begins to increase. When the pressure reaches the setting of sequence valve 2, the valve opens to allow fluid to pass to cylinder 1. As a result, cylinder 1 retracts. Fluid from the blind end of cylinder 1 is returned directly to the reservoir through the directional control valve.

This circuit is commonly used to illustrate the use of sequence valves. The use of sequence valves such as these, rather than the back pressure check valves shown in the previous circuit, results in a more easily adjusted and accurate system. The level of performance of circuits such as this depends on the valves used. Sequence valves are available in both direct- and pilot-operated designs, producing considerable variation in performance. Pilot-operated valves produce greater accuracy and repeatability. This results in a more efficient system. A review of the sequence valve section in Chapter 10 may be helpful at this time.

Synchronization of Linear Actuators

Synchronization is when two or more actuators move in unison. Synchronizing the movement of two cylinders in a hydraulic circuit is a fairly common design problem. When first confronted with synchronization, it is often considered a minor problem that can be solved by the use of simple flow control valves or a mechanical connection between the cylinder rods. In actual practice, it often becomes a challenge due to uneven loading between the actuators, internal leakage, and friction variations. Even if the cylinders are equal in size, slight variations that occur during each stroke can grow into major problems in systems that cycle many times in each hour of operation.

Figure 13-17 shows a flow control arrangement used to synchronize the extension stroke of two cylinders. The circuit has a pressure-compensated flow control valve placed in a meter-out flow configuration for each cylinder. The flow

Figure 13-17. Obtaining accurate synchronization of actuators requires a consistent rate of fluid flow into the inlet of each actuator. One method involves using meter-out flow control circuits containing pressure-compensated flow control valves.

control valves can be fine tuned to obtain identical extension times with each cylinder. The check valves allow fluid to bypass the flow control valves during cylinder retraction. The cylinders are returned to identical starting positions for the beginning of each extension stroke. This assists in maintaining synchronization.

> **Note:** The circuit shown in **Figure 13-17** allows cylinders of different sizes to be used even when identical extension rates are required.

An effective flow-dividing device can be created by mechanically connecting the output shafts of two hydraulic motors. **Figure 13-18** shows this device applied to a circuit that requires cylinder synchronization. The mechanical leakage forces the motors to turn at the same speed. This produces equal fluid discharge from each motor. The equal discharge rates produce equal extension rates in the two cylinders. The circuit shown also synchronizes the retraction of the cylinders. The motors meter an equal rate of fluid out of the blind end of each cylinder. If synchronization is not needed on cylinder retraction, insert check valves in lines around the motors. This bypasses the metering effect of the motors.

The performance of these systems must be monitored to assure the circuit is maintaining the required extension rate. Internal leakage, unexpected load variations, and fluid viscosity changes can cause variations in those systems that depend on fluid flow rates to produce synchronization of movement.

Design and Operation of Circuits with Other Functions

The previous sections present a small selection of pressure, flow, and motion control methods. There is a large number of additional circuits. For example, additional examples include accumulator, locking, motor-control, pressure-intensification, pump-control, and safety-related circuits. The following three sections describe additional circuits representing important concepts of system operation. These selected circuits represent only a small percentage of the methods that can be used to fill a specific design need in hydraulic powered equipment.

Figure 13-18. Hydraulic motors with mechanically connected output shafts can be used to produce two equal flows. Output flow from the motors assures synchronized movement of the two cylinders.

Hydraulic Motor Control

In hydraulic motor circuits, system pressure establishes torque output and flow rate controls speed. In addition to these two basic factors, the design of a hydraulic motor circuit can be used to produce other operating characteristics. For example, it may be desirable to allow the motor to freewheel while coasting to a stop. On the other hand, it may be necessary to brake the motor and associated load to a controlled stop. If these factors are not considered, high shock pressures and physical component damage may occur.

Figure 13-19 shows a basic hydraulic motor circuit. It allows the motor to be powered by the prime mover, but the machine operator can allow the motor to freewheel or be braked to a stop by the system relief valve. When the directional control valve is in the center position (position 2), the majority of pump output is returned to the reservoir through the open center. The hydraulic motor can freewheel by taking fluid from the motor-inlet side and directing any fluid coming from the outlet side to the reservoir.

When the directional control valve is shifted to position 1, the motor rotates. Fluid from the motor outlet is returned directly to the reservoir. When the load stalls the motor, system pressure increases to the relief valve setting. Pump output is then returned to the reservoir through the relief valve.

Figure 13-19. The inertia of the load on a hydraulic motor can present special problems. This relatively simple circuit can allow a motor to be powered, freewheel, or braked to a stop.

Any time during operation, the system operator can allow the motor to freewheel by shifting the directional control valve to position 2. If the inertia of the load causes the motor to continue to rotate, the motor can be braked by shifting the directional control valve to position 3. This forces the motor to pump fluid over the relief valve. The result is a braking action that quickly slows and then stops motor rotation.

Figure 13-20 shows an advanced hydraulic motor circuit that provides more flexible control than the previous circuit. Two pilot-operated pressure control valves are used to provide control. They serve as a relief valve and a counterbalance valve. The directional control valve controls the vent chambers of the pressure control valves.

During system startup, when the directional control valve is in position 2, both the relief valve and counterbalance valve are vented. This allows all pump output to return to the reservoir at near-zero pressure and the hydraulic motor to freewheel. Shifting the directional control valve to position 3 blocks the vent of the relief valve and opens the vent of the counterbalance valve. The power unit drives the hydraulic motor at pressures up to the relief valve setting. The motor fluid discharge is returned to the reservoir at near-zero flow resistance through the open counterbalance valve.

The hydraulic motor continues to turn the load until the directional control valve is moved to either position 1 or 2. Moving the valve to position 2 vents both the relief and counterbalance valves. This allows the inertia of the load to continue to turn the motor (freewheel). Shifting the directional control valve into position 1 opens the vent of the relief valve and closes the vent of the counterbalance valve. The relief valve then opens. This allows pump output to return to the reservoir. The activated counterbalance valve creates a back pressure on the motor, braking it to a stop. Once the motor stops, the closed counterbalance valve holds the motor and load in a fixed position.

System Decompression

It is generally assumed that hydraulic fluids compress so little as not to be considered important when designing, operating, or maintaining hydraulic systems. This thinking may be appropriate when working with systems containing

Figure 13-20. The venting capability of the pressure control valves in this circuit provides remote control of the valves and accurate pressure setting for both motor power output and braking action.

only a small volume of fluid. However, in major systems containing a large volume of fluid, the energy of compressed fluid must be considered as well as the energy stored by machine members under tension. The sudden release of either pressure can cause unexpected flow or pressure surges in a system. Both of these factors are safety issues and issues concerning service life of the hydraulic system or machine.

Figure 13-21 shows a simple circuit used with a forming press that has a large-diameter cylinder containing a large volume of hydraulic fluid. The circuit includes two check valves. The first is a staged, pilot-operated check valve that acts as a *decompression valve.* This valve first decompresses the fluid in the system and then acts as the primary route to return fluid to the reservoir as the platen quickly lowers. The second check valve is a simple check valve that acts to prevent the reverse flow of fluid through the working line leading from the directional control valve to the blind end of the cylinder.

When the directional control valve is in position 2, the tandem center returns the output of the pump to the reservoir at near-zero pressure. The cylinder is locked in place by the directional control and check valves. Shifting the directional control valve to position 1 directs hydraulic fluid to the blind end of the cylinder. This opens the simple check valve in the cylinder inlet line and closes the pilot-operated check valve leading to the reservoir. When the press closes, the pressure in the system increases to the setting of the relief valve. This maintains a constant force on the platen.

When the pressure portion of the cycle is completed, the directional control valve can be returned to position 2 to reduce the load on the system power unit. This occurs because pump output is directed to the reservoir. At this point, the weight of the platen pressurizes the fluid in the blind end of the cylinder to keep both check valves closed.

When the press platen needs to be lowered, the directional control valve is shifted to position 3. This connects the piloted-check valve to the pump. As a result, pressure in the pilot line quickly increases to the point where the pilot mechanism opens a small orifice in the check valve. This small opening decompresses the fluid in the cylinder. The pressure in the blind end of the cylinder drops, allowing the pilot mechanism to completely open the check valve. Gravity acting on the heavy platen forces the fluid in the cylinder to the reservoir and rapidly retracts the cylinder.

Figure 13-21. The decompression of circuits having large fluid capacities and static loads is sometimes necessary to reduce shock pressure. The process may also be used to reduce unanticipated machine member movement, which is a safety issue.

Safety-Related Circuits

Safe operation should always be a concern for equipment designers, manufacturers, and operators. The hydraulic field is not an exception to this concern. High pressures, immense forces, high flow rates, and high rates of movement are common elements in hydraulic systems. A variety of safety circuits are used to ensure:

- Equipment can be operated without injury.

- Equipment cannot unexpectedly operate.

- Costly equipment is protected against excessive pressure and operating rates.

The *two-hand-operation circuit* is common on presses and other equipment. This circuit requires both hands of the machine operator be kept clear

of dangerous areas. The circuit as described here operates using hydraulic principles. However, versions of this circuit may use pneumatics or electricity.

The circuit shown in **Figure 13-22** uses two manually shifted, spring-offset, two-position, four-way directional control valves to shift the system directional control valve. The system directional control valve is a pilot-operated, spring-centered, three-position valve with a closed center. When the power unit is not operating, the internal elements of the control valves should all be in position 2, as shown in the circuit schematic. This positioning is established by spring offsets in the two manual valves and spring centering in the system directional control valve. The cylinder is locked into position by the closed center of the system directional control valve.

Figure 13-22. Safety circuits are used in systems for the protection of operators and equipment. This two-hand safety circuit will not function unless valves A and B are activated at the same time.

Starting the power unit pressurizes the pilot lines. This causes the system directional control valve to shift into position 3. As a result, the cylinder fully retracts. Pressing and holding valves A and B in position 1 redirects pilot pressure to shift the system directional control valve into position 1. This causes the cylinder to fully extend and hold that position until valves A and B are released. When these valves are released, the pilot pressure shifts the system directional control valve to position 3 to fully retract the cylinder.

No other combination of valve A and valve B operation will fully extend and retract the cylinder or allow the cycle to be repeated. Releasing either valve A or valve B during cylinder extension causes the centering springs in the system directional control valve to shift the valve into position 2. This stops cylinder movement. Tying down either valve A or B to free one hand will not allow a new cylinder cycle to begin. As a result, the operator must have *both* hands on valves A and B for the cylinder to operate.

Chapter Test

Answer the following questions. Write your answers on a separate sheet of paper.

1. Internal geometry and system _____ determine the force output of a hydraulic cylinder or motor.

2. Remote control of pilot-operated relief valves can be easily achieved by using the _____ feature of the valve.

3. A maximum pressure lower than the relief valve setting may be obtained by using a(n) _____ valve in a circuit.

4. In a system with an open-center directional control valve, the pressure required to shift pilot-operated valves may be obtained by inserting a(n) _____ check valve in the return line to the reservoir.

5. In the _____ flow control circuit, the flow control valve is placed between the outlet port of the actuator and the directional control valve.

6. In a meter-out flow control circuit controlling the extension of a single-acting cylinder the pressure in the rod end chamber of the cylinder may be _____ than the maximum pressure setting of the relief valve.

7. Name the three characteristics of basic flow control circuits that need to be considered when selecting a design for use in a hydraulic system.

8. Which flow control circuit is considered the most energy efficient?

9. In the high-low pump design, when is the high volume pump discharge routed to the reservoir?

10. How is rapid cylinder advance created in a regenerative-cylinder-advance circuit?

11. Sequence valves must be _____ drained. Explain why.

12. Describe two methods that may be used to synchronize the movement of two or more hydraulic cylinders.

13. What occurs when a hydraulic motor is freewheeling?

14. Define *decompression* as related to a hydraulic system.

15. Name three factors that hydraulic safety circuits ensure.

This agricultural machine is used for taking soil samples. It makes extensive use of hydraulics to adjust the height of the machine, adjust the width between the wheels, and drive the wheels. In addition, since this machine is used around food crops, it makes use of plant estolide–based hydraulic fluid. (USDA)

In pneumatic systems, energy is transmitted using a gas. This contrasts with the liquid used in hydraulic systems, electron movement in electrical circuits, and belts and gears used in mechanical systems. To a casual observer, gas may appear to be a simple and inexpensive approach to energy transmission. However, this is not the case in actual practice. Free air from the atmosphere must be conditioned before it can be used in pneumatic systems. This chapter deals with the composition of free air and the nature of the compressed air needed in a pneumatic system.

Objectives

After completing this chapter, you will be able to:

- Describe the characteristics of free air.
- Identify the problems associated with the use of unconditioned air in a pneumatic system.
- Compare the composition of compressed air in a pneumatic system with the characteristics of free air.
- Explain the terms used to describe characteristics of compressed air.
- Compare the scales used to measure the pressure of free and compressed air.
- Compare the isothermal, adiabatic, and actual modes of air compression.
- Apply the principles of the general gas law to the compression and expansion of the air used in a pneumatic system.

Chapter 14

Compressed Air
The Energy Transmitting Medium

Selected Key Terms

The following terms will be used in this chapter. As you read the text, record the meaning and importance of each. Additionally, you may use other sources, such as manufacturer literature, an encyclopedia, or the Internet, to obtain more information.

adiabatic compression
adiabatic expansion
atmosphere
dewpoint
dry air
free air
ionosphere
isothermal compression
isothermal expansion
lubricant
mesosphere
ozone layer
relative humidity
saturation
stratosphere
troposphere
water vapor

Internet Resources

http://hyperphysics.phy-astr.gsu.edu/hbase/kinetic/relhum.html
Georgia State University—Dept. of Physics and Astronomy: HyperPhysics
Explains relative humidity, dewpoint, and saturation vapor pressure. Calculation methods are shown for each of these factors.

http://resources.norgren.com/document_resources/USA/HelpfulEngineeringInfo.pdf
Norgren
Provides basic engineering information related to determining water content in compressed air systems and establishing pressure drop in pneumatic circuits.

www.en.wikipedia.org/wiki/Gas_laws
Wikipedia: The Free Encylopedia—Gas Laws
Provides a brief description of gas laws with related mathematical equations.

www.en.wikipedia.org/wiki/adiabatic_process
Wikipedia: The Free Encyclopedia—Adiabatic Process
Provides a basic definition of the adiabatic process. Several examples are given with links to other sites that provide more detailed mathematical explanations.

www.en.wikipedia.org/wiki/isothermal
Wikipedia: The Free Encyclopedia—Isothermal Process
Provides a basic definition of the isothermal process. Several examples are given with links to other sites that provide more detailed mathematical explanations.

Basic Source of System Air

The source of the air used in a pneumatic system is the *atmosphere* that blankets Earth. This blanket is a layer of gas approximately 360 miles deep. This layer of air sustains life as we know it, supporting activities we take for granted. These include acting as our personal source of oxygen as we breath, the source of weather that distributes moisture to our food crops, our protection from solar radiation, and insulation from the frigid temperatures of space. In addition, the atmosphere supplies the oxygen that supports combustion, which is our primary source of heat in many activities.

Structure of the Atmosphere

The atmosphere is made up of distinct layers separated by transition zones, **Figure 14-1.** The layer in which we live is called the *troposphere.* It extends from Earth's surface to a maximum of 10 miles above the surface. The *stratosphere* extends from the troposphere to approximately 30 miles above Earth's surface. The *ozone layer*

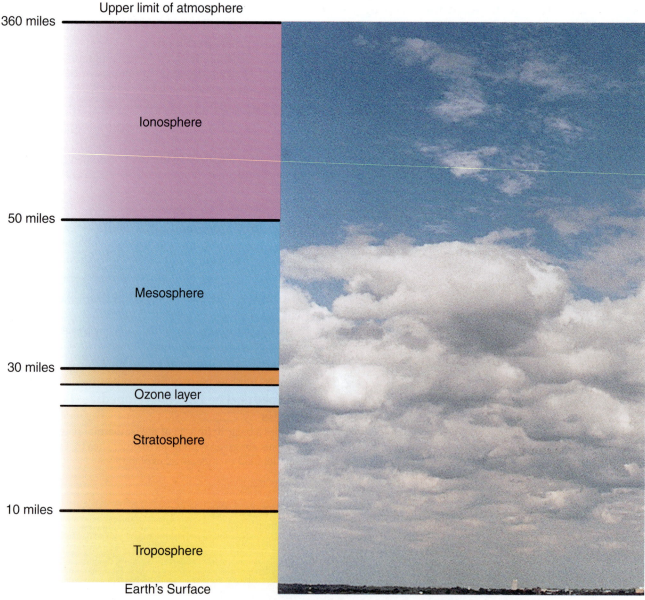

Figure 14-1. The atmosphere of Earth is made up of a series of layers. We live in the lowest level, called the troposphere.

that filters ultraviolet energy is located in the upper limits of this layer. The troposphere and stratosphere contain almost all of the gases that we normally consider *air*. The *mesosphere* begins above the stratosphere and goes to approximately 50 miles above Earth's surface. The last layer, the *ionosphere*, extends to approximately 360 miles. This is considered the limit of Earth's atmosphere.

From this general description it can be seen that, in theory, we have an enormous quantity of air available for use in pneumatic systems. However, in actual practice, careful conditioning is required to make atmospheric air usable in pneumatic systems.

Composition of Atmospheric Air

Many definitions indicate the composition of atmospheric air only by the proportion of the individual gases that make up the mixture. This may be critical for users that need one or more of the individual gases for use in a chemical process. However, when used as the energy-transmitting medium in pneumatic systems, additional factors must also be considered if the system is to operate effectively.

When only the gases are considered, air is made up of approximately 79% nitrogen, 20% oxygen and 1% other gases, **Figure 14-2.** These other gases include argon, ozone, carbon dioxide, and a number of additional gases in very small quantities. Other components include water vapor and entrapped dirt. The gas composition is relatively consistent, however, the other components vary depending on location.

Some water is always present in atmospheric air. The amount of **water vapor** varies from very little to the point of saturation. *Saturation* is reached when the maximum amount of water is held for a given temperature of air. This basic concept is an important factor. Water forms liquid water that accumulates in pneumatic system lines and other components as atmospheric air is compressed and then cooled in the system.

Dust and dirt may become suspended in atmospheric air and cause severe problems in a pneumatic system. The amount of dust and dirt depends on the location. Dust and dirt in a pneumatic system can be directly associated with internal wear of the compressor and other system components.

The weight and pressure of atmospheric air must also be considered. In daily activities, these

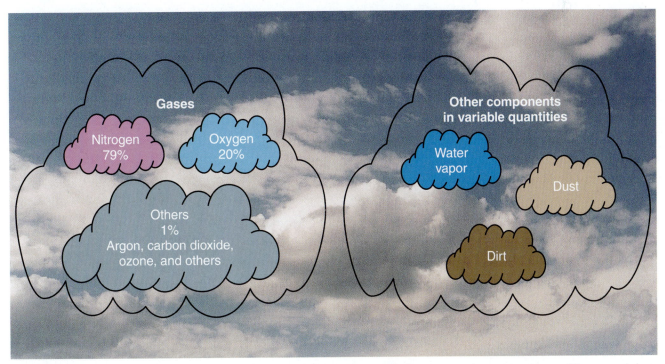

Figure 14-2. The major portion of the atmosphere is made up of nitrogen and oxygen gases. The remaining portion is other gases, water vapor, and entrapped dust and dirt.

factors are usually not a conscious concern. However, the weight of an atmosphere that is 360 miles deep is a real factor. It produces an air pressure of 14.7 pounds per square inch at sea level (14.7 psia or 0 psig). See **Figure 14-3.** Also, at 68° Fahrenheit, one cubic foot of air weighs approximately .0752 pounds.

Pneumatic System Compressed Air

Atmospheric air, typically referred to as *free air,* needs to be conditioned before distribution and use in a pneumatic system. Conditioning involves four factors to produce air that results in an efficiently operating system with minimal equipment service:

- Removal of entrapped dirt.
- Removal of water vapor.
- Removal of heat.
- Incorporating oil for component lubrication.

The location of the pneumatic system has a great deal of influence on the amount of dirt in free air. This dirt must be removed before the free air enters the compressor intake. The free air in clean environments needs minimal filtration. In contrast, a compressor at a dusty construction site needs several levels of filtration to prevent excessive wear of the compressor and air tools, **Figure 14-4.** Details of the methods used to remove dirt from intake air are found in Chapter 16.

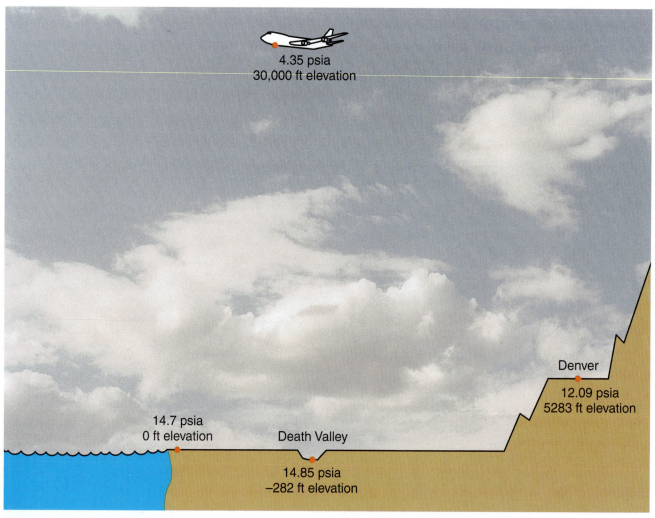

4.35 psia
30,000 ft elevation

Denver
12.09 psia
5283 ft elevation

14.7 psia
0 ft elevation

Death Valley
14.85 psia
−282 ft elevation

Figure 14-3. Atmospheric pressure varies depending on the vertical location at which the measurement is taken. The higher the location, the lower the pressure reading.

Water that accumulates in a pneumatic system can be detrimental to system components and must be removed, **Figure 14-5.** It can also affect the manufactured product if used for cleaning or in any other step of the process. Designers and operators of pneumatic systems need to be aware of air's ability to hold water vapor. They also need to understand the relationship of humidity and temperature to the formation of water in a system.

The amount of water carried in air largely depends on the temperature of the air. Hot air can hold more water than cool air. When air cannot hold additional water at a given temperature, it is saturated.

Relative humidity is the amount of water vapor actually carried by a given volume of air. This is a percentage showing the relationship between the actual amount of water in the air and the maximum amount the air can hold at the given temperature. For example, a relative humidity of

A

B

Figure 14-4. The amount of entrapped dirt in free air varies, depending on local conditions. A—The air at construction sites often needs additional filtration. B—Quarrying operations are very dusty. (Manufactured Housing Institute, Atlas Copco)

Figure 14-5. Compressed air containing an excessive amount of water vapor results in liquid water forming in distribution lines. This water may be collected by the use of water legs placed in strategic locations in the distribution lines.

Reaction of Air to Temperature, Pressure, and Volume

The concepts involved with isothermal, adiabatic, and actual compression and expansion of air begins to identify the complexities of handling air. However, these complexities may be more easily understood if two basic principles are remembered:

- A change in either air temperature, pressure, or volume results in a change in at least one, and possibly both, of the other factors.

- The relationships expressed are based on the absolute scales of temperature and pressure.

In an operating system, the variations that occur can be complex. They can be a challenge when analyzing the operation or performance of a pneumatic system. However, it must be remembered that these changes adhere to the general gas law:

$$(P1 \times V1) \div T1 = (P2 \times V2) \div T2$$

In addition to the pressure, volume, and temperature of the air, many outside variables make direct application of the general gas law impractical in an operating system. Several sources, including publications of component manufacturers and data handbooks for compressed air and gas, provide engineering data that can be applied to compensate for additional variables. The resulting calculations are improved estimates of how a component or system design will respond in an operating situation.

The following sections discuss several aspects of how air in an operating pneumatic system responds to the general gas law and the other variable factors. The discussion does not involve calculations, but is intended to illustrate the complex relationship between changes in air pressure, volume, and temperature and the influence of other system factors.

Pressure Variations

According to the general gas law, any change in the pressure of air will result in a temperature or volume change. For example, when shifting

Figure 14-9. To extend the cylinder under load, both air pressure and volume must be increased. Air temperature also changes in the process due to rapid pressure changes.

the system directional control valve, air flows from the receiver to the cylinder, **Figure 14-9.** The air entering the cylinder increases the pressure in the confined space of the cylinder chamber. The increase in cylinder air pressure is due to air from the system receiver entering the fixed volume of the stalled cylinder and the slight increase in temperature resulting from the heat of compression.

When the force generated by the air pressure acting on the piston surface is adequate to extend the cylinder, the chamber volume begins to increase. As a result, the cylinder rod moves. Air from the receiver continues to flow into the cylinder chamber at a pressure adequate to move the load. If the load remains constant, the pressure and temperature of the air in the cylinder chamber remain constant.

Volume Variations

Changes in the volume of air result in a pressure or temperature change. An example is the reduction of volume in a cylinder during the compression stroke of a piston-type compressor,

Figure 14-10. The compression stroke increases air pressure by both the reduction of volume in the cylinder chamber and the heat of compression that increases air temperature. The smaller the volume left in the cylinder at the end of the compression stroke, the higher the pressure produced and the higher the air temperature.

Temperature Variations

An increase or decrease in the temperature of air results in a pressure or volume change. For example, the pressure in the fixed space of the system receiver changes as the temperature of the air changes, **Figure 14-11.** As heated air coming from

Figure 14-10. Reducing the volume of air during the compression stroke in a piston-type compressor increases the pressure of the air. The rapid air compression also produces heat that contributes to an additional increase in pressure of the air.

Heat transfers from the receiver to the air in the room.

Figure 14-11. The rate of heat loss to the atmosphere through the external walls of the receiver influences the final pressure of the air in the unit.

the compressor enters the receiver, the existing air is warmed by both the heat of the incoming air and the heat generated by the heat of compression as the air existing in the receiver is further compressed. Both of these factors increase the pressure in the fixed-volume receiver.

Temperatures in a variety of situations can cause changes in the operation of a pneumatic system. Examples include compressor intake air temperature, compressor room temperature, and the temperature of the air surrounding the system distribution lines. The temperatures existing in each of these situations can produce variations in system performance.

Chapter Test

Answer the following questions. Write your answers on a separate sheet of paper.

1. The atmosphere blankets Earth with a layer of gases approximately _____ miles deep.

2. The air we breath and use as the source of air for pneumatic systems contains several gases. Name the gases and indicate the percentage at which they exist in the atmosphere.

3. Define *saturation*.

4. Standard atmospheric pressure is _____ pounds per square inch at sea level.

5. _____ refers to the amount of water vapor actually carried by a volume of air compared to the maximum amount it can carry at the specified temperature.

6. Where are lubricants typically introduced into a pneumatic system?

7. The process that assumes all heat is retained in the air during compression is known as _____ compression.

8. Name two basic principles that must always be considered when applying the general gas law to the compression of air in a pneumatic system.

9. Provide the formula for the general gas law.

10. According to the general gas law, what happens when the pressure of air is changed?

Compressed air used to apply paint must be carefully controlled to assure a quality product finish. (pistolseven/Shutterstock)

The compressed-air unit is the source of power for the remainder of the pneumatic system. These units range in size from small, portable compressors to huge units that provide compressed air to large manufacturing facilities. The units may be relatively simple, consisting of a prime mover, compressor, receiver, and simple pressure switch. In contrast, major installations not only include larger versions of these components, but additional components designed to provide a higher level of conditioning for cleaner and drier compressed air. This chapter introduces each of the components and then discusses compressor design in more detail to provide an understanding of compressor operation and application.

Objectives

After completing this chapter, you will be able to:

- Describe the function of a compressed-air unit.
- Name and explain the function of each of the components in a compressed-air unit.
- Identify the basic designs used in air compressor construction.
- Compare the operating characteristics of positive- and non-positive-displacement air compressors.
- Compare the operating characteristics of rotary and reciprocating air compressors.
- Describe the general construction characteristics of the various compressor types.
- Explain the operation of the various systems used to control the maximum air pressure available from the compressed-air unit.
- Identify the factors that must be considered to estimate the required output of a compressor to meet the air demands of a pneumatic system.
- Interpret performance data supplied by a compressor manufacturer.

Chapter 15

Source of Pneumatic Power
Compressed-Air Unit and Compressor

Selected Key Terms

The following terms will be used in this chapter. As you read the text, record the meaning and importance of each. Additionally, you may use other sources, such as manufacturer literature, an encyclopedia, or the Internet, to obtain more information.

air filter
capacity-limiting system
central air supply
compressed-air unit
compressor-capacity control
coolers
coupling
double-acting compressors
dryers
dynamic compressors
lobe-type compressors
non-positive-displacement
portable unit
positive displacement
prime mover
pumping chamber
pumping motion
receiver
reciprocating compressor
rotary compressor
rotary screw compressors
rotary, sliding-vane compressors
single-acting compressors
slip
staging

Internet Resources

www.atlascopco.us/usus/
Atlas Copco
Enter "screw compressors" in the product search box to review product information on the various rotary screw compressors manufactured by the company. Includes stationary and portable units.

www.dresser-rand.com/literature/turbo/2099_DATUM_C.pdf
Dresser-Rand
Describes a centrifugal compressor designed for a process application in the oil, gas, or chemical industries.

www.fs-elliott.com/pdf/brochures/PAPPLus.pdf
FS-Elliott Co., LLC
Describes a line of centrifugal compressors. Illustrated with a general description of the operation of a two-stage unit including intercoolers.

http://hydraulicspneumatics.com/fluid-power-basics/pneumatics
Hydraulics & Pneumatics Magazine
Review two articles: *Air Compressors–Part 1* and *Air Compressors–Part 2*. These articles cover the styles of compressors available in the pneumatic field, emphasis on rotary compressors. Also includes information on power and efficiency, lubricated or lubrication-free compressors, cooling methods, and capacity control.

www.ingersollrandproducts.com/am-en/products/air/small-reciprocating-air-compressors
Ingersoll-Rand Company
Provides information on reciprocating air compressors manufactured by the company. The brochure linked on the page refers to single- and two-stage air compressors.

www.jennycompressor.com/howtochoose.html
Jenny Products, Inc.
Guide to selecting an air compressor for a small business operation or home shop. Covers factors such as the need for staging, type of prime mover, and specific questions that should be answered before purchasing a compressor.

www.saylor-beall.com
Saylor-Beall Manufacturing Company
Manufacturer that produces reciprocating and screw compressor units. Information contained includes product brochures and service manuals with illustrated, informative materials covering general questions and answers.

www.senco.com/Documents/Manuals/PC1130_MAN.pdf
SENCO Brands, Inc.
Product manual for a single-stage, reciprocating compressor. Applications for these compressors range from the home shop to industrial situations.

Compressed-Air Unit

The *compressed-air unit* produces pressurized air for the pneumatic system. The design and appearance of this unit varies, depending on the use of the air and the size of the system it must operate. The unit may be a very simple compressor supplying a fraction of a cubic foot per minute for an artist's airbrush or portable air tool, **Figure 15-1.** On the other hand, it may be a large, stationary installation in an industrial plant, **Figure 15-2.** Industrial units may produce several thousand cubic feet per minute of conditioned air for the operation of a number of system workstations.

Figure 15-1. A small, portable air compressor is commonly used at home building sites to operate nailing equipment or other tools. (DeVilbiss Air Power Company)

Classification of Compressed-Air Units

A compressed-air unit is classified as a portable unit or central air supply. A *portable unit* is one that can be relatively easily moved from one location to another. Portable units are moved to the work location and provide compressed air to remote work sites, **Figure 15-3.** A *central air supply* has a permanently mounted compressor and air distribution lines. Central air supplies provide air to fixed workstations in a permanent installation.

Physical size may be a factor in selecting the classification of unit to use, but certainly cannot be used as the primary deciding feature. Physical size could be considered a major difference if the small, portable compressors that have become a popular consumer tool were the only portable compressor units, **Figure 15-4.** However, portable compressors used in the construction, mining, and quarrying fields often exceed the size of central air supplies in small manufacturing plants.

Even though the appearance of these units is different, they do include a number of elements that serve identical functions. Units under both classifications:

- Provide the prime mover energy needed to compress the air.

- Control air temperature, retained water vapor, and entrapped dirt.

- Provide a means of controlling the maximum air pressure in the system.

- Store the conditioned air until needed at the workstation.

The components in the compressor unit that provide these functions are described in the next section.

Components Parts and Functions

The compressed-air unit serves as the overall source of energy for a pneumatic system. A *prime mover* drives the compressed-air unit. Typically, the primer mover is dedicated to the operation of the compressor. The prime mover in a central air supply is most often an electric motor, although steam and turbine engines may be used in very large applications. Small, portable units are typically powered by electric motors. Internal combustion

Figure 15-2. Industrial plants use central air stations that supply compressed air to numerous workstations. (Badger Iron Works, Inc.)

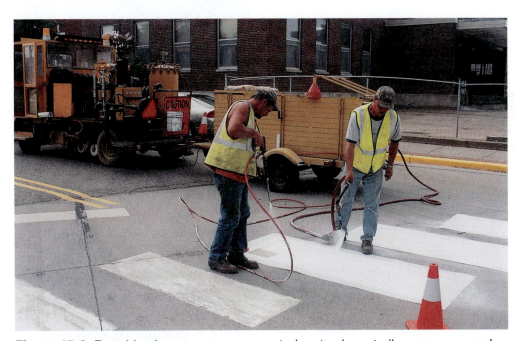

Figure 15-3. Portable air compressors mounted on trucks or trailers are commonly used at highway and other construction sites.

engines are used to power portable units for use in remote locations, such as new building sites and road construction, where electricity is not available. The energy that drives the prime mover comes from electric current for a motor or fuel for an internal combustion engine.

The prime mover is connected to the input shaft of the compressor by a mechanical *coupling.* Several different coupling designs are used, including direct, mechanical couplers, and belt drives. See **Figure 15-5.** The design used in a specific compressed-air unit is based on rated

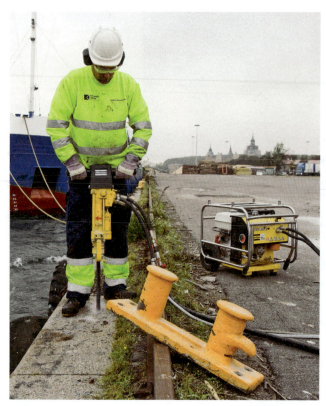

Figure 15-4. Portable units allow the compressor to be operated where the work is being performed. (Atlas Copco)

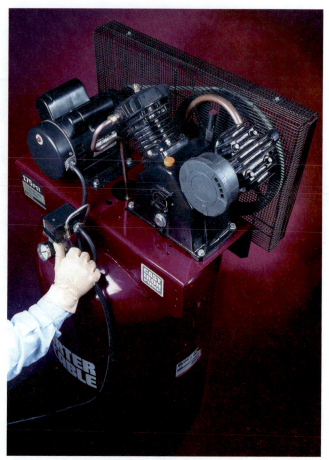

A

speed, load, and allowable shaft misalignment between the prime mover and compressor.

The *compressor* is driven by the prime mover and generates compressed air. A variety of compressor designs are available for use in pneumatic systems. The design selected for an application depends on physical size restrictions, required system air flow rates, operating pressure, and a variety of other requirements. All of the compressor designs increase the pressure of the air from atmospheric pressure to the desired operating pressure. This is done by moving additional atmospheric air into the fixed volume of the components that comprise the compressed-air unit.

Another major component of the compressed-air unit is the *receiver.* This component serves a number of purposes. First, it is the sealed, fixed-volume chamber that holds the compressed air until needed by the pneumatic system. When the system needs air for the operation of a workstation circuit, that air is drawn from the receiver. The receiver also assists in cooling the air and removing water vapor to assure the distribution of dry air to the system.

B

Figure 15-5. Belts (A) or direct couplers (B) are used to transfer energy from the system prime mover to the compressor. (DeVilbiss Air Power Company)

Another vital part of the compressed-air unit is a *capacity-limiting system*. This system is used to control the maximum air pressure produced by the unit. The maximum pressure is somewhat higher than the pressure needed to operate the actuators in the system. This higher pressure assures efficient air distribution in the system. The capacity-limiting function is commonly performed by an electrically operated pressure switch that automatically starts and stops the prime mover at preselected minimum and maximum pressures. In systems where the prime mover is an internal combustion engine, the pressure switch opens and closes a valve that vents compressor output when the preselected maximum system pressure is reached. The compressed-air unit also includes a mechanically operated *safety valve*. If the electrically operated valve should fail, the safety valve opens to vent pressure before a dangerous pressure is reached.

The compressed-air unit includes a number of additional components, depending on the size and complexity of the unit. A critical unit is an *air filter* to remove dirt from the atmospheric air. *Coolers* and *dryers* are also used to assure the delivery of clean, cool, dry air to the distribution lines and workstations of the system.

Typically, compressor manufacturers are willing to provide all of the necessary components and electrical controls as a package. This provides a compressed-air unit that meets the requirements of an installation without the trouble of assembling the system on site.

Basic Compressor Design

A number of compressor designs are used to convert the mechanical power output of the prime mover into high-pressure air for use in a pneumatic system. The capacity of these units varies significantly, as does the general appearance and structure. However, basic operating concepts are similar. This section examines the conceptual design and operation of a compressor to illustrate how air compression is accomplished.

Basic Compressor Structure

A pneumatic system compressor should be thought of as a specialized air pump. It is simply a pump designed to deliver air at an elevated pressure. The following explanation uses a basic reciprocating compressor design to illustrate both compressor structure and operation. The design has been used for years in the pneumatic field and is still common today.

The structure of a basic reciprocating compressor can be compared to an internal combustion engine minus the fuel and ignition systems. The structure of such a compressor is shown in **Figure 15-6.** The space above the pistons form *pumping chambers* that vary in size as the crankshaft rotates to move the pistons up and down.

The cylinder head of the compressor contains ports. Each port is fitted with a valve. The inlet ports are open to the atmosphere through the filtering system. The inlet valves are open as the pistons move down in the cylinder and closed as the pistons move up. The outlet ports are connected to the inlet port of the system receiver. The outlet valves are open when the pistons are moving up in the cylinder and closed as the pistons move down.

Figure 15-6. A reciprocating compressor is similar in appearance and structure to an internal combustion engine. (DeVilbiss Air Power Company)

Basic Compressor Operation

The following sections describe the basic operation of a compressor. Refer to **Figure 15-7.** There are three basic phases of operation:

- Air intake.
- Air compression.
- Air discharge.

Air intake

When the rotating crankshaft pulls the piston down, the pumping-chamber volume increases. This increased volume lowers the air pressure in the cylinder. As a result, atmospheric pressure forces air into the pumping chamber through the open inlet valve. Ideally, the air pressure in the pumping chamber will be as close to atmospheric pressure as possible during the intake phase. This maximizes the amount of air moved through the compressor. When the continued rotation of the crankshaft begins to move the piston upward, the inlet valve closes. The volume of the pumping chamber also begins to decrease.

Air compression

As the crankshaft continues to rotate, the piston is forced upward in the cylinder. This decreases the volume of the pumping chamber. The decrease in volume causes the pressure of the air in the chamber to increase. This increase in pressure indicates two important items:

- The amount of actual decrease in chamber volume.
- The amount of energy from the prime mover that has been converted to potential energy, as represented by the higher air pressure.

Air discharge

The pressurized air in the pumping chamber only moves into the system receiver when the pressure in the chamber is higher than the pressure in the receiver. When the receiver and system pressures are low, the transfer to the receiver is early in the compression stroke. When the system pressure is close to the maximum pressure setting of the compressed-air unit, the transfer is late in

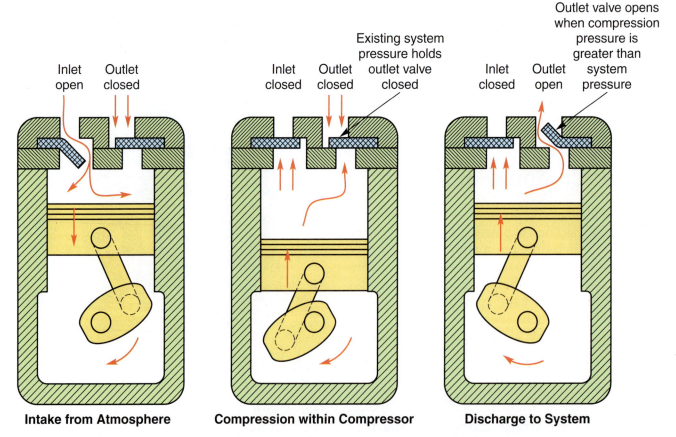

| Inlet open | Outlet closed | Inlet closed | Outlet closed | Existing system pressure holds outlet valve closed | Inlet closed | Outlet open | Outlet valve opens when compression pressure is greater than system pressure |

Intake from Atmosphere   **Compression within Compressor**   **Discharge to System**

Figure 15-7. The point at which air flows from a compressor depends on the existing pressure in the system.

the compression stroke. The response of a compressor to the air demands of a pneumatic system is determined by:

- Size of the pumping chamber of the compressor.
- Operating speed of the compressor.
- Pressure and volume of the air needed to operate the circuits of the system workstations.

Basic Compressor Classifications

All compressors used in pneumatic systems are classified using two methods. These classifications involve the design, structure, and operation of the compressors relating to:

- Displacement.
- Fundamental pumping motion.

Displacement

The *displacement* of a compressor is the volume of air displaced per unit of time. It is usually stated in cubic feet per minute (cfm). The method used to calculate displacement depends on the compressor design. Compressors are classified as either having positive displacement or nonpositive displacement.

Positive displacement

Compressors classified as *positive displacement* confine air in a chamber that can be mechanically reduced in volume to increase the pressure. When the air chamber volume is decreasing, air is transferred to the receiver of the compressed-air unit after the chamber pressure exceeds the air pressure in the system. Reciprocating, sliding vane, and rotary screw compressors are considered positive-displacement designs, **Figure 15-8**.

Non-positive displacement

Non-positive-displacement compressors compress air by using impellers or vanes rotating at high speed. This movement increases both air velocity and pressure. In the past, these designs were often called *blowers* because of their relatively low pressure output. Today, they are typically referred to as *dynamic compressors.* Both centrifugal and axial-flow dynamic compressors are considered non-positive-displacement units.

Fundamental Pumping Motion

The *pumping motion* classification relates to the basic concepts used to increase the air pressure and move air through the compressor. Compressors are rated either as reciprocating or rotary under this classification.

In the reciprocating design, a cylinder fitted with a piston produces a variable-volume chamber. A *reciprocating compressor* produces an up-and-down movement of the piston to bring in atmospheric air, compress the air, and then move it to the receiver of the compressed-air unit. The process follows the basic compressor operation described earlier. However, a number of refinements are used to increase compressor efficiency, including multiple cylinders and single- or double-acting piston designs.

In the rotary design, a rotating component inside of the compressor produces the variation needed to compress the air. One of a number of different variations is used in a *rotary compressor* to bring in the atmospheric air, complete compression, and then move the air to the system receiver.

Figure 15-8. The reciprocating compressor design is a common example of a positive-displacement compressor. (DeVilbiss Air Power Company)

In each case, the process is continuous, rather than the stop-and-go performance of a reciprocating compressor. Sliding-vane, screw, and rotary-liquid-piston compressors are examples of rotary, positive-displacement designs. In each of these designs, a moving chamber that varies in size provides a more-continuous compression than a single-stage reciprocating unit.

Non-positive-displacement, dynamic compressors also operate with a continuous, rotary motion. The high-speed rotation of both centrifugal and axial compressors causes airflow through the units. The cross-sectional area of the discharge of these compressors is much smaller than the intake, resulting in an increased pressure. These compressors generally use multiple stages that produce the desired pressure increase.

Compressor Design and Operation

System designers have a variety of air compressors to consider when developing a new pneumatic system. The compressor should be selected only after a careful analysis matches it to the performance needs of the system. Factors to consider include:

- Compressor capacity.
- Expected length of service.
- Environmental conditions.
- Initial unit and installation costs.
- Ongoing operating costs.

This section examines the basic designs and features of both positive- and non-positive-displacement compressors commonly found in pneumatic installations.

Compressor Types

The basic principles involved in the operation of compressors are similar in the fact atmospheric air must be drawn into the compressor and then reduced in volume to increase the pressure to a desired level. However, there are some differences between designs. The following sections describe the various compressor designs that may be encountered in the field.

Reciprocating-piston compressors

Reciprocating-piston compressors, or simply reciprocating compressors, are very common in pneumatic systems. They are available as both single- and double-acting units. At the low end of performance, they may be units designed for the home user. These small units usually have a single cylinder with a displacement that may be as low as 1 cubic foot per minute. System pressure may range from 40 psi to 120 psi. At the high end of performance, multi-cylinder industrial units with prime movers of several thousand horsepower produce extremely high operating pressures. See **Figure 15-9.**

Figure 15-9. Reciprocating compressors in large, industrial, central air systems may have multiple cylinders and deliver large volumes of air at very high pressures. (Atlas Copco)

Single-Acting Compressors. *Single-acting compressors* are, from a technical standpoint, the simplest of the compressors. They use cylinders, pistons, connecting rods, and crankshafts of the same type found in internal combustion engines. See **Figure 15-10.** Air is compressed in the cylinder

A

B

Figure 15-10. Multiple-cylinder, reciprocating compressors may have an inline, V-type, or other cylinder arrangement. A—Inline. B—V-type. (DeVilbiss Air Power Company)

chamber, which is on the top of the piston. The piston bottom is exposed to the crankcase, which serves to lubricate the crankshaft bearings and other bearing surfaces. Most units use pressure-operated reed valves to control airflow. Some models use cam-operated valves synchronized with the position of the piston and crankshaft.

During compressor operation, atmospheric air enters the cylinder through the inlet valve. This results from the reduced pressure in the cylinder produced by the increasing volume of the cylinder chamber as the piston moves downward. When the piston moves upward, the chamber volume is reduced. This increases the air pressure. When the pressure becomes higher than the air pressure in the compressor discharge line, the outlet valve opens. This allows the high-pressure air to move toward the system receiver. With this design, there is intake of atmospheric air each time the piston moves down and a discharge of higher-pressure air each time the piston moves upward.

Single-acting compressors are available as single- or multiple-cylinder units. Multiple-cylinder units may have from two to eight cylinders in an inline, opposed, V, W, or semiradial arrangement. The W arrangement is typically used with three-cylinder compressors. The semiradial arrangement has a single crankshaft throw with a master connecting rod.

Both air- and water-based cooling systems are used to remove the heat of compression from single-acting compressors. Air-cooled units have cooling fins on the outside of the cylinder to transfer heat to the atmosphere. The cooling air is moved across the fins by a fan that is often an integral part of the compressor flywheel. The flywheel helps dampen the natural irregular power needs of reciprocating machines. Large units that operate in dirty or extremely hot environments may require a water cooling system. This involves a cooling jacket around the compressor cylinders and an external radiator to eliminate the excessive heat. Water circulated through the cooling jacket picks up heat and then releases it to the atmosphere via the radiator.

Double-Acting Compressors. Double-acting, reciprocating compressors are normally considered heavy-duty units designed for continuous service. *Double-acting compressors* have separate cylinder compression chambers at each end of a piston. **Figure 15-11** shows the internal structure of a single-cylinder, double-acting unit.

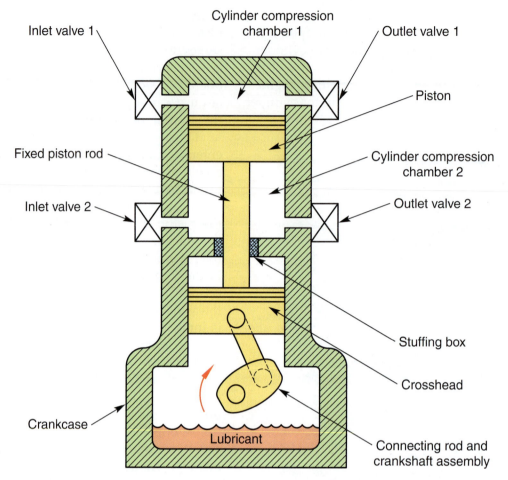

Figure 15-11. The double-acting compressor design allows air intake and air compression to simultaneously occur.

The piston is powered by a fixed piston rod that extends through a stuffing box located between the piston and a crosshead. The end of the piston rod outside of cylinder compression chamber 2 is attached to the crosshead that slides in a cylinder bore with walls parallel to the walls of the compression chambers. The crosshead is attached to the compressor crankshaft with a connecting rod. This mechanical arrangement allows the energy of the prime mover to be transferred through the crankshaft and connecting rod to the crosshead. The reciprocating motion of the crosshead allows the piston to accurately travel within the cylinder, while the stuffing box packing around the piston rod provides a positive seal for that compression chamber.

A double-acting compressor produces a maximum volume of compressed air. During each crankshaft revolution, the compressor experiences two intake strokes and two compression strokes. For example, as the piston extends during the first 180° of rotation, there is compression in compression chamber 2 and intake in compression chamber 1. During the second 180° of rotation, there is intake in chamber 2 and compression in chamber 1.

Cooling and lubrication are critical to the performance and service life of these units. They often operate on a continuous basis. However, the exact cooling system used heavily depends on the application. Water cooling is recommended in most situations. Both cylinders and cylinder heads are fitted with water jackets. Splash lubrication, pressure-feed lubrication, or a combination of both is used with these compressors. Pressure-feed lubrication is typically considered the best.

Very large and complex single-cylinder, double-acting compressors have been built. They also exist as multiple-cylinder units. The pistons may be arranged in the same manner as previously described for single-acting compressors.

Rotary, sliding-vane compressors

Rotary, sliding-vane compressors use rotary motion directly to produce an almost pulsation-free compressing action. They are positive-displacement units. The compressor basically consists of a housing forming a compression chamber with inlet and outlet ports and a slotted rotor. The rotor contains moveable vanes that use centrifugal force to maintain a seal with the walls of the compression chamber. See **Figure 15-12.** The rotor and vanes form the compression mechanism of the unit.

The slotted rotor is directly attached to the input shaft of the compressor, which is turned by the system prime mover. The rotor is offset in the compressor housing. This results in a series of varying-size compression chambers between the rotor, vanes, and walls of the housing. Intake and compression occur as the size of these chambers increases and then decreases as the rotor turns. As the chambers are increasing in size, they are exposed to the inlet port. This allows atmospheric air to enter. As the rotor continues to turn, the chambers begin to reduce in size. When the chambers are exposed to the outlet port, the pressurized air is allowed to move into the system.

The vanes of these units are subjected to friction resulting from centrifugal force pressing the vane tip against the compression-chamber wall. This pressure and the scraping action against the chamber wall results in considerable vane wear. To reduce this wear and increase sealing, oil is often injected into the compression chamber. Oil separators are placed in the compressor discharge line to remove any liquid oil that remains in the compressed air.

Rotary, sliding-vane compressors are typically water cooled. The cylindrical compression chamber housing and the housing heads contain water jackets. The oil injected to improve sealing and reduce wear also helps remove some of the heat of compression.

Rotary screw compressors

Rotary screw compressors use intermeshing, helical screws to form chambers that linearly move air through the compressor, **Figure 15-13.** The design provides a continuous, positive displacement

Figure 15-13. Rotary screw compressors have intermeshing, helical screws. The screws form sealed chambers during rotation to transport a volume of air from the atmosphere to the receiver and distribution lines. (Atlas Copco)

Contact between sliding vane and compression chamber wall

Figure 15-12. Centrifugal force keeps the vanes of the sliding-vane compressor in constant contact with the walls of the compression chamber.

of air to deliver a nonpulsating flow at all operating speeds. This group of compressor designs has become very popular in larger industrial applications because of:

- Lower initial cost for high-volume outputs.
- Low maintenance.
- Availability of sophisticated control systems.

These compressors may be dry or oil flooded. The dry version uses timing gears to turn the screw elements. These units are constructed with an extremely small clearance between the screw surfaces. This requires little or no lubrication. In the oil-flooded design, the screws are driven by direct screw contact. As a result the timing gear mechanism is not required. In both versions, when the screws are turned, a volume of atmospheric air is trapped in the open spaces between the screw threads. This trapped air is pushed along the axis of the compressor to the discharge port leading to the system receiver.

The oil-flooded compressor design uses an oil separator to remove the oil from the compressed air. The separated oil is reused after it is cooled and filtered. An oil separator is normally not needed with the dry compressor as the small screw clearance eliminates the need for lubricating oil. Both designs require a cooling system to remove the heat of compression. The oil in the oil-flooded version acts as a coolant to remove some heat, as well as lubricate and seal.

Dynamic compressors

Dynamic compressors are typically used to compress air or other gases in large applications such as petroleum refineries, chemical plants, and steel mills. In recent years, the pressure capability of the design has increased from under 40 psi to several hundred psi. Also, these designs are most often available as high-flow-output units producing thousands of cubic feet per minute.

Dynamic compressors use the rotary design and may be of centrifugal or axial design. The designs vary considerably in physical appearance, but achieve compression using basically the same theory. Kinetic energy of moving air is converted into pressure by slowing the airspeed with a fixed diffuser or stationary blades. In the centrifugal compressor, the airflow is radial from the shaft of the unit. Compression is primarily due to changes in centrifugal forces. Airflow in the axial compressor is parallel to the shaft. Compression is caused by the action of alternate rows of rotating and fixed blades.

The basic internal structure of a centrifugal-flow compressor is shown in **Figure 15-14**. The

Figure 15-14. The impeller is the only moving part in a centrifugal compressor.

compressor consists of an inlet nozzle, impeller, and volute (spiral) collector. The impeller consists of a circular disk fitted with blades that taper from a deep, center intake section to a thin discharge section on the exterior edge. See **Figure 15-15.** The impeller is surrounded by a housing that forms the volute collector. The impeller is normally directly driven by the prime mover. As the impeller turns, centrifugal force moves atmospheric air into the inlet nozzle, through the impeller, and on to the collector. Energy from the prime mover is converted into kinetic energy as the airspeed rapidly increases through the impeller. This kinetic energy is then converted into pressure as the airspeed sharply decreases as air moves into the collector and on to the system distribution lines.

Figure 15-16 shows the internal structure of a multi-stage, axial-flow compressor. The unit is actually a series of basic compressors functioning together to compress the air. Each basic compressor stage consists of a row of moving rotor blades and a row of fixed stator blades. See **Figure 15-17.** Each of these basic units moves the air parallel to the compressor shaft. The length of the rotor and stator blades decreases from the first-stage compressor to the last stage. This results in a reduced volume of air transferred from each of the basic compressors. See **Figure 15-18.**

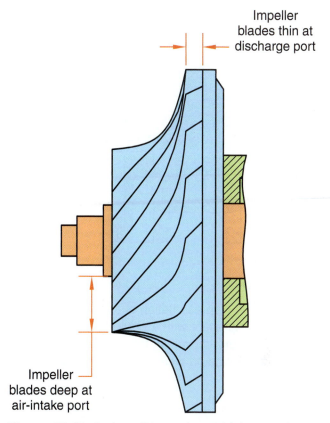

Figure 15-15. An impeller turning at high speed moves air through a centrifugal compressor. Energy from the prime mover is converted into kinetic energy in the rapidly moving air.

Figure 15-16. Most axial flow compressors are really several staged, basic compressor units working together to increase air pressure.

Figure 15-17. Representation of an axial flow, dynamic compressor with a single set of rotor blades. The rotor moves air through the stator blades to produce an increase in air pressure.

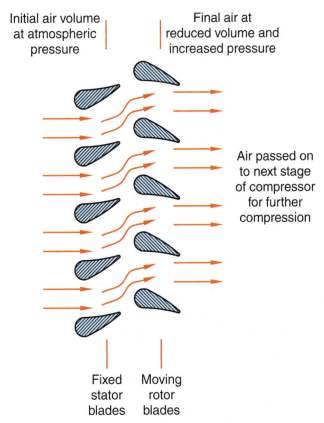

Figure 15-18. The speed of the air increases as it passes through the rows of fixed-position stator blades and rotating rotor blades. The increased airspeed increases the kinetic energy in the air, which results in increased air pressure.

In axial compressors, pressure is created in both the rotor and stator sections of the unit. As the prime mover turns the rotor blades, energy is transferred to the air as its speed increases. When the moving air impacts the fixed stator blades, it is slowed. This converts the kinetic energy into pressure. In the second stage, the rotating rotor blades of the second basic compressor again increase the airspeed. This increasing speed transfers additional energy from the prime mover to the moving air. Impact with the fixed stator blades of the second compressor again slows the air. This converts additional kinetic energy into pressure. This process repeats through each of the basic compressor units until the last unit discharges the air into the system distribution lines at the final pressure.

Dynamic compressors do not require internal lubrication for either sealing or friction control. Compressor shaft bearings are outside of the compression chamber in these units. The bearings can be externally lubricated without contributing to air contamination. Therefore, the units can provide large volumes of oil-free air without the use of an oil separator.

Lobe-type compressors

Most models of *lobe-type compressors* only develop air pressures ranging from 10 to 20 psi. Therefore, they are often referred to as *blowers*. The compressors have two impellers containing two or three lobes each. See **Figure 15-19.** The impellers

Figure 15-19. The internal structure of lobe-type compressors allows a volume of atmospheric air to be swept through the unit and added to the air already confined in the receiver and distribution lines. (Atlas Copco)

operate in an elongated chamber located in the compressor body. Inlet and discharge ports are also built into this chamber. The impellers rotate in opposite directions and are synchronized by timing gears. A constant, minute clearance is maintained between the surfaces of the impeller lobes. There is also a minute clearance between the tips of the impeller lobes and the surface of the elongated chamber.

In this design, there is no reduction of air volume in the compressor as the impellers turn. The air is simply swept from the inlet port to the discharge port. Compression is achieved when the transported volume of air is added to the volume of air already trapped in the receiver or distribution lines. Atmospheric air is trapped on both the right and left sides of the inlet port as the impellers turn. With further rotation of the impellers, both of these air volumes are added to the pressurized system air when they are transported to the discharge port. While discharge is occurring, additional atmospheric air enters the inlet port. This fills the next chambers that will sweep air through the compressor.

The minute clearances involved between the impeller lobes and the walls allow air leakage between the discharge and intake port areas. This leakage is called *slip* and reduces the efficiency of these compressors. Because of the lost operating efficiency caused by this slippage, compressors of this type need to be operated at speeds close to, but not exceeding, the maximum recommended by the manufacturer. Some design variations use oil or other liquid injection to improve sealing and help cool the air. When this is done, a separator is required to remove the liquids to assure the delivery of clean, dry air to the system.

Compressor Staging

It is more effective to compress air to high pressures in multiple stages, rather than in a single step. This is related to a number of factors:

- Discharge temperature of the air.
- Compression-ratio limitations of the compressor.
- Balancing of loads on compressor bearings.
- Power requirements of the prime mover needed for the installation.

Staging basically involves a number of compressor units to compress air to the desired pressure by increasing pressure in small increments.

Figure 15-20 illustrates a two-stage, reciprocating-piston compressor. The compressor contains two cylinder and piston assemblies connected to a crankshaft containing one throw (journal) for each cylinder. Each cylinder contains inlet and outlet valves. The inlet of the large-displacement, first-stage cylinder is open to the atmosphere. The outlet of that cylinder is connected to the inlet of the small-displacement, second-stage cylinder by a transfer line. This line transports air between the compressor stages, but also acts as a cooler to remove heat from the air. The outlet of the second-stage cylinder is connected to the system receiver and distribution lines. The throws on the compressor crankshaft not only move the pistons, but synchronize their movement. This allows the pistons of the two stages to smoothly move air through the unit.

Figure 15-20. This two-stage, reciprocating compressor contains two cylinders connected in series. Other staged designs use individual compressors arranged in series to produce similar results. (DeVilbiss Air Power Company)

the compressor intake to provide air that is as clean, dry, and as cool as possible for the location. Intakes should be:

- Located away from dust-producing conditions.

- Shielded from the weather to eliminate the entrance of rainwater.

- Protected from direct exposure to sunlight. Direct sunlight can significantly increase the intake-air temperature.

Dirt Removal

Atmospheric air contains dirt, even in the best of conditions. However, in situations such as road or building construction, the level of airborne dirt may be extremely high, **Figure 16-3**. Preventing dirt from entering the pneumatic system is essential to the service life and performance of the total system. Dirt allowed to enter a system produces unnecessary component wear. It also leads to the accumulation of dirt in lines and sensing orifices

A

B

Figure 16-3. Dust and moisture in atmospheric air must be removed to protect pneumatic components from excessive wear (A) and system-powered processes from contamination (B). (Atlas Copco, DeVilbiss Air Power Company)

of control valves. These conditions can quickly reduce the efficiency of compressors and workstation circuits.

The *intake-line filter* removes dust and dirt from atmospheric air before the air is allowed into the compressor inlet, **Figure 16-4.** The type of filter depends on atmospheric conditions, the size and type of compressor, and the final use of the compressed air. Many different filter designs are used with pneumatic systems. However, filters can generally be placed in one of three categories:

- Dry.
- Oil wetted.
- Oil bath.

Dry filters are available in a large number of configurations using paper, plastic, cloth, ceramic, or metal elements. The filter element remains dry as it traps dirt.

Oil-wetted filters use metal-packed, woven wire or other material held in a metal frame. A light oil coating is used on this material to trap the dust and dirt entering the filter.

Oil-bath filters have a filter medium that is constantly bathed in oil. The oil captures the dirt and transports it to a sump where it settles and is removed during periodic service. This style of filter was popular in older compressor systems and is still used in some applications today.

Filters require regular servicing, regardless of their design. A filter that is clogged with dirt reduces system efficiency by limiting the volume of air intake. Limiting intake volume reduces the output volume of the compressor and can cause overheating of the unit. A regular schedule should be developed for service of the intake filter based on the environment surrounding the pneumatic system and the duration of compressor operation.

Controlling Air Temperature

The temperature of both the intake air and compressed air are critical considerations when working with a pneumatic system. According to the general gas law, any change in the temperature of a gas is directly reflected in gas pressure or volume change. Therefore, maintaining a consistent air temperature throughout a system would be a real advantage. Ideally, the temperature of intake air should be low, the temperature of the air should not increase during compression, and the compressed air distributed to workstations should be near the air temperature where the workstation is located, **Figure 16-5.**

Air temperature relates to both the energy consumed by the prime mover and the water vapor brought into the system with the atmospheric air. Compressing air produces heat. This increases the temperature and pressure of the air in the confined space of the pneumatic system. This warmed, compressed air retains the water vapor present in the atmospheric air brought into the compressor.

Two conditions result when this warm, humid, high-pressure air cools as it moves through the system distribution lines. The first of these conditions is a reduction in the pressure of the air as it cools. The amount of pressure reduction is based on the general gas law. The second condition relates to the dew point of the compressed air. As the temperature of the compressed air drops to the

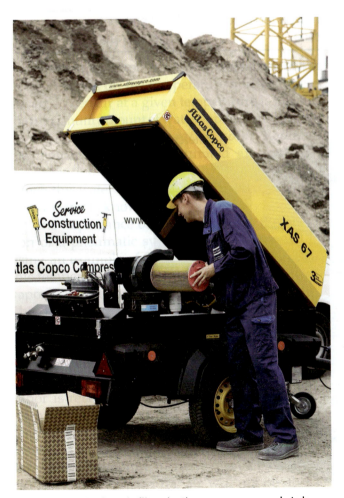

Figure 16-4. An air filter in the compressor intake line is a first step in the production of clean compressed air. (Atlas Copco)

abrasions. The lines should be kept straight and free from excessive kinks, although this is difficult in a work situation. Pressure drops in a twisted hose are as real as any pressure drop caused by excessive fittings in a fixed distribution system.

Sizing of Distribution System Lines

The size of the pipe used in the distribution system should be based on the air needed by system actuators. However, determining this volume of air becomes a problem as consumption varies considerably during the operation of most systems. This variation is based on the fact that a distribution system serves multiple workstations. In addition, each of these workstations contains actuators that intermittently operate under varying loads. These multiple variables make application of a simple formula for pipe sizing impractical. The solution recommended by many experienced pneumatic system staff is to design new systems based on information from similar, proven, existing systems.

This process begins by calculating the total air consumption of the actuators planned for the new system. Individual consumption data are available from manufacturer literature or engineering data handbooks. System air consumption can be determined by totaling the air consumption for each actuator to be included in the system. This figure will be greater than the actual volume of air used by the operating system, as few of the system actuators operate on a continuous basis or at full load, **Figure 16-12**. In most systems, the actual amount of air needed is considerably less than the combined consumption figure.

Next, multiply the estimated air consumption by two additional factors. First, multiply by a *time factor* that is the percentage of time actuators function during system operation. A fairly accurate time factor can be established by averaging the percentages of operating times of actuators based on time and motion studies using similar operations in existing equipment. Then, multiply by a *load factor* that is an estimated percentage of maximum actuator load delivered during operation. The load factor is more difficult to establish and requires the assistance of experienced pneumatic system staff.

Once the final estimated air consumption has been determined, the pipe sizes can be selected for the distribution system. Selection is based on the

Figure 16-12. Estimating air consumption for use in sizing distribution lines for a new system is difficult as actuator operating times and load factors vary. This is emphasized by the many variables at typical workstations. (DeVilbiss Air Power Company)

anticipated airflow rates and pressure drop caused by the lines and fittings in the system.

Other Distribution System Considerations

A number of additional factors may affect the distribution of air in a pneumatic system. Two common factors encountered are the presence of liquid water in the distribution lines and excessive pressure drop caused by high, short-term air usage in sections of the system.

The formation of liquid water is the result of the continued cooling of air in the lines. As previously discussed, the components in the compressor station remove a large portion of the water vapor that enters the system with the atmospheric air. However, some water vapor still remains in the compressed air. Liquid water forms if the temperature of the compressed air in the lines drops below the dew point. This situation may become a serious problem when distribution lines are exposed to low temperatures, such as during cooler seasons of the year.

The liquid water that forms in the distribution lines can be removed by designing the lines with a slope. A pitch (slope) of 1″ per 10′ of line allows the water to accumulate at selected low points in the system. *Drop lines* from these low points lead to water traps where the water is removed, either manually or with an automatically operated device. See **Figure 16-13.**

Figure 16-13. The distribution lines in a permanent installation should be sloped to allow drainage of liquid water that has condensed in the lines. Water legs (dirt legs) provide a point for the accumulation and removal of this water. (DeVilbiss Air Power Company)

Figure 16-22. In a pilot-operated pressure regulator, the control spring and adjustment screw are replaced with a sealed air chamber. Varying the air pressure in this chamber allows the regulator to be adjusted to the desired operating pressure.

Figure 16-23. The small, remotely located regulator is used to adjust the main system regulator by controlling pilot chamber pressure. (IMI Norgren, Inc.)

Regulator selection

Factors that must be considered when selecting a regulator for a system are:

- Choice of a direct-operated; basic, diaphragm-chamber; relieving-type; balanced–poppet valve; or pilot-operated style, depending on system size and performance demands.

- Pressure range of the unit, which considers the pressure in the input distribution line compared to the desired output pressure at the workstation.

- Airflow range of the unit.

- Conductor connection size.

Manufacturer information and specification sheets supply these types of detail for the various models produced.

Air Line Lubrication at the Workstation

A lubricator is the third of the components in the FRL unit. A *lubricator* provides the oil needed to improve the performance and service life of pneumatic system valves, actuators, and air-powered tools. The lubricator must accurately meter the oil to provide component lubrication. Providing inadequate lubrication can result in rapid component wear. Adding excessive amounts of oil is wasteful and can contaminate work areas around component exhausts.

Lubricator design and operation

Figure 16-24 shows the internal structure of one type of lubricator. The major external parts are a cast body containing inlet and outlet ports and a clear glass or plastic reservoir. Some designs have a metal reservoir. The internal elements are:

- Air passageway leading from the inlet port to the reservoir. This passageway contains a venturi.

- Oil siphon tube connecting the reservoir and the venturi in the air passageway.

- Adjustable oil-drip orifice.

- Passageway leading from the reservoir to the outlet port.

- Spring-loaded bypass valve that directly connects the inlet and outlet ports.

During operation, system air passes through the lubricator inlet port and passageway leading to the reservoir. The venturi in the passageway creates a reduced pressure. This causes oil to move through the siphon tube and enter the airstream through the oil-drip orifice. The rapidly moving air breaks up the oil drops into a mist that passes with the system air into the reservoir air space. The air space reduces turbulence in the air/oil mist. This allows larger droplets of oil to return to the oil supply. The oil mist then moves to the outlet port and on to the workstation air supply line. The amount of oil added to the air is controlled by adjusting the size of the oil-drip orifice.

During periods of low air use, all of the system air passes through the venturi of the lubricator. During periods of high air use, the higher pressure drop produced by the venturi causes the spring-loaded bypass valve to open. This allows a portion of the system air to pass directly into the supply line. The unlubricated air that bypasses the venturi is mixed with lubricated air before being passed out of the lubricator outlet port. This air contains less oil than normal, but does provide adequate component lubrication during brief periods of high-air-volume operation.

Figure 16-24. A lubricator atomizes a measured amount of oil and mixes it with the compressed air used at system workstations. (IMI Norgren, Inc.)

Lubricator selection

Factors that must be considered when selecting a lubricator for a system are:

- Conductor connection size.

- Oil reservoir capacity

- System pressure drop created by the volume of air that must pass through the unit to operate the components of the workstation.

Manufacturer information and specification sheets supply these types of detail for the various models produced.

Distribution System Conductors and Fittings

Moving compressed air from the compressor to the point of use at the workstation is a critical part of any pneumatic system. Any reduction in pressure between those points represents a loss of energy, which cannot be recovered. Previous sections of this chapter discuss general styles of distribution systems and their layout. This section is devoted to information about the pipe, hose, and fittings that form these systems.

The volume of products and variety of available designs can make the expansion of an existing system or the development of a new system confusing. The following materials provide general guidelines and suggestions for the selection of appropriate conductors and fittings required for installation.

Rigid Conductors

Pipe is the most common *rigid conductor* used for major compressed-air-distribution systems. Steel, copper, and ridged-plastic tubing may be found in smaller systems and on air-powered machines. Some of these materials may not be recommended for all installations, but they may provide adequate service in some installations. However, the conductor must be correctly rated for pressure and protected from external abrasives and activities that could crush the lines.

When constructing or modifying a distribution system, pressure losses caused by system lines should receive careful consideration. The sizing of conductors can be done only after the volume of air and the location of the workstations have been established. The objective is to hold pressure losses between the receiver at the compressor station and the workstation to less than 10% of the receiver pressure. Both the conductors and the associated fittings must be considered, **Figure 16-25.**

Tables are available in engineering handbooks that provide data on the air pressure loss in rigid conductors transferring a specific volume of compressed air expressed in cubic feet per minute (cfm). These pressure-loss figures are typically given as the loss per 100 feet of conductor of a specific nominal size and schedule rating. Tables are also available indicating the pressure losses caused

Figure 16-25. The fittings used in distribution lines produce a pressure loss. Pressure loss for each fitting must be included as part of the total loss in the system. (IMI Norgren, Inc.)

by the common fittings used in distribution lines. The pressure loss caused by these fittings is given in feet of comparable-sized straight conductor.

Determining the correct size of conductor for various sections of the distribution system requires the development of a map of the air needed, in cfm, at the junction points of the main distribution lines and each workstation drop line. This map can be constructed by starting at the point that is the greatest distance from the compressor station receiver and adding the air needed by each station. These figures must be increased by 10% to cover air leakage in the system. An estimated air volume should be added to these figures to cover the air needed for future system expansion. The resulting flow rates can then be used to select a pipe size that will deliver air at a pressure drop lower than the recommended level.

Adequately sized distribution lines are critical to the operation of a pneumatic system. The cost of the conductors is only a small part of the total installation cost. Therefore, when an estimated flow rate falls between conductor sizes, the *larger* size should be installed. The larger conductor will increase the performance of the current system and provide increased capacity for future system expansion.

Flexible Conductors

Hoses and semirigid plastic tubing are examples of *flexible conductors* used extensively in pneumatic systems, **Figure 16-26.** Hose applications include temporary distribution lines for portable-compressor setups, connections between moveable machine members, and connections between the FRL unit of a workstation and air-powered tools. Semirigid plastic tubing is often used in small equipment as the primary conductor between components. It is also used as sensing and control lines in many other compressed-air applications.

Hose is constructed with a minimum of three layers of materials:

- Inner tube providing a smooth surface to minimize airflow resistance.
- Middle section containing woven reinforcing materials to withstand system pressure.
- Outer layer to protect the hose from damage resulting from handling and the abrasive and corrosive materials encountered in the work environment.

Manufacturers provide a selection of hose types to meet various system working conditions. Many of these are built to meet standards established by industry standardizing groups.

Ideally, hoses should be no longer or larger in diameter than necessary to supply the compressed air needed at a work location, **Figure 16-27.** This ensures the delivery of air with minimal pressure drop. It also ensures minimal hose and installation costs. However, in work conditions that are often less than ideal, this may not be possible. These points should be considered when working with hose to minimize installation costs and pressure drops:

- Hoses should be no longer than necessary.
- Hoses should be no larger in diameter than necessary.
- A minimum number of fittings and couplings should be used.
- Layout lines to eliminate kinks and reduce the number of bends.

Figure 16-26. A wide variety of hose and semirigid tubing is available for use in pneumatic systems. (Atlas Copco)

Figure 16-27. Hoses should be no longer or larger in diameter than necessary to supply the compressed air needed at a work location.

Chapter Test

Answer the following questions. Write your answers on a separate sheet of paper.

1. The first factor contributing to the quality of compressed air that can be produced is the condition of the _____ at the system air intake.

2. Name five materials that may be used to make the element in a dry filter for use in pneumatic systems.

3. Name two things that occur when warm, humid, high-pressure air cools in the distribution lines of a pneumatic system.

4. _____ is the temperature at which water vapor in the air begins to condense to form liquid water.

5. Name five places where liquid water may form in the pneumatic system.

6. In addition to air storage, the receiver dampens air _____ caused by compressor operation.

7. The _____ compressed-air-distribution system distributes air from a single location to all workstations in a facility.

8. Which design of air distribution system is generally considered to provide maximum airflow with a minimum of flow resistance?

9. List the three key variables that make the application of a simple formula for the calculation of distribution line sizes impractical.

10. Identify the two factors that are critical to selecting the actual sizes of the lines used in a air distribution system.

11. What can be used to supply compressed air to a section of the distribution system that contains equipment using a large volume of air on an intermittent basis without increasing the size of the system?

12. Name the two ways in which a two-stage filter traps and removes unwanted particles.

13. Why is it necessary to use a pressure regulator in the final air preparation unit at a system workstation?

14. What is a disadvantage of the direct-operated pressure regulator and what can this cause?

15. The pressure setting of regulators using a flexible diaphragm is set by controlling the _____ applied to the diaphragm.

16. What is the advantage of the relieving-type pressure regulator?

17. The force produced by the pilot-chamber air pressure replaces the force generated by the _____ of the basic regulator design.

18. In typical pneumatic system lubricators, oil is atomized by adding oil to an airstream through a(n) _____ orifice.

19. What is the most common rigid conductor for major pneumatic systems?

20. List four factors that should be considered to minimize installation costs and reduce pressure losses in applications using hose for the distribution of compressed air.

Air for use in a pneumatic system must be cleaned of dirt and other contaminants. Filters come in various sizes and shapes to meet the needs at hand. When replacing filters, be sure to use the filter specified by the manufacturer or a compatible filter. (Atlas Copco)

Pneumatic systems must convert the potential energy contained in compressed air to force and movement. Without effective energy conversion, the storage, transmission, and control advantages of pneumatic systems cannot be realized. Actuators are used in pneumatic systems to produce the linear and rotary motion basic to the operation of industrial equipment. The actuators are designed to effectively produce the required force, torque, and speed of movement needed in an application by using the characteristics of compressed air. Actuators include basic cylinders and motors, as well as specialized devices. This chapter reviews the basic structure of actuators discussed in Chapter 9, which covers the details of hydraulic actuators, and examines features special to pneumatic applications.

Objectives

After completing this chapter, you will be able to:

- Describe the construction features of basic, pneumatic linear and rotary actuators.
- Compare the design and operation of pneumatic cylinders.
- Compare the design and operation of pneumatic motors.
- Explain the performance characteristics used to rate the operation of pneumatic motors.
- Describe the basic design and operation of specialized pneumatic tools commonly used in consumer and industrial applications.
- Size pneumatic cylinders and motors to meet the force and speed requirements of a basic work application.
- Interpret manufacturer specifications for basic, pneumatic cylinders, motors, and power tools.

Chapter 17

Work Performers of Pneumatic Systems
Cylinders, Motors, and Other Devices

Selected Key Terms

The following terms will be used in this chapter. As you read the text, record the meaning and importance of each. Additionally, you may use other sources, such as manufacturer literature, an encyclopedia, or the Internet, to obtain more information.

absolute pressure ratio
air nozzles
blowguns
chipping hammers
corrosion resistance
double-acting cylinders
force
gripper
impact wrenches
linear actuators
material agitation
material transfer
metal rolling
nail drivers
nutsetters
paving breaker
piston motors
pneumatic reciprocating motors
rammers
riveting hammers
rock drills
rotary actuators
scaling hammers
spraying
tampers
turbine motors
vane motors

Internet Resources

www.atlascopco.us/usus/products/navigationbyproduct/
Atlas Copco
Provides links to the wide selection of products offered by this company. Review information on several of the air-powered tools available.

www.bimba.com/en/Library/#technicaltips
Bimba Manufacturing Company
Lists library resources. Technical Tips includes listings of several factors users must understand and regularly follow to maximize cylinder performance and operating life.

www.apexpowertools.com/catalog/pdffiles/SP-104_EN.pdf
Apex Tool Group
Provides information on the construction and performance of vane, axial-piston, and radial-piston pneumatic motors. Includes basic air motor selection procedures and formulas for one brand of motor.

Pneumatic Cylinders

A pneumatic cylinder is the system component that converts the energy in compressed air to linear force and movement, **Figure 17-1.** Cylinders are often called *linear actuators.* The force generated by the actuator is determined by the area of a cylinder piston or diaphragm and the air pressure in the system. The rate of movement is determined by the volume of the actuator chamber per inch of stroke and the amount of air allowed to enter the actuator. The maximum distance that a linear actuator can move is determined by the physical length of the unit.

Controlling air pressure and flow rates can produce a wide range of actuator force outputs and operating speeds. However, the compressibility of air affects the output of both force and speed. Operation of a pneumatic actuator is a practical example of the general gas law.

Basic Cylinder

A basic pneumatic cylinder involves a piston and piston rod assembly operating within a close-fitting, cylindrical tube called the barrel. One end of the barrel is closed. An opening in the opposite end of the barrel allows the piston rod to extend outside of the component. Seals are used to prevent leakage to the exterior of the cylinder. Seals also prevent leakage around the piston that separates the barrel into two sealed chambers. Ports located near the ends of the cylinder allow air to be forced into and vented from each of the sealed chambers. **Figure 17-2** illustrates the construction of such a basic cylinder.

The basic structure of pneumatic cylinders is very similar to that of hydraulic cylinders. Chapter 9 examines the features and structure of hydraulic cylinders in detail. That chapter should be reviewed for details of cylinders. Features are similar for hydraulic and pneumatic cylinders. The remainder of this chapter is devoted to features and operating concepts primarily limited to cylinders used in pneumatic applications.

Construction Materials

Cylinders must be designed to withstand harsh operating conditions. Both hydraulic and pneumatic cylinders are subjected to corrosion

Figure 17-1. Cylinders are used to generate force to produce linear movement in a variety of industrial equipment. (Parker Hannifin)

Figure 17-2. A pneumatic cylinder uses relatively simple construction to convert the potential energy of compressed air into linear force and movement. (IMI Norgren, Inc.)

and abrasion that can produce rapid wear. However, in pneumatic applications, water vapor produces corrosive conditions that require construction materials different than those needed in hydraulic applications. Another condition that is different between the two systems is operating pressure. Pneumatic applications typically operate at lower pressures than hydraulic systems. Therefore, pneumatic actuators do not need the same structural strength that must be built into hydraulic actuators.

Increased *corrosion resistance* is provided by:

- Use of resistant materials for the construction of component parts.

- Plating of parts subjected to high corrosion.

Brass is a corrosion-resistant material that has been used to construct component parts. Chromium-plated steel is commonly used for piston rods and the bore surface of cylinder barrels. Stainless steel and anodized aluminum are also used to provide corrosion resistance.

Aluminum parts, including machined heads and caps and extruded barrels, are used on some models of pneumatic cylinders to reduce the weight. These materials easily provide the strength required for the low forces encountered in most pneumatic applications.

In most pneumatic cylinders, *resilient materials* are used for piston seals. The style of the piston seal ranges from standard, commercially available designs to special forms designed especially for a given product. Resilient materials are also used between the cylinder rod and cylinder head and between the barrel and cylinder head and cap, **Figure 17-3.**

Some manufacturers produce a product line of nonlubricated pneumatic cylinders. These operate without adding oil to the compressed air of the system. Nonlubricated cylinders are used in applications requiring oil-free air, such as in the food industry. Lubrication is provided by coating the surfaces of the cylinder bore, piston, and bearing surfaces with lubricant that adheres, but is not scraped off by the seals.

Pneumatic, linear actuators are also produced with a diaphragm instead of a piston. The diaphragm is attached to the rod, which extends from the body the same as in a conventional cylinder. This actuator design typically has a short stroke and is single acting with a spring return.

Classification of Cylinders

A number of pneumatic cylinder designs are available. These designs can be classified in much the same way as hydraulic cylinders, as described in Chapter 9. Some variations exist that must be considered when selecting a cylinder for use with compressed air. However, like hydraulic cylinders, pneumatic cylinders may be classified based on basic operating principles and construction type.

Operating principles

Pneumatic systems use both single- and double-acting cylinders. The operating principles of these designs are basically the same as for hydraulic cylinders. *Single-acting cylinders* can generate force only in one direction. The force required to return the piston and rod to their original position is provided by an internal spring or an external weight. See **Figure 17-4.** *Double-acting cylinders* can generate force during both extension and retraction of the piston and rod assembly. As in hydraulic cylinders, there is a difference in the effective area of each side of the piston. This is due to the cross-sectional area of the piston rod. As a result, extension and retraction strokes have different force outputs.

Resilient seals

Figure 17-3. Resilient materials are typically used in pneumatic cylinders to assure a positive seal between the bore of the cylinder and the piston. (IMI Norgren, Inc.)

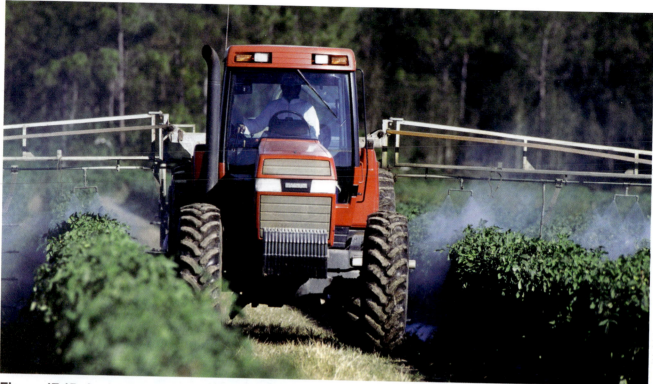

Figure 17-15. Applying insecticide or herbicide is an example of spraying. When compressed air is used for spraying, it is a process function, rather than a producer of force and movement. (USDA)

Material agitation

A number of processes require continuous *material agitation* to prevent the settling of solids or to maintain a uniform distribution of materials in a mixture. In some situations, compressed air can be directly used to continuously agitate the materials. A factor that must be considered when thinking about this type of application is the fact that air can cause oxidation. This may interfere with the primary process. In other applications, such as sewage treatment, agitation with compressed air not only keeps the material mixed, but also supplies oxygen to the bacteria converting the waste materials, **Figure 17-16.**

Material transfer

Compressed air is often used to expedite the movement of liquid, powder, or granular materials. Various techniques of *material transfer* are used in industries as diverse as food processors and cement production. Compressed air is used for the unloading and loading of railroad cars and trucks. Materials are often transferred to processing areas and then returned to storage using air-powered equipment.

Other Mechanical Equipment

A wide variety of pneumatic tools and equipment have been developed to serve various industrial situations. Many of these have been refined and modified to fit many situations beyond their original application. The following sections discuss a number of air-powered items that are commonly found in industrial or consumer applications.

Nozzles

Air nozzles convert the potential energy of static, compressed air into kinetic energy of rapidly moving air. They are commonly used with pneumatic systems. Most people who have worked with even the simplest of compressed air systems have probably used a nozzle as a cleaning device. Nozzles may look simple, but the theory of their design can be complex. They can be used with turbines to produce rotor motion or ejectors to produce vacuum or mix fluids.

Nozzles in their simplest forms are either the convergent or convergent-divergent designs shown in **Figure 17-17.** Varying the length of the curved portion of these designs provides a means

Figure 17-16. The agitation of sewage in a waste-treatment facility is often done with compressed air.

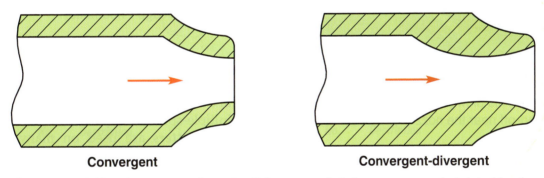

Convergent   **Convergent-divergent**

Figure 17-17. Nozzles convert the potential energy of static, compressed air into kinetic energy in rapidly moving air.

to control the velocity and pressure of the air as it passes through the nozzle. On the other hand, in a simple ejector, a vacuum is formed at the nozzle. This vacuum draws in and mixes a second fluid with the air as it moves through a diffuser section at the outlet of the device.

Figure 17-18 shows simple *blowguns* used in many shops for cleaning. They use the principles of a nozzle in their operation. These may be simple devices, but, from a safety standpoint, they need to be treated with a great deal of respect. Eye protection is needed due to flying chips and particles from cleaning operations. Also, the air pressure

may be high enough to inject air through the skin. This can occur if the nozzle is brought into direct contact with the skin. The blowgun shown in **Figure 17-19** eliminates this risk by including a simple bleed-off just before the nozzle discharge. This bleed-off prevents the buildup of high-air pressure if the nozzle is blocked.

> **Caution:** Never point a blowgun at anyone. Blowguns are capable of causing serious injuries, especially to the eyes and ears.

Figure 17-22. Pneumatic nutsetters apply multiple fasteners at one time and are commonly used on assembly lines. (Atlas Copco)

Figure 17-23. Pneumatic nail drivers are used in building construction to drive nails and other fasteners.

The drivers contain a cylinder, piston, piston rod that impacts the fastener, return air chamber, and valve to trigger the driving action. In addition, a mechanism to hold and deliver the fasteners and a safety mechanism are incorporated into the driver, **Figure 17-24.** The safety mechanism prevents the driving action from being triggered unless the tip of the driver is in firm contact with the material to be fastened. This mechanism is essential to reduce the possibility of fasteners becoming projectiles, which can cause serious injury.

When the driver is connected to the compressed-air line, the piston is in the retracted position. The piston rod is aligned with a fastener ready to be driven. Pressing the tip of the driver against the material to be fastened releases the safety catch. At that point, pulling the trigger of the driver releases compressed air into the cylinder above the piston. This causes rapid piston movement. The rapidly moving piston and piston rod impact the fastener, driving it into place. The movement of the piston also compresses the air on the rod side of the piston. This develops pressure in the return-air chamber.

When the trigger is released, the compressed-air supply is blocked and the high-pressure air used to drive the piston is released to the atmosphere. The pressure in the return-air chamber is then used to retract the piston and piston rod to their starting position. When the driver is raised from the material being fastened, the safety mechanism is reset and the driver is ready for driving another fastener at the next location.

Figure 17-24. A pneumatic nail driver uses the impact of a rapidly moving piston to drive brads, staples, and nails. A—Idle position. B—The tool is on the material surface so the safety is released. When the trigger is pulled, the fastener is driven into the material.

Rock drills

Pneumatic *rock drills* are often used to make rock cuts in quarries, mines, and highway construction. The result of this drilling is often visible along highways constructed in hilly and mountainous areas. The sheer rock faces may show a series of bore marks spaced about every 2 1/2 feet. These bore holes allow placement of explosive charges that break the rock between the holes to produce a relatively smooth face. The

9. Some _____ air motor designs can operate at speeds approaching 100,000 rpm.

10. An air motor should be sized to operate on what fraction of the workstation pressure to allow for the extra power needed for starting loads and unexpected overloads?

11. Identify the two factors that determine the impact force of a percussive, reciprocating air motor.

12. Name six tools that use reciprocating motors to power their operation.

13. A(n) _____ is used in an assembly line to run-down and torque multiple fasteners in a single operation.

14. List four applications of compressed air in typical rock-drilling equipment.

15. Robots often make use of pneumatic _____ to pick up and place materials.

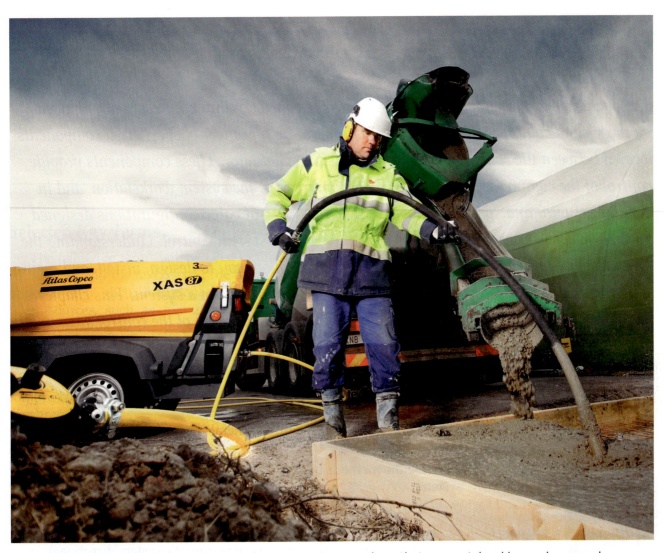

Compressed air is used to operate an agitator to assure no air pockets are retained in newly poured concrete. (Atlas Copco)

A *diaphragm* is often used in pneumatic valves as a control element, **Figure 18-5.** These are made from a relatively thin, flexible material. They are often used in place of pistons to sense pressure and operate other control elements. Control chambers can be located on one or both sides of the diaphragm. A biasing spring can be used to limit diaphragm movement until selected pressures have been reached. Diaphragms function well as control elements in the relatively low operating pressure of a pneumatic system.

A *poppet valve* is used as a control element in a number of pneumatic directional-control applications. Poppet valve assemblies consist of a seating surface located in the valve body and a valve face located on a moveable poppet. In pneumatic valve applications, a resilient material is generally used on the poppet face, rather than the traditional, machined-metal face and seat. This design variation provides a positive seal at a reduced production cost. An advantage of the poppet-valve design is a large-diameter poppet can allow maximum flow through a valve with minimal movement of the control element.

A precision-machined *sliding plate* is another means used to control airflow through pneumatic directional control valves. The plates contain a series of channels and ports that are used to direct airflow. Moving the plates in relation to each other connects the channels and ports in various combinations to produce different airflow patterns. Some designs use rotary movement of the plates to change the airflow combinations, **Figure 18-6.** Other designs use linear travel to operate the valve.

Positioning Control Valve Internal Elements

Pneumatic pressure, direction, and flow control valves are operated by varying the position of their various control elements. Forces both internal and external to the valve are used to position the control elements.

Internal force refers to forces generated by air pressure in the internal passageways of the valve or by internal valve components. Springs and air pressure are the typical sources of internal force. Springs are used to position piston, spool, and diaphragm control elements, **Figure 18-7.** They may also be used to resist the movement of these elements until a selected pressure is reached. Air pressure sensed in valve passageways or control chambers is also used to move elements.

External force refers to forces applied by sources outside of the valve. Manually applied force, pilot air pressure, and electromagnets are external forces commonly used to position the internal control elements of valves. Manually applied force may be as simple as an operator shifting the handle of a directional control valve. Air pressure is often sensed at a remote location in the circuit and transmitted by lines to a control valve. Electromagnets operate solenoids, **Figure 18-8.** Solenoids

Figure 18-5. A flexible diaphragm is commonly used as a control element in a pneumatic pressure regulator.

Diaphragm chamber

Diaphragm

Figure 18-6. Precision-machined sliding plates are often used as the control element in rotary-style pneumatic directional control valves.

Figure 18-7. Internal springs are commonly used to establish the normal position of the spool in pneumatic directional control valves.

Figure 18-8. Solenoids are used to shift the position of the control spools in many pneumatic directional control valves.

are commonly used in both simple control circuits and complex, computer-controlled systems.

Additional details relating to operation of control elements can be found in Chapter 10, which covers hydraulic system control valves. The initial sections of that chapter discuss in more detail the basic features of control elements. Many of the features of hydraulic and pneumatic components are similar. However, it must always

be remembered that air is compressible and the liquid of hydraulic system is, for all practical purposes, not compressible.

Pressure Control Methods and Devices

Several methods are used to control air pressure in a pneumatic system. The method used depends on several factors. These factors include the section of the system being controlled, required accuracy of control, effect on efficient system operation, and safety of both system components and machine operators. Three system sections can be identified that use different pressure control methods:

- System distribution lines.
- Workstations providing compressed air to various system circuits.
- Portions of circuits that require pressures above or below the level set by the workstation controls.

System Distribution Line Pressure Control

The pressure in the distribution lines of a pneumatic system can be controlled by two basic methods. The first method, called a bypass control system, bleeds excess air from the compressor to the atmosphere to maintain a constant, maximum pressure in the distribution line. The second method controls the capacity of the compressor to maintain distribution line pressures from slightly above the required pressure to a predetermined maximum pressure, **Figure 18-9**.

In a *bypass control system*, the compressor continuously operates at a constant speed. When the desired pressure is reached in the distribution line, a relief-type valve discharges excess compressor output to the atmosphere. The system is generally considered inefficient, except in specific situations.

Compressor-capacity control is the most common means of controlling distribution line pressure. It is accomplished by using control over compressor starting-stopping, inlet valve unloading, compressor speed, or, in the case of dynamic compressors, the size of the inlet opening. Each

Figure 18-9. A pressure switch operates the start-stop compressor-capacity control system commonly used with small to medium compressors.

of these methods is discussed in the Compressor-Capacity Control section of Chapter 15. That section provides the details of compressor-capacity control and explains how distribution line pressures are maintained.

Several designs are used when the compressor capacity method is used to control distribution line pressure. Each of these designs allows the compressor to deliver only sufficient compressed air to maintain the desired distribution line pressure. Air pressure varies in the distribution lines when these designs are used. However, it will not drop below the level needed for effective operation of the circuits of the overall pneumatic system. These methods are considered more efficient for most pneumatic installations.

Workstation Pressure Control

A second pressure-control point is at system workstations. The pressure regulator of the filter/pressure regulator/lubricator (FRL) unit further controls air pressure. This additional pressure control is necessary for two reasons. First, the distribution line air pressure is higher than needed for the operation of working circuits. Second, the air pressure in the distribution lines varies considerably because of the inherent characteristics of the compressor-capacity control system.

Therefore, pressure regulators located at workstations are necessary to provide air at a consistent pressure that is below distribution line pressure, **Figure 18-10.**

Air pressure regulators maintain a desired output pressure for a workstation as long as the distribution system pressure does not drop below the selected pressure. Several different regulator designs and features are available for use in pneumatic systems. The regulator selected for a system depends on the accuracy and type of control needed by the system. Designs include direct operated; basic, diaphragm chamber; relieving; balanced poppet; and pilot operated. Many manufacturers produce air line regulators, **Figure 18-11.** Some designs include a combination of design features.

Each of these regulator features is discussed in the Pressure Regulation at the Workstation section of Chapter 16. That section provides the details of design, construction, and operation of air pressure regulators.

Figure 18-10. Workstation pressure control is provided by a pressure regulator that is often part of a filter/pressure regulator/lubricator (FRL) unit. (IMI Norgren, Inc.)

Figure 18-11. Properly sized pressure regulators maintain a uniform workstation operating pressure even though pressure in the system distribution line varies.

Pressure Control within System Circuits

Pneumatic circuits may need higher or lower air pressure than that supplied by the workstation pressure regulator. Lower pressures can be easily provided by the placement of additional regulators. These additional regulators easily reduce pressure in the subcircuit to the desired level. **Figure 18-12** shows a circuit in which a low pressure is available to operate an air motor. The higher setting of the workstation regulator controls the higher pressure available to the remainder of the circuit.

When the required pressure is higher than that available from the system distribution line, it may be obtained by using a **pressure booster.** Boosters, often called *pressure intensifiers,* may operate using an air-to-air or air-to-oil designs. Air-to-oil designs are primarily used when very high pressures are needed. Both types are limited to situations in which only a small volume of high-pressure air is needed for circuit operation.

Figure 18-13 shows a circuit containing an air-to-air booster. The unit is basically a reciprocating pump with two cylinder chambers. The piston in the larger cylinder chamber is powered by air coming from the system distribution line. The force produced by this air pressure drives the second,

To circuit section operated at maximum workstation pressure

Motor operated at a pressure below the workstation setting

Figure 18-12. A pressure lower than that selected as the workstation pressure may be easily provided by the addition of a second pressure regulator.

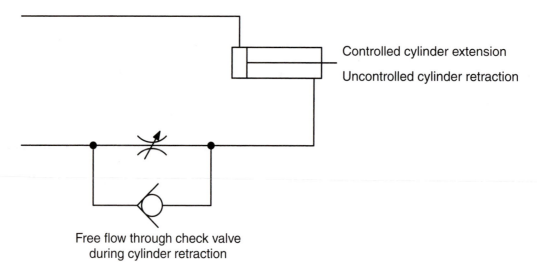

Controlled cylinder extension

Uncontrolled cylinder retraction

Free flow through check valve
during cylinder retraction

Figure 18-15. Check valves are often used to allow free reverse airflow around flow control valves.

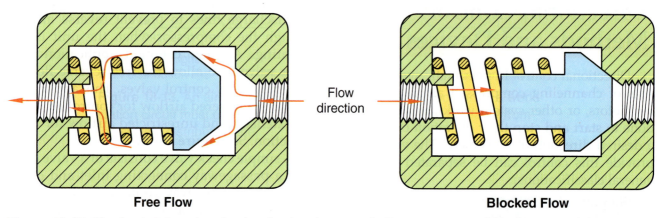

Flow direction

Free Flow

Blocked Flow

Figure 18-16. The basic internal parts of a check valve are a ball or poppet, machined seat, and, in some designs, a spring.

Three-Way Valves

Three-way directional control valves are used to direct compressed air to single-acting cylinders for extension. When the valve is shifted to allow cylinder retraction, the air returning from the cylinder is allowed to directly pass to the atmosphere through the valve exhaust port. See **Figure 18-17.**

Three-way valves consist of a valve body with passageways connecting a pressure port, actuator port, exhaust port, and sliding spool. The sliding spool operates in a bore machined into the valve body. It can be shifted to connect the ports in various combinations. One of these combinations powers the extension of single-acting cylinders. Another combination exhausts air from the cylinder to allow retraction.

Four-Way Valves

Four-way directional control valves are used to direct the extension and retraction of double-acting cylinders, select the direction of air motor rotation, or operate other workstation circuit configurations requiring alternate compressed air delivery. A basic four-way valve has a minimum of four external port connections. These include a port connected to the pressurized-air supply, inlet

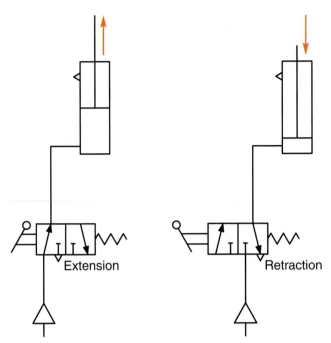

Figure 18-17. One application of a three-way directional control valve is to operate a single-acting cylinder.

Four-port configuration

Five-port configuration

P = pressure
E = exhaust
A1 = actuator connection 1
A2 = actuator connection 2

Figure 18-18. Pneumatic four-way directional control valves are available having either four or five port connections.

and outlet ports connected to the actuator, and an exhaust port leading directly to the atmosphere.

Basic operation of four-way valves

Shifting a four-way valve in one direction connects the air supply to one port of the actuator. This allows pressurized air to move the actuator in one direction. In this shifted position, the valve also connects the second port of the actuator to the exhaust port. Air is returned to the atmosphere through the exhaust port. Shifting the valve in the opposite direction connects the air supply and exhaust ports to the opposite ports of the actuator. These connections pressurize the opposite side of the actuator, causing it to move in the opposite direction. Air returned from the actuator is again exhausted to the atmosphere through the exhaust port.

Four-way valves are also available using a design that has five ports. A *five-port directional control valve* provides the same basic operation as the four-port-design described above. The primary difference is the fact that a separate exhaust port is provided for each valve position. The symbols shown in **Figure 18-18** illustrate the airflow paths of four-way directional control valves with the four-port and five-port designs.

Internal sealing in four-way valves

The control element in a four-way, pneumatic directional control valve is generally a sliding plate or spool. Movement of these elements connects internal passageways in the valve body, thus routing airflow to produce the desired actuator movement. In order to efficiently operate using these control elements, the valves must route the compressed air through the valve without leakage around internal parts and between the internal passageways. Both machined surfaces and resilient materials are used in these valves to provide this tight seal.

Manufacturers use a number of seal variations to provide an airtight seal. Seals made of resilient materials are most common in pneumatic valves. This is primarily due to the fact that the close tolerance required to prevent air lose with precision-fit parts is both expensive to create and difficult to maintain. Even so, some pneumatic directional control valves use a precision-fit spool in combination with the lubricating oil added to the system as the only sealing method. However, two resilient seal designs are more commonly used in pneumatic valves to provide the seal between the spool and bore.

The first of the resilient seal designs is sometimes called a *packed-spool design.* It uses resilient seals mounted on the spool lands. This provides the positive seal needed between the spool and the machined metal bore located in the valve body. See **Figure 18-19.**

The second of the resilient seal designs is called the *packed-bore design.* It uses resilient seals placed in grooves machined into the surface of the bore in which the spool operates. See **Figure 18-20.** This provides the positive seal.

Control element positions of four-way valves

The above description of four-way directional valves is based on a two-position valve. The first of these positions activates the actuator in one direction. Shifting the valve to the second position reverses the movement. However, many pneumatic circuits more effectively function when a third valve position is provided to allow additional control. In three-position valves, the third position is typically called the center position, **Figure 18-21.** The center position can be used simply to stop actuator movement or provide some addition operating characteristic.

Center position configurations

There is a variety of center position configurations. Three of these can be considered basic designs that are readily available and commonly used in pneumatic equipment. These three are:

- Blocked center.
- Open center.
- Pressure center.

Figures 18-22, 18-23, and **18-24** show these center designs as used in a five-port, three-position directional control valve operating a double-acting cylinder.

Figure 18-19. The packed-spool design of directional control valve has resilient seals on the lands of the valve spool. (IMI Norgren, Inc.)

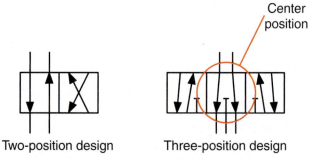

Figure 18-21. The center position of a three-position, four-way directional control valve provides additional circuit operating options.

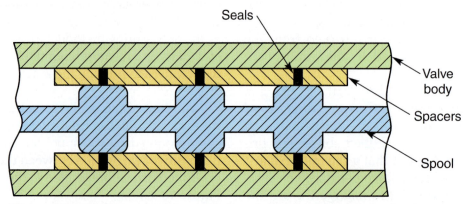

Figure 18-20. The packed-bore design of directional control valve has resilient seals on the surface of the valve bore.

Figure 18-22. The blocked-center configuration blocks all air movement through a three-position, four-way directional control valve.

Figure 18-23. The open-center configuration of a three-position, four-way directional control valve blocks the pressure port and opens the actuator ports to exhaust.

Figure 18-24. The pressure-center configuration of a three-position, four-way directional control valve connects the pressure port to both actuator ports.

The blocked-center configuration blocks the pressure port supplying compressed air from the system distribution line. See **Figure 18-22.** Both lines leading to the circuit actuator are also blocked. This design is often called a *closed center* as the movement of air is blocked at all port connections. No compressed air can enter the circuit and actuator movement is restricted. The amount of restriction is based on the compressibility of the air in the lines connecting the actuator and the directional control valve.

The open-center configuration is often called the *exhausted-center design.* In this configuration, the pressure port is blocked. Both ports connected to the actuator are connected to exhaust. See **Figure 18-23.** This allows the actuator to float, depending on the external forces applied to it.

The pressure-center design connects the pressure port of the valve to both lines leading to the actuator. See **Figure 18-24.** When the directional control valve is in the center position, equal air pressure is applied to both sides of the cylinder piston. Because of the difference of force generated on the rod and blind ends of the piston, the actuator will rapidly extend. This regenerative operating concept produces a faster extension speed, but reduces the useable force. However, the pressure-center design can also be used in other ways, such as supplying pressurized air to more than one actuator or subcircuit.

Flow Control Methods and Devices

Controlling the flow of compressed air in a pneumatic system controls the operating speed of cylinders, air motors, and other actuators. Control is achieved by using flow control valves that contain an orifice. The orifice meters the amount of air that can enter a circuit. Because of the compressibility of air, accurate speed control is difficult in pneumatic systems. The accuracy of a pneumatic flow control valve is influenced by system load changes that cause:

- Variations in system pressure.

- Changes in internal system friction, which cause load variations.

- Increases or decreases in the temperature of the compressed air, which cause variations in system pressure.

Types of Flow Control Valves

Fixed-size orifices and needle valves are commonly used to control actuator speeds in pneumatic systems. Both of these devices are described in the Valve Control Elements section of this chapter. However, the flow control valve most commonly found in a pneumatic circuit is a component that combines a needle valve and check valve in the same valve body. This *combination valve* uses the needle to accurately adjust the rate of airflow in one direction, while the check valve allows unrestricted return flow. **Figure 18-25** illustrates the internal structure of a flow control valve that combines these two features.

It is not recommended that shut-off devices, such as ball, globe, and gate valves, be used as metering devices in pneumatic systems. The internal structure and method of valve adjustment of these valve designs do not allow consistently accurate metering of the airflow.

Figure 18-25. Pneumatic system flow control valves are often needle valves combined with a check valve that allows free flow in the reverse direction.

Check valve poppet

Free flow

Controlled flow

Placing Flow Control Valves in Circuits

Flow control valves can be placed to create either meter-in or meter-out circuits. In a *meter-in circuit*, the valve is placed in the inlet line of an actuator. In a *meter-out circuit*, the valve is placed in the outlet line. See **Figure 18-26.** However, a meter-out circuit is the method preferred by system designers.

Meter-out flow control is preferred because of the compressibility of air. Backpressure that develops in the line between the directional control valve and actuator in a meter-out circuit results in stable line pressure and actuator movement. If a meter-in circuit is used, the air in this line will compress and expand during varying actuator loads. These changes will result in poor actuator speed control.

In pneumatic meter-out flow control circuits, the flow control valve can be placed between the actuator and the directional control valve or on the exhaust ports of the directional control valve. When a four-port, four-way directional control valve is used, the flow control valve must be placed between the actuator and the directional control valve. In this arrangement, the flow control valve should include an integral check valve to allow unrestricted return flow when the direction of the actuator is reversed. When a five-port, four-way directional control valve is used, place the flow control valve in the lines leading from the direction control valve exhaust ports. A check valve is not needed in this case as airflow through the valve is only toward the exhaust opening. **Figure 18-27** shows schematics of circuits using four- and five-port directional control valves.

Special-Purpose Control Valves and Other Devices

Several special-purpose valves and devices are used in pneumatic circuits to provide increased control and efficiency. A number of these are exclusively produced by a single company. Some of these are available from most manufacturers. The following sections discuss several of the special-purpose valves and devices often found in pneumatic circuits.

Figure 18-26. The preferred flow-control circuit for pneumatic systems is the meter-out design.

Figure 18-27. When using a four-port directional control valve, place the flow control valve between the directional control valve and the actuator. The preferred location of the flow control valve when using a five-port directional control valve is on the exhaust port of the directional control valve.

Quick-Exhaust Valves

A *quick-exhaust valve* is used to increase the rate at which air is exhausted from a cylinder. Increasing the exhaust rate allows the cylinder to extend or retract as rapidly as possible. **Figure 18-28** shows a circuit that includes a four-way directional control valve, double-acting cylinder, and quick-exhaust valve. The quick-exhaust valve is located in the line between the cylinder port on the blind end and the directional control valve.

When the directional control valve is shifted to retract the cylinder, the pressure and movement of the air coming from the blind end opens the quick-exhaust valve. This air then quickly escapes through the valve to the atmosphere. As a result, backpressure in the line is reduced to below the level required to force the air through the four-way valve. Due to this lower resistance to air movement, the cylinder retracts at a higher rate of speed. When the directional control valve is shifted to extend the cylinder, the quick-exhaust valve shifts and directs air to the cylinder to produce a normal extension speed.

Figure 18-29 shows the internal structure of a typical quick-exhaust valve. The valve body contains three ports. The first port is connected to an inlet line coming from the circuit directional control valve. The second port is connected to a line leading to one cylinder port. The third port leads directly to the atmosphere. A flexible disk freely floats in a cavity connected to the three ports. When the inlet port is pressurized, the disk moves to cover the exhaust port. This directs compressed air to the cylinder port. When air is being forced out of the cylinder, the disk moves to cover the inlet line port. This allows air from the cylinder line to quickly exhaust to the atmosphere. Very low air pressure and airflow are required to shift the disk.

Shuttle Valves

Shuttle valves are used to automatically select the higher of two air pressures and connect that source to the outlet port of the valve. The valve is basically a quick-exhaust valve used in a different application. The inlet and exhaust ports of the valve are connected to two different pressure sources. The third port is connected to the actuator.

Activating only one of the pressure sources shifts the shuttle valve and connects that pressure source to the actuator. If both pressure sources are active at the same time, the higher pressure source automatically shifts the shuttle to connect the higher-pressure line to the actuator. If the relationship of the pressures changes during operation, the shuttle valve shifts to keep the higher pressure connected to the actuator. See **Figure 18-30**.

Quick-exhaust valve

Quick-exhaust valve reduces cylinder-retraction time

Figure 18-28. Quick-exhaust valves allow the rapid release of large quantities of exhaust air from a cylinder.

Figure 18-29. A common model of quick-exhaust valve uses a floating, flexible disk to block and open ports.

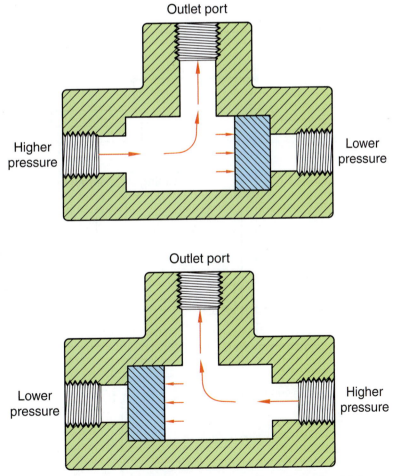

Figure 18-30. A shuttle valve automatically selects the higher of two inlet pressure sources.

Timing-Volume Reservoirs

Timing-volume reservoirs are simply air chambers used to time the shifting of pilot-controlled pneumatic valves. The component is usually a tube fitted with end caps containing ports for connection to the pilot-control portion of a circuit. The end caps also contain a means for mounting the device. One manufacturer has reservoirs available in sizes ranging from 5 to 225 cubic inches.

One port of the reservoir is connected to the pilot actuator of a control valve. The other port is connected to a pressure source. Air is bled into the reservoir through a needle valve. The rate of airflow and the volume of the reservoir determine the time it takes to increase the pressure in the reservoir to a level sufficient to shift the piloted valve.

When the control valve shifts, the air in the reservoir is exhausted. The reservoir can then repeat the function during the next cycle of the circuit.

Figure 18-31 shows a schematic of a circuit using a timing-volume reservoir to time the shifting of a control valve.

Exhaust Mufflers

Mufflers are commonly used to reduce to an acceptable level the noise of air exhausting from pneumatic control valves and tools. In addition, these devices prevent dirt and other foreign materials from entering the system. Mufflers are fitted to the exhaust ports. Muffler designs range from compact, sintered-bronze units to designs that use sheet metal outer shells filled with plastic or metal materials. See **Figure 18-32.** Sizes are readily available for 1/8" to 1 1/2" diameter ports with pipe threads.

In clean mufflers, the resistance to airflow is relatively low. However, as a muffler is used,

Timing-volume reservoir

Figure 18-31. Timing-volume reservoirs allow control of the sequence of a circuit using pilot-activated directional control valves.

Figure 18-32. A variety of mufflers are used on the exhaust ports of components to reduce the noise of component operation. (Norgren, Inc.)

resistance to airflow can increase to excessive levels. This increase is due to system lubricating oil, condensed water, and rust and scale particles that pass through the muffler with the exhausted air. These materials slowly build up in the muffler and eventually restrict airflow. Mufflers should be periodically inspected. If there appears to be an excessive build up of dirt or oil, the muffler should be cleaned or replaced. The design of most mufflers allows them to be easily removed and cleaned.

Selecting and Sizing Pneumatic Components

Manufacturers of pneumatic components provide considerable detail about the units they produce. This information is available from both company catalogs and websites. In many situations, these sources provide the detail required to select appropriate features and correctly size units for a circuit.

Information Available Concerning Component Features

Catalog and website information includes a general description of the components; the operating temperature, flow, and pressure ranges; and schematic symbols for variations. In addition, line drawings of the units provide physical dimensions of the exterior, as well as port and mounting-hole sizes.

Pressure drop and flow rate data are also displayed in these descriptions, if that information is important for selection. The illustrations shown in **Figure 18-33** are examples of part of the information provided in catalog and data sheets.

Determining Required Valve Size

Selecting the size of a control valve for a pneumatic system is a relatively difficult task. This is due to the many variables involved in system operation. However, selection of a properly sized control valve for an application is critical for effective operation of a system. Proper sizing is also important because selecting a valve that has a larger capacity than necessary increases both construction and operating costs for the system. It also consumes extra physical space, which can be important in some situations.

Several approaches described by different manufacturers can be used to determine a valve capacity appropriate for an application. Most of these approaches begin with a relatively simple formula. The data required for this calculation (for a circuit containing a cylinder) typically include:

- Cylinder piston area.
- Cylinder stroke.
- System compression.
- System pressure drop.
- Cylinder stroke time.

The approaches to this calculation are similar in design, but use variations of formulas and data from tables provided by the manufacturer. Due to

A

B

Figure 18-33. Manufacturer catalogs provide considerable technical information about components, including operating details, physical dimensions, pressure ratings, and air-flow rates (often provided in graph form). A—Dimensions of critical features for a switch. B—Pressure drop and flow for a filter.

this, specific calculations are beyond the scope of this text. However, it will be beneficial if you select a manufacturer and study its approach to valve sizing. Printed brochures and Internet-based calculators are often provided by companies.

Chapter Test

Answer the following questions. Write your answers on a separate sheet of paper.

1. List the three factors control valves regulate to produce a desired performance in a pneumatic circuit.

2. The valve control element that consists of a precision-machined hole used to control the rate of airflow is called a fixed _____.

3. List four sources of internal and external force that are used to operate the control elements in pneumatic control valves.

4. What is the most common method used to control the maximum operating pressure in the distribution line of a pneumatic system?

5. List five designs of air pressure regulators.

6. A(n) _____ valve is a direct-operated valve that fully opens as soon as the maximum allowable system operating pressure is reached.

7. The only shut-off valve that is recommended for metering airflow in a pneumatic system is the _____ valve.

8. What happens to the compressed air used to operate an actuator after it is returned to the directional control valve?

9. The _____-center configuration of four-way pneumatic directional control valve has an exhaust port located at both the extension and retraction positions of the valve.

10. Name the two basic functions of the center position of a three-position, four-way, directional control valve.

11. The two most common basic flow control devices used in a pneumatic system are fixed-sized orifices and _____ valves.

12. The _____ flow-control circuit is the preferred method to use for controlling the operating speed of cylinders in pneumatic circuits.

13. A(n) _____ valve increases the rate at which air is exhausted from a cylinder.

14. Describe the function of a shuttle valve.

15. List five primary types of information available about pneumatic components found in manufacturer catalogs.

The ability to read and interpret a schematic diagram of a complex pneumatic system begins with familiarity of the operation of basic circuit segments. Basic segments are often found in a variety of pneumatic systems. Even though a circuit or system may look complex, a careful analysis of the valves, lines, and actuators should reveal a number of component groups used to control air pressure, airflow rate, and direction of airflow. This chapter presents examples of basic circuits. These examples were selected to illustrate how various, basic components can be combined to produce desired actuator force, operating speed, operating motion, or other function.

Objectives

After completing this chapter, you will be able to:

- Describe basic circuits that are frequently used to assemble complex pneumatic systems.
- Identify the placement and explain the function of workstation air-preparation components.
- Compare the design and operation of basic pressure-control circuits used in pneumatic systems.
- Explain the design and operation of control circuits used to regulate actuator operating speeds in pneumatic systems.
- Describe the operation of circuits using four- and five-port pneumatic four-way directional control valves.
- Explain the design and operation of basic motion-control circuits for pneumatic cylinders.
- Explain the design and operation of pneumatic quick-exhaust valves.
- Describe a two-handed safety circuit used to provide operator safety.

Chapter 19

Applying Pneumatic Power
Typical Circuits and Systems

Selected Key Terms

The following terms will be used in this chapter. As you read the text, record the meaning and importance of each. Additionally, you may use other sources, such as manufacturer literature, an encyclopedia, or the Internet, to obtain more information.

automatic reciprocation
automatic return
boosters
composite symbol
logic functions
memory circuit
multiple-pressure circuit
quick-exhaust valve
remote pressure adjustment
time-delay circuit
trio unit
two-hand safety circuit

Internet Resources

www.boschrexroth.com/various/utilities/mediadirectory/index.jsp?oid=254241&language=en-GB

Bosch Rexroth AG

The Rexroth media directory page. Use the search function to select the "Pneumatics" category (or a more specific topic of interest), and then view the extensive array of related technical articles and bulletins.

www.fairchildproducts.com/industries.php

Fairchild Industrial Products Company

The page provides a number of pneumatic applications for the company's pneumatic products. The information is classified by the industries where the products are being used.

www.hydraulicspneumatics.com/other-technologies/Chapter-19-moving-part-air-logic-controls

Hydraulics & Pneumatics Magazine

An extensive chapter on air logic controls. The descriptions of air logic elements include illustrations and circuit examples. Among the illustrations are logic symbols, standard pneumatic symbols, and graphic component cutaways.

www.fluidpowerjournal.com/html/articles.html

International Fluid Power Society: Fluid Power Journal

A listing of professional papers, including some which support the use of pneumatic control of fluid power systems. This site is an example of materials provided by a professional organization.

Basic Circuits and System Analysis and Design

The assembly of valves, lines, and actuators in a pneumatic system can become confusing. This is especially true for individuals beginning work in the field. Designing, installing, and maintaining the pneumatic system of a machine can be a challenge even to experienced fluid power personnel. However, this challenge can be considerably reduced if the systems are considered groups of basic components and subcircuits working together to produce a final desired action. Being familiar with basic component groups and circuits can be very helpful when working on pneumatic systems.

The majority of the basic pneumatic component groups and circuits can be divided into five function groups:

- Workstation air preparation.
- Pressure control.
- Flow control.
- Motion control.
- Other miscellaneous functions.

A typical pneumatic circuit includes segments from several of these group merged together. The basic components and circuits selected for the final version of any circuit depend on the functions and level of sophistication demanded by the machine powered or controlled by the circuit.

The following sections illustrate several pneumatic circuit segments that can be used to produce various types of system operation. Each of these circuit segments are briefly explained in the text and shown in an accompanying figure. It must be emphasized that the segments shown here are only a few common examples of the many possibilities.

Workstation Air-Preparation Components

The compressor and system distribution lines leading to various workstations are designed to deliver clean, high-pressure air. This air must be further conditioned to:

- Reduce the pressure to the level needed for circuit operation.
- Remove remaining contaminants.
- Add a lubricant, if needed.

The Final Preparation of Air at the Workstation section in Chapter 16 discusses these individual processes in detail. This section illustrates examples of the placement of the basic conditioning components in operating circuits to assure clean, lubricated air at the required pressure level.

The circuit shown in **Figure 19-1** shows the placement of the air filter, pressure regulator, and lubricator in a basic pneumatic circuit. These components are usually assembled into a close-fitting group often considered a single component. This FRL assembly may be referred to as the *trio unit* in some literature. The component enclosure symbol surrounding the filter, regulator, and lubricator symbols in this circuit indicates the close relationship of the components.

Component enclosure

Figure 19-1. The final preparation of air in a pneumatic system occurs at the workstation. The air is filtered, the pressure is adjusted to the requirements of the working circuit, and needed lubricants are added.

In an operating circuit, the conditioning units are located between the system distribution line and the circuit. Incoming air from the distribution line is first filtered to remove dirt and condensed water. It is then routed to the regulator, which reduces the air pressure to the level needed to operate the circuit. The air then passes through the lubricator where atomized oil is added to the airstream. This provides lubrication for the system components. The clean, pressure-controlled, lubricated air is then distributed throughout the workstation circuit.

Figure 19-2 shows a circuit designed to provide additional lubrication to a specific system component. Because of a remote location or internal design characteristic, some components may not receive adequate lubrication from the FRL unit lubricator. The design shown in the figure has an additional lubricator located in a line parallel to the normal supply line for the cylinder. The parallel line must be correctly sized to assure appropriate airflow through the secondary lubricator.

The diagram shown in **Figure 19-2** uses the *composite symbol* to designate the FRL unit. This symbol is typically used instead of three separate symbols in an enclosure. The symbol is much less complex, but indicates the three conditioning elements. This symbol is used in previous chapters. It will be used in the remainder of the circuits shown in this chapter unless individual filters, pressure regulators, or lubricators are needed for the operation of the circuit.

Controlling System Air Pressure

The pressure of the air delivered to a pneumatic cylinder or motor determines the force and torque output of the units. The maximum air pressure in an operating circuit is controlled by the regulator that is part of the final air-preparation components at the workstation. However, it is also important to have the capability to control pressure beyond limiting the maximum pressure. Any portion or phase of a circuit operating below the maximum system pressure can reduce the volume of atmospheric air that must be compressed. Reducing the volume of high-pressure air needed increases system efficiency and reduces the operating costs of the pneumatic system. This section illustrates several basic pressure control circuits that can be used to control circuit pressure. These circuits either:

- Control maximum circuit pressure.
- Provide multiple levels of pressure.
- Allow remote adjustment of pressure level.

Controlling Maximum Circuit Pressure

Figure 19-3 shows a circuit that provides different maximum pressure levels to operate three different circuit functions. One pressure level is provided for cylinder extension. A second pressure

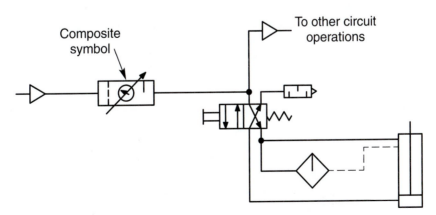

Figure 19-2. This circuit shows one method that may be used to add lubricant to remote or difficult-to-lubricate actuators. The three air-preparation components are indicated by a standardized composite symbol.

is set at a level only high enough to retract the cylinder. The third pressure is based on the level required to shift the pilot-operated directional control valve.

The highest pressure of this circuit is controlled by the setting of the workstation pressure regulator. When the circuit is pressurized, the normal position of valve C causes pilot pressure to shift valve D. This retracts the cylinder. The setting of the pressure regulator identified as valve A controls the maximum air pressure available for retraction. When valve C is manually shifted, it directs pilot pressure to shift valve D to extend the cylinder using maximum circuit pressure. The pressure regulator identified as valve B controls the maximum pressure available to operate the pilot portion of the circuit.

Providing Multiple Pressures

The circuit shown in **Figure 19-4** is basic *multiple-pressure circuit.* It provides high pressure to extend a cylinder and a reduced pressure for retraction. Another feature of this design is the use of a three-way directional control valve to provide both extension and retraction of a double-acting cylinder. Three-way directional control valves are typically lower in cost than similar four-way directional control valves.

Figure 19-3. Several operating pressures may be obtained within a single pneumatic circuit by the installation of additional pressure regulators.

Figure 19-4. The difference in effective piston area and a relieving-type pressure regulator allow this cylinder to be operated using different extension and retraction pressures.

The key to the operation of this circuit is the control valve identified as valve A. This valve must be a relieving-type pressure regulator. It acts as a relief valve when the downstream pressure exceeds the pressure setting of the valve. Valve B is a three-way directional control valve that controls the extension of the cylinder. Cylinder retraction automatically occurs whenever the three-way valve is returned to the normal position.

When the circuit is pressurized, the cylinder is returned to the retracted position by low-pressure air controlled by valve A. Air from the blind end of the cylinder is exhausted through valve B. When valve B is shifted, the blind end of the cylinder is exposed to maximum circuit pressure. The force generated by this pressure acting on the blind side of the piston is greater than the force generated by the lower pressure on the rod end. Therefore, the cylinder will extend. The low-pressure air from the rod end is exhausted through the vent of valve A.

Two limiting factors must be considered with this design. First, the effective force is reduced due to the opposing pressures in the cylinder. Second, the venting capacity of the pressure regulator (valve A) must be adequate to exhaust air from the rod end of the cylinder.

Allowing for Remote Pressure Adjustment

Figure 19-5 shows a circuit that allows *remote pressure adjustment.* A pilot-operated pressure regulator is used to change the setting of the system operating pressure. A pilot-operated pressure regulator uses air pressure in a chamber above the regulator diaphragm, instead of a spring, to establish the operating pressure of the unit. The design and operation of pilot-operated regulators is discussed in detail in Chapter 16.

Valve A is the main pressure regulator for the circuit. The pressure setting of this pilot-operated regulator is established by the second, small-capacity regulator identified as valve B. This design allows valve B to be placed in a convenient location. Also, the primary pressure regulator can be placed close to the operating components. This results in minimum line lengths, which increases system response and efficiency.

> **Note:** The circuit shown in **Figure 19-5** demonstrates how the filtration, pressure regulation, and lubrication functions can be located at separate points in a circuit, rather than combined in the FRL unit at one location.

Controlling System Airflow Rate

The rate at which compressed air is allowed to flow into a pneumatic cylinder or motor determines the operating speed of the actuator. The

Figure 19-5. Pilot-operated pressure regulators allow the adjustment of circuit pressure from a remote location. This allows the primary regulator to be placed close to other circuit components, while an adjustment valve is placed in a more convenient or safer location.

When this circuit is idle, the normal position of the directional control valve holds the cylinder in the retracted position. When the directional control valve is shifted to extend the cylinder, pressurized air goes directly to the blind end of the cylinder. As the cylinder begins to extend, the exhausted air passes through the directional control valve and on to the needle valve A. This valve meters the exhaust air, which indirectly controls the volume of air that can enter the cylinder. The end result is control of cylinder extension speed. During retraction, the directional control valve passes pressurized air directly to the rod end of the cylinder. Exhaust air from the blind end passes through the directional control valve and on to needle valve B. This results in the control of cylinder retraction speed.

Both this circuit and the previous circuit allow independent adjustment of extension and retraction speeds. An advantage of this design over the previous circuit is elimination of the integral check valves used to bypass the needle valves. What is lost is the close proximity of the flow control valves to the actuator.

One Retraction and Two Extension Speeds

The circuit shown in **Figure 19-9** provides two adjustable extension speeds and one adjustable retraction speed. The design allows rapid advance of a machine tool to a location where a lower speed is needed for a forming or machining operation. The retraction speed of the cylinder is also adjustable.

The low extension speed is activated by a mechanically operated, normally open, two-way valve. A solenoid valve can also be used. When the system is initially pressurized, the cylinder moves to the retracted position. Shifting the five-port, four-way directional control valve directs airflow to the blind end of the cylinder. Air from the rod end is exhausted through the normally open, two-way valve; to the four-way directional control valve; and on to needle valve A. The rate of exhaust through this needle valve establishes the higher speed of cylinder extension.

When the ramp on the cylinder rod shifts the mechanically operated, two-way valve, air is forced

Figure 19-9. Multiple operating speeds can be achieved during cylinder extension or retraction using a mechanically or solenoid-operated, two-way valve. This circuit provides two different cylinder rod extension speeds.

through needle valve B. This valve reduces the airflow to slow the cylinder speed for the forming or machining operation. Air exhausted from valve B passes to the atmosphere through the four-way directional control valve and valve A with little restriction.

When the four-way direction control valve is shifted to retract the cylinder, pressurized air passes though the check valve. This allows reverse flow around valve B. When the ramp moves off of the mechanically operated, two-way valve, the valve is shifted by the internal spring. The valve then provides an unrestricted route for the air to enter the rod end to aid in cylinder retraction. The air exhausted from the blind end is routed through the four-way directional control valve to valve C. Valve C is an adjustable needle valve that controls cylinder retraction speed.

Controlling Actuator Motion

Pneumatic systems must provide a wide variety of motion-control functions. These functions range from simple manual controls for extending and retracting a cylinder to complex circuits that produce multiple movements. Often these multiple movements are automatically performed after a machine operator manually shifts a valve to start the sequence.

This section examines two circuits that provide automatic reciprocation of a cylinder. The principles involved in these simple designs can be expanded to more complex circuits to provide the sequencing of two or more actuators. The circuits shown use mechanically operated, three-way valves controlling pilot-operated directional controls. A combination of limit switches and solenoid-operated valves are also commonly used to control these circuits.

Single-Cycle Operation

Figure 19-10 shows a circuit that provides for *automatic return* of the cylinder after it extends. To initiate the sequence, the operator momentarily depresses a three-way directional control valve (valve A). This action directs pilot pressure to the five-port, four-way directional control valve (valve B). As a result, valve B shifts to extend the cylinder. When the ramp on the cylinder rod shifts valve C, pilot pressure is supplied to shift the four-way directional control to the opposite position. This returns the cylinder to the original, retracted position. The cylinder is held in the retracted position until the machine operator again depresses valve A to repeat the cycle.

Continuous Operation

The circuit shown in **Figure 19-11** provides continuous, *automatic reciprocation* of a cylinder.

Figure 19-10. The automatic operation of portions of a circuit cycle often increases the efficiency of a machine. This circuit extends and retracts a cylinder after the control valve is activated.

Figure 19-11. Automatic cylinder reciprocation is possible with this circuit. The cylinder continuously extends and retracts as long as the system is pressurized and valve A is activated.

This continues as long as the system is pressurized and the operator holds the manually operated valve in the shifted position. The diagram shows the circuit before it is pressurized. The cylinder is in the fully retracted position with the extension phase of the cycle about to begin.

Depressing valve A pressurizes the system and starts the sequence. Air passes through valves A and B. Valve B provides pressure to the extension pilot of the five-port, four-way directional control valve. The four-way valve shifts to direct air to the blind end of the cylinder. This air begins to extend the cylinder. As the cylinder extends, the ramp on the rod moves off of the activation mechanism on valve B. This allows the valve to shift and vent the extension pilot line to the atmosphere. The cylinder continues to extend until the ramp depresses the activation mechanism on valve C.

When valve C is shifted, air is directed to the retraction pilot of the four-way directional control valve. The pilot shifts the valve to direct air to the rod end of the cylinder. The cylinder begins to retract. The ramp moves off of valve C, which

allows the retraction pilot line to vent. The cylinder continues to retract until the ramp depresses the activation mechanism on valve B. This allows system air pressure to again activate the extension pilot, beginning another sequence. The cylinder continues to reciprocate as long as valve A is held in the shifted position to keep the system pressurized.

Circuits for Other System Functions

A large number of additional basic circuits are often illustrated in materials prepared by component manufacturers, equipment manufacturers, trade magazines, and professional associations. The circuits shown in this section are included to illustrate the variety of specialized components and circuit segments available to support the operation of pneumatic equipment.

Booster Circuit

A pressure higher than that produced by the system compressor is sometimes needed for the operation of a section of a pneumatic circuit. This higher pressure can be provided by specialized air-to-air or air-to-oil *boosters*. Many boosters are basically small-volume, air-powered reciprocating pumps. They produce intensified pressure for use in a small section of a system.

Figure 19-12 illustrates the circuit and operating concepts involved in an air-to-air booster. The circuit shows components in position to begin the process. When the system is pressurized, air flows through valve A, holding the five-port, four-way directional control valve in the position to retract the control rod and extend the piston of the booster pump. As the control rod retracts, valve A shifts and vents the pilot pressure. As the piston of the booster pump extends, air moves out of the pump cylinder. This closes check valve C and opens check valve D. The air moved by the booster pump increases the pressure in the line leading to the circuit section requiring high pressure.

When the power cylinder approaches the end of its extension stroke, the ramp on the control rod shifts valve B. Pilot pressure is then routed through that valve to shift the four-way directional control valve to the opposite position. This causes the control rod to extend and the booster pump piston to retract. System air enters the pump through valve C. Valve D closes because of the higher pressure in the high-pressure section of the circuit.

As the control rod extends, valve B shifts and vents the pilot pressure. As the rod approaches the end of its extension stroke, the ramp shifts valve A. Air flowing through this valve shifts the pilot-operated directional control valve to automatically repeat the cycle.

Quick-Exhaust Valve Circuit

A *quick-exhaust valve* is used to reduce back pressure during cylinder extension or retraction to allow maximum operating speeds. The valve should be close-fit to the ports of the actuator to obtain maximum performance.

Figure 19-13 shows a basic cylinder circuit using a quick-exhaust valve. The circuit is shown in the normal position with the cylinder retracted. System air is routed through the directional control valve to the quick-exhaust valve. Pilot

Figure 19-12. This is an air-to-air pressure booster circuit. Boosters are used in pneumatic circuits to provide air pressures higher than the pressure available from the distribution line.

Figure 19-13 Quick-exhaust valves reduce back pressure in the exhaust line of a cylinder by routing exhaust air directly to the atmosphere. Reducing back pressure allows the cylinder to reach maximum operating speed.

pressure shifts the valve to provide a route to the rod end of the cylinder. This forces the cylinder to fully retract. During this phase of operation, the quick-exhaust valve serves only as a route for the passage of air and does not contribute to circuit performance.

When the directional control valve is shifted to extend the cylinder, air flows from the directional control valve to the blind end of the cylinder. As the cylinder extends, the air exhausted from the rod end is routed to the quick-exhaust valve. The back pressure in the exhaust line caused by resistance to airflow through the line, the directional control valve, and the quick-exhaust valve itself causes the quick-exhaust valve to shift. This provides the exhausted air a short, direct route to atmosphere.

Logic Function Circuits

Pneumatic components can be used to perform basic logic functions to provide system control. *Logic functions* include AND, OR, and NOR functions such as used in microprocessors. In pneumatic systems, the functions are completed by small moving parts in air-operated valves. These valves are not in the circuit to help produce motion and force. They are used to sense airflow and pressure to provide the desired functions. A number of manufacturers produce lines of miniature components that operate at this level.

Figure 19-14 illustrates a *memory circuit* that holds a valve position once the manually operated

valve initiates the operation. The circuit uses three three-way directional control valves and one shuttle valve.

The control function is initiated by manually depressing valve A. Air passes through that valve and shifts shuttle valve C. This directs air to the pilot of the system control valve (valve D). That valve shifts to pass pressurized air to the system. It also passes air through valve B to the second pressure port of the shuttle valve. When the start valve (valve A) is released, the shuttle valve shifts, transferring the source of the pilot pressure for valve D to the air line of the operating circuit (through valve B).

The circuit holds valve D open until the stop valve (valve B) is shifted. When valve B is shifted, the pilot pressure for valve D is vented. This allows the internal spring in valve D to shift the valve. Airflow and pressure to the operating circuit are blocked.

Time-Delay Circuits

Circuit actuators often need to be momentarily held at a point in their operation to allow the completion of a task. **Figure 19-15** shows a *time-delay circuit* that allows automatic timing of cylinder movement. This design uses simple needle valves, timing-volume reservoirs, and piloted-operated directional control valves to provide a desired sequence of operation and time delays.

Pilot-operated, five-port, four-way directional control valves are used to operate the two cylinders.

Figure 19-14. Pneumatic control valves can be assembled into circuits that perform logic functions. This is a memory circuit that holds the system control valve in the activated position using only a momentary activation of a start valve.

Figure 19-15. Timed delays in a sequence of cylinder movements can be produced using flow control valves coupled with simple timing reservoirs. These control segments time the activation of pilot-operated directional control valves.

The sequence in this circuit is extension of cylinder 1 followed by extension of cylinder 2 after a time delay. When the task performed by cylinder 2 is completed, that actuator can be immediately retracted followed by cylinder 1 after another time delay.

When this circuit is pressurized, both cylinder 1 and cylinder 2 are moved to the retracted position. Circuit operation is initiated by the operator activating valve A. This pressurizes the control portion of the circuit. Air flows through the check valve in valve B and on to the timing reservoir and the pilot control of valve C. The pressure quickly climbs to pressurize the timing chamber and activate the pilot. This shifts valve C to extend cylinder 1.

During the same period of time, the check valve in valve D closes. This forces the control air to be metered through the needle valve. This metered air flows to the second timing reservoir, which slowly fills. After a period of time, the pressure in this reservoir increases to the point where the pilot mechanism shifts valve E. Shifting valve E extends cylinder 2. This performs the second task of the operation.

When the tasks performed by the cylinders are completed, the machine operator allows valve A to return to its normal position. This vents the control portion of the circuit to the atmosphere. The pilot air that shifted valve E can easily vent through the check valve in valve D. This allows the spring mechanism in valve E to quickly shift the valve. As a result, cylinder 2 quickly returns to the retracted position. The retraction of cylinder 1 is delayed as the pilot air in the timing chamber must bleed through the needle valve in valve B.

When this pilot air has been vented to the atmosphere, valve C shifts to return cylinder 1 to the retracted position.

The length of the time delays is set by adjusting the needle valves. Different timing-volume reservoirs are also available, which allows a large range of time delays.

Safety Circuits

The safe operation of machines is always a concern. Operators may be injured when they become distracted or take chances. Pneumatic control circuits are often used in an attempt to improve workplace safety.

Figure 19-16 shows a pneumatic safety circuit often used to reduce hand injures. This circuit requires the use of both hands to activate the operation of a press or other machine that can cause serious hand injuries. A *two-hand safety circuit* generally prevents complete or partial machine operation if either of the hand controls is tied down in an attempt to override the system.

When this circuit is at rest, valve A pressurizes pilot A of the system valve (valve C). This shifts

Figure 19-16. Machines often require both hands to activate the start cycle. These safety circuits will not function if an operator attempts to bypass the system by tying down either of the hand controls.

valve C, allowing the single-acting cylinder to vent to the atmosphere. The press platen in this example fully lowers.

Depressing only valve A blocks system pressure and vents pilot A to the atmosphere. Depressing only valve B pressurizes pilot B, which attempts to shift valve C. However, the valve will not shift because both pilots A and B are pressurized. This prevents movement of the valve control element. Simultaneously depressing valves A and B vents pilot A to the atmosphere and pressurizes pilot B. This allows the system control valve (valve C) to be shifted.

The system will not operate beyond a part of a cycle if a machine operator attempts to bypass the safety system by tying down one of the control valves. Tying down valve A and attempting to operate the system by activating only valve B results in the cylinder extending. However, there is no method to pressurize pilot A, which is needed to shift valve C to retract the cylinder. Tying down valve B results in a similar situation. When valve A is activated in this second situation, the air at pilot A is vented. This allows the pressurized air coming from valve B to shift valve C. The cylinder extends, but cannot be lowered. Valve B, which is tied down, maintains a constant pressure on pilot B. This prevents further shifting of valve C.

Chapter Test

Answer the following questions. Write your answers on a separate sheet of paper.

1. The challenge of understanding the operation of a pneumatic system can be reduced if it is considered groups basic of _____ and _____ working together to produce a desired result.

2. The three air-preparation components located at the workstation are often close fitted into a unit referred to as the _____.

3. Whenever additional pressure regulators are used downstream from the air-preparation unit, they can regulate pressures to a level _____ the pressure setting of the workstation unit.

4. A(n) _____-operated pressure regulator can be used when pneumatic system pressure must be controlled from a remote location.

5. Which type of valves are integrated into needle valves to allow unrestricted return airflow?

6. Where are flow control valves often placed when used with five-port, four-way directional control valves?

7. Quick-exhaust valves are designed to increase the operating speed of cylinders. How is this achieved?

8. What is a logic function?

9. What is the purpose of a time-delay circuit?

10. In most two-hand safety circuits, what happens if one of the controls is tied down?

Glossary

A

Absolute pressure: A pressure scale where the zero point represents a perfect vacuum.

Absolute rating: A filter rating that states the largest pore opening, indicating that all particles larger than the stated size will be trapped.

Absolute viscosity: The resistance of a fluid to the relative movement of its molecules. Measured in *pounds per foot second* in the US customary system and by *poise* in the metric system.

Absolute zero: The temperature at which all molecular movement stops.

Absorbent filter: A filter that cleans a fluid using a medium to trap contaminants and then holds them by mechanical means.

Accumulator: A hydraulic component in which system liquid is stored under pressure using a mechanical device, compressed gas, or weight to maintain pressure.

Actuator: A component used to convert the energy in hydraulic fluid or compressed air into mechanical linear or rotary motion.

Adapter: Machine or system parts used to align components or change sizes of shafts and distribution lines.

Adapter fitting: A fitting used to connect components having dissimilar diameters or thread types.

Additive: A chemical compound added to a hydraulic fluid to modify its characteristics and improve system performance.

Adiabatic: The general term used to identify the process in which no heat is transferred when a change occurs in the volume of a gas.

Adiabatic compression: The process in which no heat is transferred as the volume of a gas is decreased.

Adiabatic expansion: The process in which no heat is transferred as the volume of a gas is increased.

Adsorbent filter: A filter that cleans a fluid using a medium that holds contaminants by molecular adhesion.

Aftercooler: A pneumatic system heat exchanger that cools compressed air as it leaves the compressor; used to both cool air and liquefy water vapors.

Air consumption: The volume of air used to operate a pneumatic actuator or system; usually stated in *standard cfm.*

Air distribution system: The network of lines used to distribute compressed air from the receiver to the various workstations in a pneumatic system.

Air-drying equipment: Devices used to dry air after compression by cooling, centrifugal force, absorption, adsorption, or a combination of these methods.

Air filter: A component designed to remove solid particles, moisture, and/or lubricant from pneumatic system air.

Air-line respirators: Self-contained breathing devices that provide a breathable atmosphere to workers in conditions where safe, clean air is not available.

Air nozzle: A device located on the end of an air line and containing an outlet orifice designed to maximize kinetic energy from the controlled release of pressurized system air. Often used in a blow gun.

Air receiver: A tank used for the storage of compressed air in a pneumatic system. Also assists in the control of air temperature and water vapor.

Alternating current: Electrical current that reverses direction on a regular cycle due to a change in voltage polarity. The common international standards are 50 or 60 cycles per second (or hertz).

American National Standards Institute (ANSI): The primary standards coordinating group in the United States. The group consists of numerous professional, technical, trade, labor, consumer, and governmental organizations and agencies. It is also the United States member of the International Organization for Standardization (ISO).

American Petroleum Institute (API): An organization consisting of corporations from the petroleum and allied industries. The group includes petroleum producers, refiners, marketers, and transporters. Activities include providing information to the general public as well as advocacy, research, and technical services for members.

American Society for Testing and Materials (ASTM): A group that actively works to establish standards for materials, products, systems, and services. Membership is available to individuals and organizations representing education, research, government, and business.

American Society of Mechanical Engineers (ASME): A technical society organized to promote mechanical engineering and its associated arts and sciences. The group participates in international standardization activities.

Anti-foaming agent: A chemical compound added to a hydraulic fluid to reduce or eliminate the formation of foam during system operation.

Anti-wear agent: A chemical compound added to hydraulic fluid to help reduce wear on bearing surfaces during hydraulic system operation.

API gravity: A comparison of the weight of a given volume of a substance to the weight of an equal volume of distilled water at a specified temperature.

Application information: Information supplied by a manufacturer describing conditions and limits of operation in which a product may be expected to successfully operate.

Archimedes: The individual credited with the discovery of the principle of buoyancy. The discovery was made in the third century BC.

Armstrong, Lord William: The individual credited with the development of the accumulator, which made a significant contribution to the development of the early fluid power industry.

Atmosphere: 1. The mix of gases that surround Earth between its surface and space. 2. The pressure created by a column of gas extending between space and the surface of Earth at sea level. 3. The pressure 14.7 psia.

Atoms: The smallest parts of an element that can take part in a chemical reaction without changing the characteristics of the material.

Automatic reciprocation: A hydraulic or pneumatic circuit design that produces continuous extension and retraction movement of a linear actuator (as long as the system is activated).

Automatic return: A hydraulic or pneumatic circuit design that, when activated, produces a single extension and retraction stroke on a linear actuator.

Auxiliary air receiver: A secondary air storage tank used in a pneumatic circuit to supply additional compressed air to a high-demand, intermittently operating actuator or other system function.

Axial piston pump: A hydraulic pump using a design in which the reciprocating motion of the pistons is parallel to the power input shaft.

B

Back injury: An injury to the skeletal, muscular, or nervous-system elements of the back associated with lifting or other work-related activity.

Back-pressure check valve: A spring-loaded check valve used with some pilot-controlled hydraulic systems to maintain sufficient pressure during the idle portion of the cycle to shift directional control valves.

Baffle: A plate or other device placed in a hydraulic reservoir to direct oil flow between the system return lines and the pump inlet line.

Balanced-piston valve: Hydraulic valves that pressurize both the control and system side of a primary control piston. Desired valve action is obtained using metering orifices and/or small, spring-loaded poppets to adjust pressure on the control side of the piston.

Balanced-poppet valve regulator: A pneumatic system regulator using a design in which the action of the valve control poppet is dampened by subjecting both ends of the poppet to valve outlet pressure.

Balanced-vane pump: A hydraulic vane pump using a design with two inlet and outlet chambers that balance the load forces on the bearing of the pump rotor shaft.

Ball valve: A valve primarily used as an on-off valve. The primary valve component is a ball with a bored passageway. The valve requires only 1/4 turn to go from fully open to fully closed.

Barrel: The component containing the cylinders of an axial piston hydraulic pump.

Basic machine type: One of the simple machines: lever, inclined plane, wedge, wheel and axle, screw, and pulley.

Bent-axis design: The axial piston hydraulic pump design in which the angle between the input shaft and the barrel of the pump determines the length of the piston stroke.

Bernoulli, Daniel: A scientist of the 18th century credited with laying the foundation of hydrodynamics.

Bernoulli's theorem: A basic scientific law stating that in a flowing frictionless fluid a reduction in the cross-sectional area of the conductor will result in an increase in velocity and a decrease in pressure of the fluid at the point of restriction.

Biodegradable fluid: A hydraulic fluid formulated to degrade in nature to reduce environmental damage from spillage.

Bleed-off circuit: A hydraulic speed-control circuit in which the required flow rate to the actuator is set by returning unneeded pump output directly to the reservoir through a flow control valve.

Blow gun: A device containing an air nozzle and designed for cleaning parts and equipment with a high-velocity airstream.

Bonnet: The upper portion of a gate, globe, or needle valve that holds the adjustment mechanism and protects the valve from damage.

Booster: A hydraulic system component used to increase the pressure in a portion of the system to higher than the pump discharge pressure. Often called a pressure intensifier.

Boyle, Robert: A scientist of the 17th century identified as one of the first to study the characteristics of gases.

Boyle's law: One of the basic gas laws that states the volume of a gas, held at a constant temperature, varies inversely with the pressure.

Braided wire: Fine wire braided into a flexible tube and used as a layer of reinforcement in some high-pressure hydraulic and pneumatic hoses.

Brake valve: A pressure control valve used in some hydraulic motor circuits to slow the motor to a controlled stop against the rotational inertia of the external load.

Bramah, Joseph: An English engineer of the late 18th and early 19th centuries credited with the invention of the cup seal, which was a major contribution to the success of early hydraulic systems.

Brazed-plate heat exchange: A heat exchanger formed from metal plates brazed together into a compact unit; used to cool or heat the fluid in a system.

Btu: Abbreviation for British thermal unit, which is equal to the heat required to raise the temperature of one pound of water by one degree Fahrenheit.

Buoyancy: The force vertically exerted on a body by the fluid in which it is partially or wholly submerged.

Burst pressure: A rating for pressure vessels and conductors indicating the pressure at which the component failed during the prescribed test procedure.

Bypass flow control valve: A flow control valve that uses an integral control port to return excess fluid directly to the reservoir.

C

Calorie: The heat energy required to raise one gram of water one degree Celsius at atmospheric pressure.

Cam ring: A positioning ring found in variable-displacement pumps, both vane and radial piston, that allows adjustment of the displacement of the pumping chambers.

Cam-type radial-piston motor: A hydraulic motor in which the force generated by the pistons is applied against the slope of a cam located on the motor output shaft. This action produces motor rotation and torque.

Cap: The end closure of a hydraulic or pneumatic cylinder that is opposite of the end from which the rod extends. Often called the cylinder blind end.

Capacity-limiting system: A system used to control the maximum air pressure produced by the pneumatic system compressor.

Compressor start-stop, inlet valve unloading, and other methods can control capacity.

Capillary viscometer: A test instrument containing a capillary tube calibrated to provide information adequate to determine the viscosity of fluid.

Catalyst: A substance that increases the rate of a chemical reaction without being consumed in the process.

Cavitation: The formation of gas within a liquid stream that occurs when pressure drops below the vapor pressure of the liquid; may result in excessive noise and pump damage. The condition in a hydraulic pump can result from restricted inlet flow or excessive oil viscosity.

Central air supply: An air-distribution system consisting of a single compressor station with compressed air delivered to workstations located throughout a facility.

Centralized grid: An air-distribution system consisting of a network of branch lines extending from a central trunk line. The branch lines supply workstations located throughout a facility.

Centrifugal pump: A pump using centrifugal force to create flow. A rotating impeller draws fluid into the pump where the fluid is moved by centrifugal force to the outer edges of the impeller and directed to the pump discharge.

Charles, Jacques: A French scientist of the late 18th and early 19th centuries who is credited with developing a scientific law relating to the effect of temperature on the volume of a gas.

Charles' law: A basic gas law that states the volume of a gas, held at a constant pressure, varies directly with the absolute temperature.

Check valve: A valve that normally allows fluid flow in only one direction.

Chipping hammer: A hand-held, pneumatic, reciprocating tool designed to chip material from concrete, masonry, plaster, or other substances.

Circuit: A group of electrical, hydraulic, and/or pneumatic components assembled to produce a desired movement or sequence of movements.

Circuit diagram: A group of symbols combined to represent the components and interconnecting lines in a fluid power system.

Cleanout: A removable cover used in a hydraulic system reservoir to allow easy cleaning of the vessel.

Clevis mount: A cylinder rod and cap mounting configuration involving a C-shaped casting and a mounting pin that allows the cylinder to pivot during extension and retraction.

Closed circuit: A hydraulic circuit design in which pump output is returned directly to the pump inlet after passing through a hydraulic motor. The design is commonly used with hydrostatic drive systems.

Compact hydraulic unit: A hydraulic system design developed in the early 20th century in a move away from large, centralized systems to compact hydraulic pumps that often serve a single machine or actuator.

Component group: A group of commonly designed hydraulic or pneumatic components offered in a full range of sizes with appropriate accessories.

Composite symbol: A symbol that indicates a component containing more than one valve function. A simple example is a flow control valve with a built-in check valve.

Compound relief valve: A relief valve containing both a balancing piston and a small, direct-operated relief valve. The two designs work together to produce a valve that operates efficiently.

Compressed air unit: Commonly used to designate a pneumatic compressor station that includes a prime mover, compressor, reservoir, and pressure control components.

Compression fitting: A fitting style used with both metal and plastic tubing that obtains a seal by tightening a component of the fitting, which tightly compresses around the tubing.

Compressor: The pneumatic system device that converts the energy input of the system prime mover into high-pressure compressed air.

Compressor-capacity control: The method used to control compressor operation to obtain required air volume and pressure. Several methods may be used, depending on the demands placed on the pneumatic system.

Conduction: A basic term used with both heat and electricity relating to the transfer of heat or electricity through the material.

Conductor: The element used in fluid power systems to allow the movement of hydraulic fluid or compressed air from component to component. Hose, pipe, and tubing are the most common fluid power conductors.

Contaminant: Foreign materials carried in the liquid or air of a fluid power system: dirt, water, metallic particles, paint, and a variety of other materials that enter during system operation.

Control element: A valve component that functions to allow, direct, meter, and/or stop the flow of fluid in a control valve to obtain desired operation.

Control mechanism: A device used to control the operation of a control valve. This may include mechanically, pilot-, solenoid-, and manually operated devices.

Control valve: Any valve designed to control fluid pressure, flow rate, or actuator direction in a fluid power system.

Convection: A basic method of heat transfer involving the movement of heated air caused by density differences of air of different temperatures.

Cooler: A pneumatic system dryer that uses a refrigeration element to lower the temperature of system air for the purpose of removing moisture.

Corrosion inhibitor: A compound added to hydraulic fluid that has been formulated to resist chemical reactions caused by contaminants entering a hydraulic system.

Corrosion resistance: The ability of a material to resist damage from chemicals that may be formed when contaminants enter the fluid of a hydraulic system.

Counterbalance valve: A hydraulic system pressure control valve used as a safety device to prevent the movement of a loaded actuator without the application of additional pressure generated by the system power unit.

Coupler: A mechanical device used to directly connect the output shaft of the prime mover to a hydraulic pump or pneumatic system compressor.

Coupling: A general term used for devices that connect system components such as fluid conductors and power transmission shafts.

Cracking pressure: The pressure level required to produce the first fluid flow through a pressure control valve.

Crescent design: A low-pressure hydraulic pump using an internal gear design in which the pumping chamber is produced by a crescent-shaped element positioning the internal- and external-toothed gears.

Cup seal: A seal shaped like a cup and often used on cylinder pistons. Application of pressure to the lip edge of the seal assures a tight seal. Initial development in the late 19th century was critical to the development of industrial hydraulics.

Current: The movement of electrons through a conductor. Current is measured in amperes.

Cushioning: A design feature in fluid power cylinders that reduces fluid flow near the end of the extension or retraction stroke to decelerate piston movement, which avoids both noise and component damage.

Cutaway symbol: Represent fluid power components in schematics using miniature cutaway drawings of the basic internal structure of valves and other parts.

Cylinder: The common term for linear actuators used to generate movement and force in fluid power systems.

Cylinder barrel: The tube-shaped portion of a cylinder (linear actuator) in which the piston and rod reciprocate.

D

da Vinci, Leonardo: An Italian painter, architect, and scientist of the late 15th and early 16th centuries who made extensive contributions to the engineering fields.

De facto standard: A standard that has become generally accepted without specific government or industry endorsement.

Deceleration circuit: A speed-control circuit that involves controlled slowing of an actuator from a maximum speed to a selected feed rate.

Deceleration valve: A control valve specifically designed to allow maximum, deceleration, and feed flow rates. The valves often use mechanical activation and a check valve that allows an unrestricted return flow rate.

Decentralized grid: An air-distribution system consisting of multiple grids that serve small portions of a facility. These grid sections contain smaller compressors, but may be interconnected to provide air movement between grids to balance intermittent, high air demands.

Decompression valve: A valve designed to release pressurized hydraulic fluid in large press or other installations where high pressure may be retained in the system.

Degrees: Typically associated with temperature scales in technical fields. Care must be taken to be certain appropriate conversions are made between scales to assure usable results.

Demulsifier additive: Improves the surface tension characteristics of petroleum-based hydraulic fluids to decrease the time needed for the separation of water from the hydraulic fluid in the reservoir.

Depth-type filter: Fluid filter that uses the depth of the filter material to provide many passageways to catch and hold contaminants.

Desiccant: A material in compressed-air dryers that uses the principle of adsorption to remove moisture from the air.

Detents: Mechanical device built into a directional control valve to hold the valve spool in a selected position.

Dew point: The temperature at which the water vapors in air will begin to condense (form dew).

Diaphragm: A flexible membrane used in various fluid power components to separate adjoining chambers that need to vary in volume. Often found in hydraulic accumulators, pneumatic pressure regulators, and low-capacity pumps and compressors.

Diaphragm-chamber regulator: A pneumatic system regulator that uses a flexible diaphragm between the outlet and control chambers of the valve.

Direct current: Electrical current with a continuous, one-direction current flow.

Directional control valve: Hydraulic and pneumatic control valves that direct fluid flow to start, control direction of movement, and stop system actuators.

Direct-operated regulator: A basic design of a pneumatic system pressure regulator in which the diaphragm is directly exposed to the pressure of the outlet chamber.

Direct-operated valve: A fluid power pressure control valve in which the operating pressure of the valve is directly dependent on physical force generated by a control spring.

Displacement: The physical volume of a chamber. In fluid power applications, displacement is important in calculations involving cylinders, motors, pumps, and compressors.

Double-acting compressor: A compressor design in which air intake and compression are completed in chambers located on both the top and underside of the compressor pistons. These compressors were common in larger industrial installations, but are being replaced by current rotary designs.

Double-acting cylinder: Cylinders that may be powered both on the extension and retraction strokes.

Double-rod-end cylinder: A cylinder that has rods attached to both sides of the piston providing equal-displacement chambers that produce equal extension and retraction speeds, as well as two attachment points to the cylinder.

Draincock: A valve located at the lower end of a drop line in a pneumatic distribution system to allow removal of condensed water. A plug-type valve is often used for this application.

Drain return line: A line used to return to the reservoir fluid drained from externally-drained components.

Drop line: A pipe coming from the top of the air distribution lines to a system workstation. Tapping the line into the top of the distribution line prevents condensed water from entering the drop line.

Dry air: Pneumatic system air with a saturated vapor pressure well above the ambient temperature.

Dryer: The pneumatic system component designed to remove water vapor from the compressed air. The unit is usually located in the compressor station area and may use refrigeration, chemical, or mechanical means to reduce the water content of the air.

Dry filter: A pneumatic system filter that uses a dry element to remove dirt and debris from air at the compressor inlet and other system locations.

Drying: The process of removing moisture from pneumatic system air.

Dryseal standard pipe threads: A type of pipe thread that provides a positive seal, preventing leakage of hydraulic oil through a threaded pipe connection.

Dual pump: A design in which two pumps are built into a single case. The pumps may be rated at different pressures and flow outputs to provide additional flexibility.

Dust mask: Personal protective device designed to remove dust and other airborne particles from the air an individual breathes; generally not suitable for removing fine particles and vapors.

Dynamic compressor: A device that compresses air or other gas using rotating vanes or impellers. These moving components increase pressure by converting the energy in the high velocity air to pressure.

Dynamic seal: A seal between moving parts. These seal designs are subject to wear because the seal must contact the moving surfaces.

E

Earmuffs: Devices designed to protect the hearing of the user by reducing the decibel level of sound reaching the inner ear. The device is made from sound-deadening materials that snuggly fit over the ears.

Earplugs: Devices designed to protect the hearing of the user by reducing the decibel level of sound reaching the inner ear. The device is inserted into the outer ear canal.

Effective area: The area of a surface that contributes to the force generated by the pressure of a fluid acting on the surface.

Effective piston area: The area of a piston that contributes to the force generated by system pressure. For example, the effective area of a cylinder piston during retraction is the area of the piston minus the cross-sectional area of the piston rod.

Electrical conductor: A material that will carry electricity from one point to another. Copper is used in most basic electrical circuits.

Electrical control: Control of a variety of fluid power component and system functions ranging from simple manual on/off electrical switches to complex servomechanisms.

Electrical insulator: A material that does not allow the transmission of electricity. Ceramic, rubber, and plastic are examples of insulators.

Electrical potential: The difference in electrical charge between two points in an electrical circuit; measured in volts.

Electromagnetic induction: The generation of electrical current flow in a wire by the movement of magnetic lines of force across that wire.

Electron: One of the three basic parts of an atom. These negatively-charged portions of the atom orbit the nucleus containing neutrons and protons.

Emergency shower: A shower bath located in a work area designed to allow a worker to immediately flood their body and clothing with water if exposed to corrosive or toxic chemicals.

Emulsion: A stabilized mixture of oil and water that typically has a milky appearance. An example is fire-resistant hydraulic fluid, which is classified as oil-in-water or water-in-oil.

Energy: The ability or potential to do work.

Energy conversion device: A device that has the ability to convert energy from potential to kinetic or kinetic to potential.

Entrained air: Air that has been churned into the fluid in a hydraulic system. Can shorten the life of the fluid by increasing oxidation or reduce system performance by producing spongy actuator operation.

Equipment maintenance: Service and repair work applied to equipment to assure expected performance and service life.

External force: Force applied to a component or system from an outside source.

External-gear pump: A hydraulic pump design that uses gears with external teeth to move fluid through the pumping chamber of the unit.

Extreme-pressure agent: An additive blended into many hydraulic fluids to prevent direct contact of the surfaces of bearings and gear teeth when operated under heavy load conditions.

Eye protection device: Protective wear ranging from eyeglasses to full face shields that are designed to prevent eye injuries from caustic fluids, flying debris, or harmful light rays.

Eyewash station: A small sink located in a work area and designed to allow workers to flood their eyes with water if exposed to corrosive or toxic materials.

F

Face mask: A protective device such as a clear plastic shield designed to protect the face from flying debris while operating machinery or tools.

Feedback control: A control system that uses information from system output to vary the input function to maximize performance.

Ferrule: The ring used on a hydraulic hose fitting to secure the hose to the threaded portion of the fitting.

Film: A thin layer or sheet of material. In fluid power systems, this may be a layer of oil or a solid sheet of material used to separate materials or protect the surface of component parts.

Filter: A fluid power component that is used to remove solid contaminants from both air and liquid. The term is often used to designate the complete component including the filter element, case, and mounting hardware.

Filter-element-condition indicator: A device located on many filter designs that indicates the condition of the filter by measuring the pressure drop across the filter element.

Filtration system: Typically refers to the placement of the filter in the system. Filter placement includes pump inlet, return line, and working line.

Finned conductor: A simple form of heat exchanger consisting of a metal tube surrounded by thin metal disks that dispense heat to the surrounding air. Often used as an intercooler between the cylinders of two-stage compressors.

Fire point: The lowest temperature at which a volatile substance vaporizes rapidly enough to produce an air-vapor mixture that will continuously burn when ignited.

First aid kit: A case containing a variety of basic first aid items suitable for emergency situations. These kits should be readily available in work areas and need to be maintained on a regular schedule.

First class lever: A lever that functions with the fulcrum between the load and effort arms.

First Law of Thermodynamics: States that energy cannot be created or destroyed during a process, although it can be changed from one form to another.

Fitting: A small accessory part with standard dimensions used to assist the assembly of major component parts into systems. An example is parts used to facilitate the connection of hydraulic hose to valves and actuators.

Fixed-centerline mount: A cylinder-mounting design in which the load carried by the cylinder rod and piston is supported at the centerline of the cylinder barrel, which is fixed to a machine member.

Fixed-delivery pump: Any hydraulic pump design that has a nonadjustable fluid delivery rate.

Fixed-displacement motor: Any hydraulic motor that does not allow the displacement per motor revolution to be adjusted. Varying the fluid flow rate through the motor controls the speed of these designs.

Fixed-noncenterline mount: A cylinder-mounting design in which the load carried by the cylinder rod and piston is transferred through the cylinder barrel to off-center mounts fixed to a machine member.

Fixed orifice: A hole of fixed size in a component part, allowing passage of a measured amount of fluid.

Flared fitting: A fitting design used with metal tubing that involves flaring the end of the tube. A mechanically tight seal is obtained using a fitting with a matching angle and a compression nut. Fittings are available with 37° and 45° flares.

Flash point: The lowest temperature at which vapors from a volatile substance will momentarily ignite when exposed to an open flame.

Flexible conductor: Fluid transmission lines in hydraulic and pneumatic systems that permit at least limited flexing or bending. These may range from flexible pipe joints that offer limited movement to hose and plastic tubing that offer almost unlimited movement.

Flexible hose: Hoses used in both hydraulic and pneumatic systems that can operate under full system pressure while providing maximum freedom of movement.

Flow control valve: A component designed to provide control of the rate of flow of air or liquid in hydraulic and pneumatic systems.

Fluid: A substance such as air, water, or oil that easily flows and tends to assume the shape of the container in which it is stored.

Fluid compressibility: The degree to which a set volume of fluid will be reduced when subjected to an outside force. Gases are considered compressible, while liquids are considered noncompressible.

Fluid conditioning: The methods used to provide a fluid that is clean and maintained at an acceptable operating temperature in a pneumatic or hydraulic system.

Fluid conductor: The lines used in a fluid power system to route the fluid from component to component in order to obtain the desired operation.

Fluid filter: The system component designed to remove contaminants from fluid power systems. The term typically refers to the filter element, case, and any mounting accessories.

Fluid flow weight: One of the basic factors involved in the hydraulic horsepower formula. Equal to the weight of the fluid output of the pump.

Fluid power: The transferring, controlling, and converting of energy using hydraulic and pneumatic systems.

Fluid Power Distributors Association (FPDA): An organization of companies primarily involved in the distribution of fluid power products. The organization promotes the industry through trade shows, brochures, and numerous educational programs.

Fluid Power Foundation (FPF): An organization that coordinates, supports and promotes education in fluid power. The foundation works closely with the NFPA, the FPDA, and the IFPS.

Fluid turbulence: Fluid flow through a line in which the movement of the fluid is irregular, causing increased flow resistance.

Flux: The lines of force that surround a permanent magnet or electromagnet.

Force: An influence on a body that causes it to accelerate (move).

Four-ball method: An ASTM test procedure designed to determine the wear-prevention characteristics of hydraulic fluids. The test involves rotating a steel ball under load against three stationary balls covered by the test fluid.

Four-way directional control valve: A directional control valve providing four separate fluid-flow patterns, allowing a cylinder to be powered during both extension and retraction.

Four-way valve: A term commonly used to indicate a four-way directional control valve.

Free air: Air displaying the characteristics of the atmospheric air at a specific location.

Friction: The force that retards or resists the movement of two touching materials.

FRL unit: A component group providing air filtration, pressure regulation, and lubrication for a specific pneumatic system workstation.

Fulcrum: The support point of a lever about which the load and effort arms pivot.

Full-flow filtration: A hydraulic fluid filtration system that filters all pump output before returning it to the system reservoir.

Full-flow pressure: The pressure required to force all system fluid through a pressure control valve at a given setting.

G

Gas-charged accumulator: A component containing a sealed chamber pressurized with an inert gas and used to store hydraulic fluid under pressure.

Gate valve: A shut-off valve that provides straight-line flow between the valve inlet and outlet ports. Provides minimal pressure drop when fully open.

Gauge pressure: A pressure scale where the zero point is equal to atmospheric pressure.

Gay-Lussac's law: States principles regarding the changing of a gas from one pressure and temperature condition to another while holding the volume constant.

Gear pump: A hydraulic pump design that uses gears to move fluid through the pumping chamber of the component. Gears may have internal or external teeth.

General consent standard: A standard that has not been developed by a standardizing organization, but is generally followed by an affected group.

General gas law: A general equation that expresses the relationships between pressure, volume, and temperature when applied to a gas ($P_1 \times V_1 \div T_1 = P_2 \times V_2 \div T_2$).

General hydraulic horsepower: A computed figure used in hydraulic systems based on fluid flow and pressure. Equal to fluid flow weight in pounds per minute multiplied by the pressure in feet of head divided by 33,000 foot pounds per minute.

General specification: Descriptions of fluid power components or systems providing general details of size, power, and operation.

Gerotor design: A common internal-gear pump or motor design. The pumping chamber of these units has two gear-shaped elements that form varying-size chambers as they rotate.

Globe valve: A shut-off valve that consists of a movable disk and a stationary seat. The passageway through the valve is offset, which causes fluid turbulence as the flow rate increases.

Graphic symbols: The most common symbols used to represent component features, system construction, and system operation in fluid power diagrams. Consist of a series of lines and shapes that represent specific components or features.

Gripper: An actuator commonly used with robots to handle parts in machining or assembly operations.

Ground: An electrical conducting path between an electrical circuit and Earth or another conducting body serving in place of Earth.

H

Head: The height of a column of water or other liquid necessary to develop a stated pressure.

Hearing-protection device: Protective wear, including earplugs and earmuffs, designed to prevent injury to the inner ear that could result in hearing loss.

Heat: Kinetic energy indicated by rapid molecular movement in a substance. Fluid power systems generate considerable heat during operation.

Heat exchanger: A component part in a fluid power system that adds heat to or subtracts heat from the system to keep the operating temperature within the desired range.

Heat insulator: Substances that do not readily transfer heat. Substances that contain microscopic chambers filled with gas, such as foam plastic, are generally considered good heat insulators.

Helical gear: A gear with teeth set at an angle to the axis of the gear. The design runs quietly, but has the disadvantage of producing a side thrust.

Hero: An individual who during the second century BC contributed to the development of basic scientific principles related to fluids.

Herringbone gear: A gear tooth design used in some gear pumps that is the equivalent of two helical gears placed side by side with the teeth forming a V pattern. The tooth pattern produces quiet operation with no side thrust.

High-low pump circuit: A circuit using two fixed-displacement pumps. The pumps are connected, with appropriate pressure control valves, to provide a high-volume flow at low pressure and a low-volume flow when high pressure is required to operate a system.

Horsepower: A standard unit of power equal to 550 foot-pounds of work per second or 33,000 foot-pounds per minute.

Hose: A flexible conductor for carrying hydraulic fluid or compressed air to actuators.

Hose-end fitting: Metal or plastic parts designed to provide a strong and leakproof connection between the end of a hose and a system component or rigid conductor.

Housekeeping: A general term used to indicate the process of keeping a workplace clean, orderly, and free from clutter.

Humidity: Atmospheric water vapor content expressed using several different measures.

HWCT: High-water-content hydraulic fluids that are primarily water with 2% to 5% soluble chemicals.

Hydraulic accumulator: A hydraulic system component capable of storing liquid under pressure. Uses mechanical or compressed-gas elements to maintain system pressure.

Hydraulic fuse: Devices used to prevent excess pressure or uncontrolled fluid loss if a line should rupture.

Hydraulic intensifier: A component that increases the pressure in a selected section of the system over that of the system relief valve. The flow rate in the intensified pressure section is reduced in order to achieve the increased pressure.

Hydraulic motor: A hydraulic system actuator that produces rotary motion. Circuits may be designed to produce both clockwise and counterclockwise rotation.

Hydraulics: The study and technical application of liquids in motion, especially oils and other blended liquids used in industrial systems.

Hydrostatic drive: A fluid power drive system using a hydraulic pump and motor to transmit the power of a prime mover to the input of a machine. Available in either open- or closed-circuit designs.

Hydrostatic transmission: A hydrostatic drive in which the pump, motor, or both have variable displacement. Drives with these features can be manually or automatically adjusted to control, torque, speed, and power.

I

Immersion heat exchanger: A component that heats or cools the fluid of a hydraulic system using water, electrical-resistance heaters, or steam. Usually located in the reservoir of the system.

Impact wrench: A portable hand tool, often operated by compressed air, used to install or remove bolts or nuts with a series of rapid-impact, rotary blows.

Impeller: The rotating component of a dynamic compressor or impeller pump. Energy is increased in the fluid as centrifugal force causes flow through the unit.

Inclined plane: One of the six basic machines. The inclined plane is a slope used to reduce the forced needed to lift an object.

Industrial Revolution: A period extending from the 1700s through the 1800s in which rapid change occurred in industry, including the development of many fluid power concepts, components, and systems.

Inertia: The tendency of all objects to remain either at rest or moving in the same direction, unless acted on by an outside force.

Inlet line: The conductor that provides system fluid to a component.

Inline design: A design of hydraulic axial piston pump in which the pistons are parallel with the axis of the power input shaft of the pump.

Installation drawing: Detailed drawings of components showing outside dimensions suitable for incorporating the components into equipment.

Intake-line filters: These filters are available in two forms: sump strainers (course filtration) and suction filters (fine filtration; located outside of the sump).

Integral construction: When two or more valves share the same body casting. An example is a sequence valve with an integral check valve.

Intercooler: A heat exchanger located between the stages of a two-stage pneumatic compressor. The intercooler reduces the temperature of the compressed air, thus increasing the performance of the compressor.

Internal force: The forces generated by internal valve components or by fluid pressure in the internal passageways of a valve. These forces are used to position valve elements to obtain a desired operation.

Internal gear pump: A hydraulic pump design that uses an internal-tooth gear operated by an external-tooth gear to move fluid through the pumping chamber. Common internal gear pumps use gerotor or crescent designs.

International Electrotechnical Commission (IEC): An international standardizing group that has the primary responsibility for electrical and electronic engineering. ANSI is the US representative on this commission.

International Fluid Power Society (IFPS): A group of individuals concerned with fluid power in a wide variety of technical fields. Operates a certification process requiring both a written and hands-on test.

International Organization for Standardization (ISO): An international agency for standardization located in Geneva, Switzerland. It includes the recognized national standards bodies of approximately 110 countries of the world.

Inverted emulsion: An emulsion in which oil surrounds finely divided water droplets. This principle is used in some forms of water-and-oil fire-resistant hydraulic fluids.

Ionosphere: The uppermost layer of the atmosphere, which extends from approximately 50 miles to 360 miles above Earth's surface.

Isothermal: A process completed without a change in temperature.

Isothermal compression: Compression in which any heat produced by compressing the gas is removed, producing a final pressure based only on the decrease in volume.

Isothermal expansion: Expansion of a gas in which heat is added to produce a final pressure based only on the increase in volume.

J

Jet pump: A pump that uses a variation of Bernoulli's principle to move fluid. The design forces a liquid through a nozzle placed in a venturi to produce a low-pressure area, which draws additional fluid from the primary fluid source.

K

Kinematic viscosity: A precise indicator of the viscosity of a liquid. The rating is based on the time required for a fixed amount of a fluid to flow through a calibrated viscometer under a fixed pressure and temperature.

Kinetic energy: The energy a body possesses because of its motion or velocity.

L

Laminar: Fluids arranged in distinct, thin layers.

Laminar flow: Nonturbulent flow in layers of a viscous fluid through a conductor.

Land: The full diameter portions of the control spool in a directional control valve. The annular clearance between lands directs fluid flow. The lands also block passageways through the valves to obtain desired operating characteristics.

Latent heat: The amount of heat absorbed or released by a substance without changing temperature during a change of state.

Latent heat of fusion: The amount of latent heat involved when a substance is changed from a liquid to a solid or a solid to a liquid.

Latent heat of vaporization: The amount of latent heat involved when a substance is changed from a liquid to a vapor or a vapor to a liquid.

Legal regulation standard: A standard established by government legislation. These standards typically involve safety issues.

Lever: One of the simple machines. Consists of a rigid bar pivoted about a fulcrum. Used to multiply a force or movement.

Lever arm: The part of a lever that extends from the fulcrum to the point at which force is applied.

Limited-rotation actuator: An actuator design that primarily produces rotational movement of one revolution or less. Various designs are available using a rack and pinion, vane, or helical shaft.

Linear actuator: A term often used to indicate a hydraulic or pneumatic cylinder. It converts fluid pressure and flow into linear mechanical force and movement.

Load arm: The part of a level that extends from the fulcrum to the location of the load.

Load factor: An estimate of expected actuator loads used to determine the air consumption needed to size air-distribution lines for a pneumatic system.

Lobe pump: A hydraulic pump that uses specially-designed elements with lobes to provide a pumping chamber. Liquid moves through the unit very much like an external-gear pump design.

Lobe-type compressor: Usually provide only low-pressure air. Construction involves specially-designed elements with lobes that provide a sealed pumping chamber. Air is swept from the compressor inlet to the outlet as the lobes rotate. Often referred to as blowers.

Lock-out device: A safety device designed so a machine cannot be accidentally turned on while maintenance work is being performed.

Logic function: A general term indicating several basic electronic or pneumatic circuits that perform problem-solving tasks.

Loop system: A hydraulic system design that allows continuous circulation of fluid through a motor and pump using only a low-volume system reservoir.

Lubricant: A substance used to reduce friction between moving surfaces, such as those found in bearings.

Lubrication: A method to reduce friction and minimize wear of parts moving in close proximity. Usually involves oil or other substance as a lubricant.

Lubricator: A device designed to apply a lubricant to a bearing surface. In pneumatic systems, the lubricator is a component of the FRL unit located at the workstation.

Lubricity: The ability of a lubricant to form a strong film between two bearing surfaces and adhere to those surfaces to reduce friction and wear.

M

Magnetic poles: The areas at the ends of a magnet that produce magnetic lines of force. The lines of force attract ferrous metals or may be used to generate electricity.

Magnetism: The phenomena produced by the magnetic lines of force surrounding natural or electromagnets.

Manifold: A fluid power component designed to provide branch connections to several conductors or valves to facilitate construction of circuits or systems.

Material agitation: The use of compressed-air systems to mix or compact materials in process, packaging, or other industries.

Material transfer: The use of compressed-air systems to move powder or small-particle materials through tubes from one process or storage point to another.

Maudslay, Henry: The individual generally considered to be the father of the machine tool industry. Maudslay lived during the late 1700s and early 1800s and co-developed the cup seal, which was critical for the success of hydraulic systems.

Maximum operating pressure: The highest pressure recommended for operation of a hydraulic or pneumatic system.

Mechanical advantage: The ratio of the force produced by a machine to the applied input force.

Mechanical control: Mechanical mechanisms to operate control valves, including levers and ramps (inclined planes).

Mechanical coupler: A mechanical device that connects the prime mover to the hydraulic pump or pneumatic compressor.

Mechanical efficiency: The theoretical horsepower needed to operate a pump or compressor divided by the actual horsepower required.

Memory circuit: A pneumatic control circuit that will hold an actuator in a selected position after only a momentary input signal.

Mesosphere: The portion of the atmosphere that extends from the stratosphere to the ionosphere.

Metal rolling: A metal-forming technique that is used by some manufacturers to attach the barrel to the cap and rod ends of the cylinder.

Meter-in circuit: A fluid power flow-control circuit in which fluid is metered into an actuator to control operating speed.

Meter-out circuit: A fluid power flow-control circuit in which fluid is metered out of an actuator to control operating speed.

Micron: A unit of measure equal to one millionth of a meter. Commonly used to designate the size of particles that filter media will remove.

Mill cylinder: A hydraulic cylinder constructed of heavy steel for use in industries such as foundries and steel mills.

Moisture separator: A device sometimes teamed with the compressor aftercooler to increase the effectiveness of the moisture-removal process during pneumatic system air compression.

Molecule: All substances are structured of molecules composed of atoms. Atoms are composed of electrons, neutrons, and protons.

Motion-converting mechanism: A device that converts rotary motion into linear motion or linear motion to rotary motion.

Motor: An actuator that primarily produces continuous, rotary motion. These components may be powered by electricity, hydraulics, or pneumatics.

Motor braking circuit: A fluid power motor circuit that uses back pressure to slow a hydraulic motor to a full stop without excessive system shock pressures.

Muffler: A device designed to reduce the level of noise produced at the air intake of a compressor or the exhaust of pneumatic control valves and motors.

Multiple maximum system operating pressures: Fluid power systems designed to allow a number of maximum operating pressures by using remote pressure valves to control the operation of a primary hydraulic relief valve or pneumatic regulator.

Multiple-pressure circuit: A circuit that provides for selection of several different, preset system pressure levels.

Muscular control: A valve-control element that depends on manually applied force to shift the valve into a desired operating position.

N

Nail driver: A portable pneumatic tool used to drive nails.

National Fluid Power Association (NFPA): A nonprofit trade association and standards developer serving manufacturers of fluid power products, including both hydraulics and pneumatics.

Needle valve: A common flow control and shutoff valve used in both hydraulic and pneumatic circuits. The valve contains a fixed orifice and an adjustable screw with a tapered end that is aligned with the center of the orifice.

Neutron: One of the three basic parts of an atom. Neutrons have neither a negative nor positive charge. Neutrons and protons form the nucleus of the atom.

Nominal rating: A rating that involves an approximation of performance rather than an exact rate.

Nominal sizing: A sizing system in which the actual dimension is approximately the same as the size stated. The measured dimension is uniform, but does not exactly match the stated size.

Nonintegral construction: Construction in which only a single valve is housed in a body casting rather than placing two or more basic valve functions in a single housing.

Non-positive-displacement pump: A pump that does not have a variable-volume pumping chamber. An impeller or other device is used to move the fluid. The inertia of that fluid movement produces pressure when flow is resisted.

Normally closed: Refers to a pressure control valve in which the internal elements are closed when the system is shut down, blocking fluid passage through the valve.

Normally open: Refers to a pressure control valve in which the internal elements are open when the system is shut down, allowing fluid free passage through the valve.

Normal operating position: The position that the internal elements of a control valve assume when a fluid power system is shut down.

Nucleus: The center portion of an atom, which is made up of protons and neutrons.

Nut setter: A specialized pneumatic impact wrench designed for assembly line work. The device installs nuts on bolts and torques them to a specified tightness. Available in single- and multiple-driver heads.

O

Occupational Safety and Health Administration (OSHA): Federal agency that establishes requirements for safety, including safe and healthy working conditions and the encouragement of employees to follow safe working practices.

Off-line filtration: A hydraulic filtration method that continuously pumps fluid out of the reservoir, through a filter, and back to the reservoir.

Ohm's law: Describes the fundamental relationship between voltage, current, and resistance. This is expressed as I (amps) = E (volts) ÷ R (ohms).

Oil-bath filter: An air intake filter design used in pneumatic systems that involves a filter medium constantly bathed in oil. The oil catches the dust and dirt and transports it to a sump where it is removed during periodic service.

Oil-in-water emulsion: High-water-content, fire-resistant hydraulic fluids using approximately 5% soluble oil and 95% water. Other names that may be applied are *high water content fluids (HWCF)* and *5-95 fluids.*

Oil-wetted filter: An air intake filter design using a filter medium made from a wire-screen form packed with oil-coated metal material that removes dust, dirt, and other debris.

One-piece cylinder: Usually describes cylinders that have permanently attached ends (head and cap). They are considered throw-away items as they cannot be disassembled for service.

Open circuit: A hydraulic circuit that uses the layout of a basic hydraulic motor circuit with a directional control valve to control motor direction and a reservoir to hold surplus fluid.

Orbiting gerotor motor: A variation of the gerotor motor that uses the internal-toothed gear of the gerotor set as a fixed gear. The external-toothed gear orbits following the internal-toothed gear. This produces higher torque/lower speed output.

Orifice: A precision-drilled hole in a component that is used to meter a desired flow rate through a flow control valve or the control section of other valve types.

Overall efficiency: The calculated efficiency of a hydraulic pump considering all energy losses. Overall efficiency is equal to the volumetric efficiency of the pump multiplied by the mechanical efficiency divided by 100.

Oxidation inhibitors: Materials added to hydraulic fluids to slow the complex chemical reactions that oxidize the fluid.

Ozone layer: A layer of ozone that is located in the upper limits of the stratosphere. The layer acts as a filter to reduce the amount of ultraviolet energy that reaches Earth's surface.

P

Packed-bore design: A design used with pneumatic directional control valves in which resilient seals are placed in annular grooves machined into the spool bore of the valve body.

Packed slip joint: A fitting recommended for use around the pump inlet line as it passes through the cover of the hydraulic system reservoir. The fitting absorbs system vibration as well as reduces water and dirt entering the system.

Packed-spool design: A design used with pneumatic directional control valves in which resilient seals are placed in annular grooves machined into the land surfaces of the valve spool.

Parallel circuit: An electrical or fluid power circuit that simultaneously provides multiple paths for the current or fluid to follow as it moves through a circuit.

Pascal, Blaise: A scientist of the early 17th century who is credited with producing proof of many basic fluid principles.

Pascal's law: A basic scientific law that states pressure applied to a confined, nonflowing fluid is transmitted undiminished to all points in the fluid.

Paving breaker: A pneumatic percussion tool used for breaking pavement, compacted soil, and a variety of other materials. The unit may be handled by an individual operator or mounted on mobile equipment.

Performance data: Information published by tool and equipment manufacturers specifying size and operating characteristics of a product.

Phosphate ester: A synthetic hydraulic fluid. It has a high flash point and provides excellent fire resistance.

Pictorial symbol: A symbol often used to depict a component in a fluid power circuit. The symbols often represent the external shape of components, but no standardized shapes have been identified.

Pilot-operated regulator: A pneumatic system pressure regulator that is controlled by pilot pressure from a remote location.

Pilot pressure: The fluid pressure needed by pilot-control elements to obtain expected component performance.

Pinion gear: The external-toothed gear on the output shaft of a hydraulic limited-rotation actuator that uses a rack and pinion design.

Pintle: A fixed pin that acts as the bearing for the rotating cylinder block in a radial-piston pump. The inlet and outlet ports for the pump cylinders are also machined into the pintle pin.

Pipe: Rigid tube made from steel with adequate wall thickness to be threaded. Often used as a conductor in fluid power systems.

Piston: A sliding cylinder that reciprocates in a bored circular hole in the body of a component. In a fluid power system, the piston is either moved by or moving against fluid pressure as it reciprocates.

Piston motor: A fluid power rotary actuator using reciprocating pistons and motion-converting mechanisms to produce rotational movement and torque.

Piston pump: A fluid power pump that converts the power and rotary motion of a prime mover into fluid flow, using reciprocating pistons and a motion-converting mechanism.

Piston rod wiper ring: A ring seal that scrapes the rod of a cylinder as it is retracted to prevent water, dirt, and other contaminants from being drawn into the cylinder.

Piston shoe: A component attached to the piston end in some piston pump and motor designs. The shoe is held in constant contact with the swash plate during operation.

Pivoting-centerline mount: A clevis or trunnion mounting that allows the cylinder to follow an arc as it powers a machine member. The load remains concentrated on the centerline of the cylinder.

Pneumatic reciprocating motors: Air-powered motors that produce a repeated back-and-forth, linear motion. For example, this type of motor powers a jackhammer used in street maintenance.

Pneumatics: The study and technical application of air or other gas in motion.

Polyglycol: A fluid similar in chemical make-up to automotive antifreeze. Used in fire-resistant, water-glycol hydraulic fluid.

Poppet valve: One of several designs used in fluid power control components in which a face on the valve seals against a seat located in the body.

Positive displacement: A pump or rotary actuator design with a positive seal between the inlet and outlet, which produces a constant volume of fluid for each revolution.

Positive-displacement pump: A pump that produces a fixed volume of fluid output for each revolution. Some designs are adjustable.

Potential energy: Energy that is due to position and not motion. For example, crude oil, compressed springs, and charged hydraulic accumulators contain potential energy.

Pour point: The lowest temperature at which a fluid will flow as defined by a standardized test procedure.

Pour point depressant: An additive used in hydraulic fluids to lower the pour point to permit cold-weather operation of hydraulic equipment.

Power: The rate at which work is performed or energy expended. Power = force × distance ÷ time.

Power unit: The unit in a hydraulic system that provides energy for the system, moves fluid through the system, provides a safe maximum limit of system pressure, and maintains desired system temperature and fluid cleanliness.

Precharging: The process of setting the pressure on the gas side of a gas-charged accumulator after
the pressure of the hydraulic side has been reduced to 0 psi.

Pressure: Force per unit area, which is usually expressed in *pounds per square inch (psi)* or *bar*.

Pressure balancing: A design technique used to reduce or balance uneven loading of bearings and other pump components to reduce pump wear and increase efficiency.

Pressure booster: A fluid power component that can increase pressure in a selected section of

a circuit above the general operating pressure of a system. A higher pressure is obtained by using two different-sized pistons with a low pressure acting on the larger piston.

Pressure compensation: An element in a flow control valve that maintains a constant pressure drop across the metering orifice of the valve resulting in consistent flow rates through the valve even with varying system loads.

Pressure controls: A variety of components designed to limit maximum system pressure (relief), sequence actuator movement (sequence), restrain movement (counterbalance), unload pump output (unloading), and provide reduced pressure (pressure reducing).

Pressure control valves: Control valves in a fluid power circuit designed to protect the system from damage from excessive pressure, provide motion control, and protect operators from unexpected actuator movement.

Pressure override: The difference between the cracking pressure and the full-flow pressure of a valve.

Pressure rating: The maximum operating pressure recommended by the manufacturer for a fluid power component or system.

Pressure-reducing valve: A hydraulic pressure control valve designed to maintain the pressure in a portion of a system at a lower level than the pressure set by the system relief valve.

Pressure regulator: A pneumatic system component used to control air pressure at the system workstations. The valve maintains a constant, reduced pressure for use by tools and equipment operated at the station.

Pressure safety valve: A valve used to protect a system from excessive pressure caused by valve failure or improper settings. Versions of these valves are used in both hydraulic and pneumatic systems.

Pressure surges: Unexpected increases in pressure in hydraulic systems caused by the inertia of fluid flow when valves close or loads suddenly increase or decrease.

Prime mover: The source of energy for any fluid power system. Commonly used prime movers are electric motors and internal combustion engines.

Prime mover horsepower: Rating used to indicate the power requirement of a fluid power system prime mover. Size calculations must include peak flow/pressure demands and the efficiency of the system pump.

Propeller pump: A non-positive-displacement pump consisting of a rotating, propeller-shaped pumping element located in a fluid line. Used primarily as a fluid transfer pump in hydraulic systems.

Proportional filtration: A hydraulic filtration system that directs on a continuous basis a measured amount of system fluid to the filter and then returns it directly to the reservoir.

Proton: One of the three basic parts of an atom. Neutrons and protons form the nucleus of the atom. The proton carries a positive electrical charge.

Pulley: One of the simple machines, which allows the direction of a force to be changed. When used as a movable pulley or used in multiple sets, a mechanical advantage is provided.

Pump: A hydraulic component turned by the prime mover that produces fluid flow, which transmits energy through the system.

Pumping chamber: The portion of a positive-displacement pump or compressor that creates the conditions necessary to draw in and then move fluid into the system. The chamber enlarges during the intake phase and reduces in size during the output portion of the cycle.

Pumping motion: The motion involved to change the volume of the pumping chamber, which acts to move fluid through a pump. Pumps are classified as either rotary or reciprocating.

Pump unloading control: A pump and unloading valve combination that allows the system pump to unload at near zero pressure when additional fluid is not needed to continue system operation. The combination reduces energy consumption without affecting system performance.

Q

Quick exhaust valve: A pneumatic control valve used to increase the rate air can be exhausted from a cylinder. Increasing the exhaust rate allows faster extension or retraction speeds.

R

Rack gear: A bar containing gear teeth on one side that is meshed with a pinion gear. Used in some limited-rotation actuators to rotate a pinion gear attached to the output shaft of the actuator.

Radial-piston pump: A positive-displacement piston pump design in which the pistons are located in a cylinder block or housing that is perpendicular to the power input shaft.

Radiation: Energy transmitted by waves through space. Earth's atmosphere serves as protection from radiation coming from space.

Radiator: A heat exchanger used in both hydraulic and pneumatic systems to transfer heat from the system to atmospheric air.

Ram: A single-acting linear actuator in which the diameter of the rod is basically the same diameter as the bore of the actuator barrel.

Rammer: A pneumatic, portable, reciprocating tool used to compact sand in foundry molds and other applications.

Rapid-advance-to-work circuit: A fluid power circuit design that allows a tool or other machine element to rapidly move to a preset position, slow to a point where machining is to occur, and then move at an accurate cutting speed. May also allow rapid retraction of the work elements.

Rated capacity: The maximum capacity for which a component is designed. This rating may include details related to fluid flow, pressure, and other factors.

Reaction ring: An internal component of a rotating-cylinder, radial-piston pump design. The ring is adjustable to change the stroke length of the pistons.

Receiver: A tank in a pneumatic system, located close to the compressor, that stores and assists in conditioning compressed air.

Reciprocating compressor: A common compressor design using a cylinder, piston, crankshaft, and valves similar to an internal combustion engine. The reciprocating action of the piston brings air into the cylinder, where it is compressed and then moved into the system.

Reciprocating-plunger pump: An early reciprocating pump design used for both high-volume and high-pressure applications. They are currently used primarily for very-high-pressure, low-volume applications.

Reciprocating pump: One of a variety of pump designs used to move fluid. The drive mechanisms vary, but a reciprocating piston moves the fluid against system actuator resistance.

Reduced-pressure section: A section of a fluid power circuit that functions at a pressure lower than the primary relief valve or pressure regulator setting.

Regenerative-cylinder-advance circuit: A hydraulic circuit design in which fluid discharged from the rod end of a cylinder combines with pump delivery and is directed to the cylinder blind end. The design results in increased cylinder extension speed, but reduced force capability.

Relative humidity: The relationship between the actual amount of water in the air and the maximum amount the air could hold at a given temperature.

Relief valve: A pressure control valve that bypasses excess hydraulic system pump delivery to the reservoir in order to limit system pressure to a selected maximum level.

Relieving-type regulator: A pneumatic system pressure regulator that contains a venting orifice. The vent prevents system pressure from exceeding the initial setting of the regulator, even when downstream conditions cause an increase in air pressure.

Remote location: A location for control components that is beyond the distance usually used for placement of components on machines.

Remote pressure adjustment: Adjustment of a pressure control valve from a location some distance from the valve installation.

Replenishment circuit: A circuit used with closed-loop hydraulic systems that provides makeup fluid to replace any fluid lost from leakage during system operation.

Reservoir: A component in a hydraulic system that holds the system fluid not currently in use in the pump, control components, actuators, and lines.

Resilient material: A material that has the ability to return to its original shape after being compressed. This type of material is often used for the sealing surfaces on valve faces or seats in pneumatic components.

Resistance: Opposition to the movement of electron flow in wires, fluid flow in fluid conductors, or a load placed on a fluid power actuator.

Restrictive check valve: A specialized hydraulic check valve that allows free flow in one direction and restricted flow in the opposite direction.

Restrictor flow control valve: A flow control valve design that meters the flow passing through the valve, while forcing excess pump output fluid to return to the reservoir through the system relief valve.

Return line: The hydraulic system fluid conductor that returns system fluid from an actuator to the directional valve and back to the reservoir as the system operates.

Revolving-cylinder design: A radial hydraulic pump design in which a cylinder and piston assembly rotate as a unit.

Reynolds, Osborne: A scientist of the late 1800s who did extensive development work in the fluid mechanics area. Reynolds is credited with identification of the principles of laminar and turbulent flow.

Reynolds number: A dimensionless number that is one of the basic parameters of fluid mechanics relating to fluid flow.

Rigid conductor: A fluid power conductor that is not intended to be bent. Steel pipe and some forms of tubing and plastic pipe are examples. Various fittings and adapters must be used to form these materials into required configurations.

Riveting hammer: A pneumatic, portable, reciprocating tool used to place and form the end of rivets.

Rock drill: A pneumatic tool often used to make rock cuts in quarries, mines, and highway construction. May use a drill or a percussion tool for cutting.

Rod: The rod that is attached to one side of the cylinder piston and extends through the cylinder rod end (head). The rod transfers the force generated by the piston to the system load.

Rotary actuator: A fluid power actuator used to produce torque and rotary motion. Commonly called a *hydraulic motor* or *pneumatic motor*.

Vane motor: A hydraulic or pneumatic actuator using a rotor with slots containing movable vanes to generate rotary motion and torque.

Vane pumps: Several designs of hydraulic pumps using a rotor with slots containing movable vanes. Rotating the rotor moves fluid through the pump, producing fluid flow in a hydraulic system.

Vapor pressure: The pressure at which a liquid will begin to form a vapor. Vapor pressure varies with fluid type and formulation.

Variable-displacement motor: A hydraulic motor design in which the displacement of the motor can be easily changed, allowing the speed to be changed without varying input fluid flow rate.

Variable-displacement pump: A hydraulic pump design where the displacement of the pumping chamber can be easily changed, which allows pump output to be varied.

Varnish: An undesirable material that forms on internal surfaces of hydraulic components when hydraulic fluids oxidize as they are exposed to high operating temperatures. Contributes to corrosion of metal surfaces and promotes the formation of sludge.

Velocity: The speed at which fluid moves through a line or component element. Also, the physical speed of a machine member. Expressed in terms of distance and time (feet per second).

Vent connection: An external fitting on a compound relief valve connected to the control chamber of the valve. Allows remote adjustment of that pressure to vary the maximum operating pressure of a system.

Venting: Connecting a portion of a fluid power component to the atmosphere in order to achieve or maintain atmospheric pressure in that section. Used in a reservoir to support pump inlet flow and valves for control purposes.

Venturi: A shaped restriction in a component that increases velocity and lowers the pressure of the fluid passing through the opening.

Viscosity grade: A rating indicating the viscosity of a liquid. Several different rating systems are available from standardizing groups. Each system provides consistent ratings within that system.

Viscosity-index-improver additive: An additive that may be blended with a hydraulic fluid to reduce the change occurring in viscosity as temperatures change. Especially important for systems that operate with extreme temperature changes.

Viscosity index number: A number that expresses the relative change in viscosity that can be expected for a given change in temperature of a liquid.

Viscosity: The internal resistance to flow of the molecules of a liquid.

Volumetric efficiency: A ratio of the volume of air admitted to a compressor cylinder at a given temperature and pressure to the full displacement of the cylinder.

von Guericke, Otto: A German scientist of the 17th century who conducted considerable experimentation with gases.

W

Water screw: An invention developed by Archimedes in the third century BC using a screw (simple machine) to raise water to a different elevation. It is still used today in some developing countries.

Water trap: A simple trap located at low points in pneumatic system distribution lines to collect water that has condensed from the air in the lines. Water removal may be manual or automatic.

Water vapor: Water in its gaseous form. Note: fog or mist is not water vapor, but atomized liquid water.

Water wheel: An early application of fluid power dating to ancient civilizations where moving water rotated a wheel to produce useable force.

Watt, James: An inventor of the late 18th century who extensively contributed to growth during the Industrial Revolution. Watt is typically identified as the inventor of the first practical steam engine.

Wear plate: Plates located on the end surfaces of gears and rotors in gear and vane pumps to promote sealing of the pumping chambers and to reduce wear on the components.

Wedge: One of the simple machines, consisting of two inclined planes that share a common base.

Weight: The result of the mass of a structure or body being acted on by gravity.

Weight-loaded accumulator: An accumulator design using physical weight acting on liquid in a vertical ram to store energy in a hydraulic system. The only accumulator design that provides a constant pressure as fluid is discharged.

Wheel and axle: One of the simple machines, consisting of a wheel attached to an axle. The common center of the wheel and axle is the fulcrum, allowing the device to act as a second or third class lever depending where the effort is applied.

Windmill: An early application of fluid power that some people trace to boat sails. Machines resembling our concept of a windmill appeared in early history in the Middle East. Modern windmills are still in use throughout the world.

Wiper seal: Prevents dirt and other contaminants from entering a fluid power system as the cylinder rod is retracted. Often referred to as a scraper.

Work: The application of force through a distance. Work only results when the applied force moves the object.

Appendix A

Conversions

Area

To Convert	Into	Multiply By
square centimeters	square inches	.155
square feet	square centimeters	929.0
square inches	square centimeters	6.452
square inches	square millimeters	645.2
square meters	square feet	10.76
square meters	square yards	1.196

Distance

To Convert	Into	Multiply By
centimeters	feet	.0328
centimeters	inches	.3937
centimeters	meters	.01
centimeters	millimeters	10.0
feet	centimeters	30.48
feet	inches	12.0
feet	meters	.3048
feet	millimeters	304.8
inches	centimeters	2.54
inches	feet	.0833
inches	meters	.0254
inches	millimeters	25.4
meters	feet	3.281
meters	inches	39.37
millimeters	centimeters	.1
millimeters	inches	.0394
millimeters	feet	.00328
millimeters	meters	.001
yards	meters	.9144

Energy

To Convert	Into	Multiply By
kilowatt-hours	Btu	3415.0

Flow

To Convert	Into	Multiply By
cubic feet/minute	barrels/day	256.5
cubic feet/minute	cubic feet/hour	60.0
cubic feet/minute	cubic meters/hour	1.699
cubic feet/minute	gallons/minute	7.481
cubic feet/minute	liters/minute	28.32
cubic feet/hour	gallons/minute	.1247
cubic feet/hour	liters/hour	.472
cubic feet/hour	cubic feet/minute	.0167
cubic feet/hour	cubic meters/hour	.0283
cubic meters/hour	gallons/minute	4.403
cubic meters/hour	liters/minute	16.67
cubic meters/hour	cubic feet/minute	.5886
cubic meters/hour	cubic feet/hour	35.31
cubic meters/hour	barrels/day	150.9
gallons (US)/minute	cubic feet/minute	.133
gallons (US)/minute	cubic feet/hour	8.021
gallons (US)/minute	cubic inches/minute	231.0
gallons (US)/minute	cubic meters/hour	.227
gallons (US)/minute	liters/minute	3.785
liters/minute	gallons (US)/minute	.2642

Mass

To Convert	Into	Multiply By
kilograms	pounds	2.205
ounces	grams	28.35
pounds	kilograms	.4536
tons (metric)	pounds	2205.0
tons (short)	pounds	2000.0

Power

To Convert	Into	Multiply By
horsepower	foot-pounds/minute	33,000.0
horsepower	foot-pounds/second	550.0
horsepower	watts	745.7

Pressure

To Convert	Into	Multiply By
atmospheres	bar	1.013
atmospheres	inches of mercury	29.92
atmospheres	kilograms/square centimeter	1.033
atmospheres	kilopascals (kPa)	101.3
atmospheres	psi	14.7
bar	atmospheres	.9869
bar	feet of water column (wc)	33.45
bar	kilograms/square centimeter	1.020
bar	kilopascals (kPa)	100.0
bar	psi	14.5
inches of mercury	psi	.4912
inches of water (4°C)	psi	.0361
kilograms/cubic meter	pounds/cubic foot	.06243
millimeters of mercury	psi	.0194
pounds/cubic foot	grams/cubic centimeter	.016
pounds/square inch (psi)	atmospheres	.068
pounds/square inch (psi)	bar	.069
pounds/square inch (psi)	inches of mercury	2.036
pounds/square inch (psi)	inches of water (4°C)	27.7
pounds/square inch (psi)	kilopascals (kPa)	6.895
pounds/square inch (psi)	kilograms/square centimeter	.0703

Temperature

To Convert	Into	Multiply By
Celsius	Fahrenheit	$9/5 C° + 32°$
Celsius	Kelvin	$C° + 273°$
Fahrenheit	Celsius	$5/9 (F° - 32°)$
Fahrenheit	Rankine	$F° + 460°$

Volume

To Convert	Into	Multiply By
cubic centimeters	cubic inches	.061
cubic feet	cubic meters	.028317
cubic feet	liters	28.317
cubic inches	liters	.01639
cubic meters	cubic feet	35.31
gallons (US)	liters	3.785
liters	cubic feet	.0351
liters	cubic inches	61.02
liters	cubic meters	.001
liters	gallons (US)	.2642

Appendix B

Metric-Inch Equivalents

Millimeters	Decimal Inch	Fractional Inch	Millimeters	Decimal Inch	Fractional Inch
0.100	.0039		3.969	.1563	5/32
0.100	.0039		4.000	.1575	
0.200	.0079		4.100	.1614	
0.300	.0118		4.200	.1654	
0.397	.0156	1/64	4.300	.1693	
0.400	.0157		4.366	.1719	11/64
0.500	.0197		4.400	.1732	
0.600	.0236		4.500	.1772	
0.700	.0276		4.600	.1811	
0.794	.0313	1/32	4.700	.1850	
0.800	.0315		4.763	.1875	3/16
0.900	.0354		4.800	.1890	
1.000	.0394		4.900	.1929	
1.100	.0433		5.000	.1969	
1.191	.0469	3/64	5.100	.2008	
1.200	.0472		5.159	.2031	13/64
1.300	.0512		5.200	.2047	
1.400	.0551		5.300	.2087	
1.500	.0591		5.400	.2126	
1.588	.0625	1/16	5.500	.2165	
1.600	.0630		5.556	.2188	7/32
1.700	.0669		5.600	.2205	
1.800	.0709		5.700	.2244	
1.900	.0748		5.800	.2283	
1.984	.0781	5/64	5.900	.2323	
2.000	.0787		5.953	.2344	15/64
2.100	.0827		6.000	.2362	
2.200	.0866		6.100	.2402	
2.300	.0906		6.200	.2441	
2.381	.0938	3/32	6.300	.2480	
2.400	.0945		6.350	.2500	1/4
2.500	.0984		6.400	.2520	
2.600	.1024		6.500	.2559	
2.700	.1063		6.600	.2598	
2.778	.1094	7/64	6.700	.2638	
2.800	.1102		6.747	.2656	17/64
2.900	.1142		6.800	.2677	
3.000	.1181		6.900	.2717	
3.100	.1220		7.000	.2756	
3.175	.1250	1/8	7.100	.2795	
3.200	.1260		7.144	.2813	9/32
3.300	.1299		7.200	.2835	
3.400	.1339		7.300	.2874	
3.500	.1378		7.400	.2913	
3.572	.1406	9/64	7.500	.2953	
3.600	.1417		7.541	.2969	19/64
3.700	.1457		7.600	.2992	
3.800	.1496		7.700	.3031	
3.900	.1535		7.800	.3071	
			7.900	.3110	

Millimeters	Decimal Inch	Fractional Inch	Millimeters	Decimal Inch	Fractional Inch
7.938	.3125	5/16	12.303	.4844	31/64
8.000	.3150		12.400	.4882	
8.100	.3189		12.500	.4921	
8.200	.3228		12.600	.4961	
8.300	.3268		12.700	.5000	1/2
8.334	.3281	21/64	12.800	.5039	
8.400	.3307		12.900	.5079	
8.500	.3346		13.000	.5118	
8.600	.3386		13.097	.5156	33/64
8.700	.3425		13.100	.5157	
8.731	.3438	11/32	13.200	.5197	
8.800	.3465		13.300	.5236	
8.900	.3504		13.400	.5276	
9.000	.3543		13.494	.5313	17/32
9.100	.3583		13.500	.5315	
9.128	.3594	23/64	13.600	.5354	
9.200	.3622		13.700	.5394	
9.300	.3661		13.800	.5433	
9.400	.3701		13.891	.5469	35/64
9.500	.3740		13.900	.5472	
9.525	.3750	3/8	14.000	.5512	
9.600	.3780		14.100	.5551	
9.700	.3819		14.200	.5591	
9.800	.3858		14.288	.5625	9/16
9.900	.3898		14.300	.5630	
9.922	.3902	25/64	14.400	.5669	
10.000	.3937		14.500	.5709	
10.100	.3976		14.600	.5748	
10.200	.4016		14.684	.5781	37/64
10.300	.4055		14.700	.5787	
10.319	.4063	13/32	14.800	.5827	
10.400	.4094		14.900	.5866	
10.500	.4134		15.000	.5905	
10.600	.4173		15.081	.5938	19/32
10.700	.4213		15.100	.5945	
10.716	.4219	27/64	15.200	.5984	
10.800	.4252		15.300	.6024	
10.900	.4291		15.400	.6063	
11.000	.4331		15.478	.6094	39/64
11.100	.4370		15.500	.6102	
11.113	.4375	7/16	15.600	.6142	
11.200	.4409		15.700	.6181	
11.300	.4449		15.800	.6220	
11.400	.4488		15.875	.6250	5/8
11.500	.4528		15.900	.6260	
11.509	.4531	29/64	16.000	.6299	
11.600	.4567		16.100	.6339	
11.700	.4606		16.200	.6378	
11.800	.4646		16.272	.6406	41/64
11.900	.4685		16.300	.6417	
11.906	.4688	15/32	16.400	.6457	
12.000	.4724		16.500	.6496	
12.100	.4764		16.600	.6535	
12.200	.4803				
12.300	.4843				

Millimeters	Decimal Inch	Fractional Inch	Millimeters	Decimal Inch	Fractional Inch
16.669	.6563	21/32	21.034	.8282	53/64
16.700	.6575		21.100	.8307	
16.800	.6614		21.200	.8346	
16.900	.6654		21.300	.8386	
17.000	.6693		21.400	.8425	
17.066	.6719	43/64	21.431	.8438	27/32
17.100	.6732		21.500	.8465	
17.200	.6772		21.600	.8504	
17.300	.6811		21.700	.8543	
17.400	.6850		21.800	.8583	
17.463	.6875	11/16	21.828	.8594	55/64
17.500	.6890		21.900	.8622	
17.600	.6929		22.000	.8661	
17.700	.6968		22.100	.8701	
17.800	.7008		22.200	.8740	
17.859	.7031	45/64	22.225	.8750	7/8
17.900	.7047		22.300	.8780	
18.000	.7087		22.400	.8819	
18.100	.7126		22.500	.8858	
18.200	.7165		22.600	.8898	
18.256	.7188	23/32	22.622	.8906	57/64
18.300	.7205		22.700	.8937	
18.400	.7244		22.800	.8976	
18.500	.7283		22.900	.9016	
18.600	.7323		23.000	.9055	
18.653	.7344	47/64	23.019	.9063	29/32
18.700	.7362		23.100	.9094	
18.800	.7402		23.200	.9134	
18.900	.7441		23.300	.9173	
19.000	.7480		23.400	.9213	
19.050	.7500	3/4	23.416	.9219	59/64
19.100	.7520		23.500	.9252	
19.200	.7559		23.600	.9291	
19.300	.7598		23.700	.9331	
19.400	.7638		23.800	.9370	
19.447	.7656	49/64	23.813	.9375	15/16
19.500	.7677		23.900	.9409	
19.600	.7717		24.000	.9449	
19.700	.7756		24.100	.9488	
19.800	.7795		24.200	.9528	
19.844	.7813	25/32	24.209	.9531	61/64
19.900	.7835		24.300	.9567	
20.000	.7874		24.400	.9606	
20.100	.7913		24.500	.9646	
20.200	.7953		24.600	.9685	
20.241	.7969	51/64	24.606	.9688	31/32
20.300	.7992		24.700	.9724	
20.400	.8031		24.800	.9764	
20.500	.8071		24.900	.9803	
20.600	.8110		25.000	.9843	
20.638	.8125	13/16	25.003	.9844	63/64
20.700	.8150		25.100	.9882	
20.800	.8189		25.200	.9921	
20.900	.8228		25.300	.9961	
21.000	.8268		25.400	1.0000	1

Appendix C

Dimensional Data for Drilled Orifices, Hose, and Pipe

Diameter	Orifice	Internal Area (square inches)	
		Hose	Pipe (schedule 40)
1/32	.00077		
1/16	.00307		
3/32	.0069		
1/8	.01227	.01227	
5/32	.01917		
3/16	.02761		
7/32	.03758		
1/4	.04909		.104
9/32	.06213		
5/16	.0767		
11/32	.09281		
3/8	.1104	.11	.190
13/32	.1296		
7/16	.1503		
15/32	.1726		
1/2	.1963	.196	.304
17/32	.2217		
9/16	.2485		
19/32	.2769		
5/8	.3068	.307	
21/32	.3382		
11/16	.3712		
23/32	.4057		
3/4	.4418	.442	.533
13/16	.5185		
7/8	.6013		
15/16	.6903		
1	.7854	.785	.864
1 1/4	1.2272	1.227	1.496
1 1/2	1.767		2.036
2	3.1416	3.14	3.356
2 1/2	4.9088		4.788

Appendix D

Tubing Size

Outside Diameter (OD)		Wall Thickness		Inside Diameter (ID)	
Inches	Millimeters	Inches	Millimeters	Inches	Millimeters
1/8	3.18	.028	0.71	.069	1.76
		.032	0.81	.061	1.55
		.035	0.89	.055	1.40
3/16	4.76	.032	0.81	.123	3.13
		.035	0.89	.117	2.98
1/4	6.35	.035	0.89	.180	4.57
		.049	1.25	.250	6.35
		.065	1.65	.120	3.05
5/16	7.94	.035	0.89	.243	6.16
		.049	1.25	.215	5.45
		.065	1.65	.183	4.64
3/8	9.53	.035	0.89	.305	7.75
		.049	1.25	.277	7.04
		.065	1.65	.245	6.23
1/2	12.70	.035	0.89	.430	10.92
		.049	1.25	.402	10.21
		.065	1.65	.370	9.40
		.083	2.11	.334	8.48
5/8	15.88	.035	0.89	.555	14.10
		.049	1.25	.527	13.39
		.065	1.65	.495	12.58
		.083	2.11	.459	11.66
3/4	19.05	.049	1.25	.652	16.56
		.065	1.65	.620	15.75
		.083	2.11	.584	14.83
		.109	2.77	.532	13.51
7/8	22.23	.049	1.25	.777	19.74
		.065	1.65	.745	18.93
		.083	2.11	.709	18.01
		.109	2.77	.657	16.69
1	25.40	.049	1.25	.902	22.91
		.065	1.65	.870	22.10
		.083	2.11	.834	21.18
		.109	2.77	.782	19.86
1 1/4	31.75	.049	1.25	1.152	29.26
		.065	1.65	1.120	28.45
		.083	2.11	1.084	27.53
		.109	2.77	1.032	26.21
1 1/2	38.10	.065	1.65	1.370	34.80
		.083	2.11	1.334	33.88
		.109	2.77	1.282	32.56
1 3/4	44.45	.065	1.65	1.620	41.15
		.083	2.11	1.584	40.23
		.109	2.77	1.532	38.91
		.134	3.40	1.482	37.64
2	50.80	.065	1.65	1.870	47.50
		.083	2.11	1.834	46.58
		.109	2.77	1.782	45.26
		.134	3.40	1.732	43.99

Appendix E

Pipe Size

Nominal Size	Outside Diameter	Wall Thickness		Inside Diameter		Flow Area	
		Schedule 40	Schedule 80	Schedule 40	Schedule 80	Schedule 40	Schedule 80
1/8	.41 in (1.03 cm)	.07 in (0.17 cm)	.10 in (0.24 cm)	.27 in (0.68 cm)	.22 in (0.55 cm)	.057 in² (0.366 cm²)	.036 in² (0.234 cm²)
1/4	.54 in (1.37cm)	.09 in (0.22 cm)	.12 in (0.30 cm)	.36 in (0.92 cm)	.30 in (0.76 cm)	.104 in² (0.671 cm²)	.072 in² (0.462 cm²)
3/8	.68 in (1.71 cm)	.09 in (0.22 cm)	.13 in (0.32 cm)	.49 in (1.25 cm)	.42 in (1.07 cm)	.191 in² (1.231 cm²)	.140 in² (0.906 cm²)
1/2	.84 in (2.13 cm)	.11 in (0.28 cm)	.15 in (0.37 cm)	.62 in (1.58 cm)	.55 in (1.39 cm)	.304 in² (1.959 cm²)	.234 in² (1.510 cm²)
3/4	1.05 in (2.67 cm)	.11 in (0.28 cm)	.15 in (0.37 cm)	.82 in (2.09 cm)	.74 in (1.88 cm)	.533 in² (3.439 cm²)	.432 in² (2.788 cm²)
1	1.32 in (3.34 cm)	.13 in (0.34 cm)	.18 in (0.45 cm)	1.05 in (2.66 cm)	.96 in (2.43 cm)	.864 in² (5.573 cm²)	.719 in² (4.638 cm²)
1 1/4	1.66 in (4.22 cm)	.14 in (0.36 cm)	.19 in (0.49 cm)	1.38 in (3.51 cm)	1.28 in (3.25 cm)	1.495 in² (9.645 cm²)	1.282 in² (8.272 cm²)
1 1/2	1.90 in (4.83 cm)	.15 in (0.37 cm)	.20 in (0.51 cm)	1.61 in (4.09 cm)	1.50 in (3.81 cm)	2.035 in² (13.128 cm²)	1.766 in² (11.395 cm²)
2	2.38 in (6.03 cm)	.15 in (0.37 cm)	.22 in (0.55 cm)	2.07 in (5.25 cm)	1.94 in (4.93 cm)	3.354 in² (21.638 cm²)	2.951 in² (19.041 cm²)
2 1/2	2.88 in (7.30 cm)	.20 in (0.52 cm)	.28 in (0.70 cm)	2.47 in (6.27 cm)	2.32 in (5.90 cm)	4.785 in² (30.873 cm²)	4.236 in² (27.330 cm²)
3	3.50 in (8.89 cm)	.22 in (0.55 cm)	.30 in (0.76 cm)	3.07 in (7.79 cm)	2.90 in (7.37 cm)	7.389 in² (47.670 cm²)	6.602 in² (42.592 cm²)
3 1/2	4.00 in (10.16 cm)	.23 in (0.57 cm)	.32 in (0.81 cm)	3.55 in (9.01 cm)	3.36 in (8.54 cm)	9.882 in² (63.754 cm²)	8.883 in² (57.312 cm²)
4	4.50 in (11.43 cm)	.24 in (0.60 cm)	.34 in (0.86 cm)	4.03 in (10.23 cm)	3.83 in (9.72 cm)	12.724 in² (82.089 cm²)	11.491 in² (74.136 cm²)

Appendix F

Cylinder Force

The following table can be used to quickly determine the force, measured in pounds, that a cylinder will exert at a specific gauge pressure. Friction in the cylinder is disregarded in these calculations. Friction losses should be determined using information provided by the cylinder manufacturer.

Force during retraction of double-acting cylinders can be calculated by subtracting the force generated by the cross-sectional area of the cylinder rod from the appropriate figure in this chart.

Piston Diameter (inches)	Piston Area (square inches)	Operating Pressure (psig)									
		10	20	30	40	50	60	70	80	90	100
.500	.196	1.9	3.9	5.8	7.9	9.8	11.8	13.7	15.7	17.7	19.6
.625	.307	3.1	6.1	9.2	12.3	15.3	18.4	21.5	24.5	27.6	30.7
.750	.442	4.4	8.8	13.3	17.7	22.1	26.5	30.9	35.3	39.8	44.2
.875	.601	6.0	12.0	18.0	24.1	30.1	36.1	42.1	48.1	54.1	60.1
1.000	.785	7.8	15.7	23.6	31.4	39.3	47.1	54.9	62.8	70.7	78.5
1.125	.994	9.9	19.9	29.8	39.8	49.7	59.6	69.6	79.5	89.5	99.4
1.250	1.227	12.3	24.5	36.8	49.1	61.3	73.6	85.9	98.2	110.4	122.7
1.375	1.485	14.8	29.7	44.6	59.4	74.2	89.1	103.9	118.8	133.6	148.5
1.500	1.767	17.7	35.4	53.0	70.7	88.4	106.0	123.7	141.4	159.1	176.7
1.625	2.074	20.7	41.5	62.2	82.9	103.7	124.4	145.2	165.9	186.7	207.4
1.750	2.405	24.1	48.1	72.2	96.2	120.3	144.3	168.4	192.4	216.5	240.5
1.875	2.761	27.6	55.2	82.8	110.4	138.1	165.6	193.3	220.9	248.5	276.1
2.000	3.142	31.4	62.8	94.3	125.7	157.1	188.5	219.9	251.4	282.8	314.2
2.250	3.976	39.8	79.5	119.3	159.0	198.8	238.6	278.3	318.1	357.8	397.6
2.500	4.909	49.1	98.2	147.3	196.4	245.5	294.5	343.6	392.7	441.8	490.9
2.750	5.940	59.4	118.8	178.2	237.6	297.0	356.4	415.8	475.2	534.6	594.0
3.000	7.069	70.7	141.4	212.1	282.7	353.4	424.1	494.8	565.5	636.2	706.9
3.500	9.621	96.2	192.4	288.6	384.8	481.1	577.3	673.5	769.7	865.9	962.1
4.000	12.566	125.7	251.4	377.0	502.7	628.4	754.0	879.7	1005.4	1131.1	1257.0
4.500	15.904	159.0	318.1	477.1	636.2	795.2	954.2	1113.3	1272.3	1431.4	1590.4
5.000	19.635	196.4	392.7	589.1	785.4	981.8	1178.1	1374.5	1570.8	1767.2	1968.5
5.500	23.758	237.6	475.2	712.7	950.3	1188.0	1425.5	1663.1	1900.6	2138.2	2375.8
6.000	28.274	282.7	565.5	848.2	1130.9	1413.7	1696.4	1979.1	2261.9	2544.7	2827.4

Force (pounds)

Appendix G

Free Air Consumption

Use this table to determine the cubic feet of air consumed by a pneumatic cylinder. The values in this table indicate the cubic feet of air consumed for each inch of piston stroke. For single-acting cylinders, multiply the value in the table by the piston stroke (in inches). For retraction of double-acting cylinders, subtract the cross-sectional area of the rod from the piston area. Then, select the area closest to this figure in the area column to obtain an estimate of the free air required for retraction.

Piston Diameter (inches)	Piston Area (square inches)	Operating Pressure (psig)			
		50	60	70	80
.500	.196	.0005	.0006	.0007	.0007
.625	.307	.0008	.0009	.0010	.0011
.750	.442	.0011	.0013	.0015	.0016
.875	.601	.0015	.0018	.0020	.0022
1.000	.785	.0020	.0023	.0026	.0029
1.125	.994	.0025	.0029	.0033	.0037
1.250	1.227	.0031	.0036	.0041	.0046
1.375	1.485	.0038	.0044	.0050	.0055
1.500	1.767	.0045	.0052	.0059	.0066
1.625	2.074	.0053	.0061	.0069	.0077
1.750	2.405	.0061	.0071	.0080	.0090
1.875	2.761	.0070	.0081	.0092	.0103
2.000	3.142	.0080	.0092	.0105	.0117
2.250	3.976	.0101	.0117	.0133	.0148
2.500	4.909	.0125	.0144	.0164	.0183
2.750	5.940	.0151	.0175	.0198	.0221
3.000	7.069	.0190	.0208	.0236	.0264
3.500	9.621	.0245	.0283	.0321	.0359
4.000	12.566	.0320	.0370	.0419	.0469
4.500	15.904	.0405	.0468	.0530	.0593
5.000	19.635	.0500	.0578	.0655	.0732
5.500	23.758	.0605	.0699	.0792	.0886
6.000	28.274	.0720	.0831	.0943	.1054

Air Consumed (cfm)

Appendix H

Viscosity Conversions

Approximate Viscosities	
Material	**Viscosity (centipoise)**
Water	1
Oil, SAE 10	85–140
Oil, SAE 20	140–420
Oil, SAE 30	420–650
Oil, SAE 40	650–900

Viscosity Conversions (sp. gr. = 1)				
Centipoise (cPs)	**Poise (Ps)**	**Centistokes (cSt)**	**Stokes (St)**	**Saybolt Universal (SSU)**
1	.01	1	.01	31
20	.2	20	.2	100
30	.3	30	.3	160
40	.4	40	.4	210
60	.6	60	.6	320
80	.8	80	.8	430
100	1.1	100	1.0	530
120	1.2	120	1.2	580
140	1.4	140	1.4	690
160	1.6	160	1.6	790
180	1.8	180	1.8	900
200	2.0	200	2.0	1000
220	2.2	220	2.2	1100
240	2.4	240	2.4	1200
260	2.6	260	2.6	1280
280	2.8	280	2.8	1380
300	3.0	300	3.0	1475
320	3.2	320	3.2	1530
340	3.4	340	3.4	1630
360	3.6	360	3.6	1730
380	3.8	380	3.8	1850
400	4.0	400	4.0	1950
420	4.2	420	4.2	2050
440	4.4	440	4.4	2160
460	4.6	460	4.6	2270
480	4.8	480	4.8	2380
500	5.5	500	5.5	2480
600	6.0	600	6.0	2900
700	7.0	700	7.0	3380
800	8.0	800	8.0	3880
900	9.0	900	9.0	4300
1000	10.0	1000	10.0	4600
1100	11.0	1100	11.0	5200
1200	12.0	1200	12.0	5620

Appendix I

Fluid Power Organizations

The following list identifies several national and international professional and standards organizations that are related to fluid power. These groups have many items that may be very helpful resources.

American National Standards Institute (ANSI)
25 West 43rd Street, 4th Floor
New York, NY, 10036
212-642-4900
www.ansi.org

American Petroleum Institute (API)
1220 L Street, NW
Washington, DC 20005-4070
202-682-8000
www.api.org

ASTM International
100 Barr Harbor Drive
P.O. Box C700
West Conshohocken, PA 19428-2959
610-832-9500
www.astm.org

American Society of Agricultural and Biological Engineers
2950 Niles Road
St. Joseph, MI 49085
269-429-0300
www.asabe.org

ASME International
Three Park Avenue
New York, NY 10016-5990
800-843-2763
www.asme.org

Association of Equipment Manufacturers (AEM)
6737 W. Washington Street, Suite 2400
Milwaukee, WI 53214-5647
414-272-0943
www.aem.org

British Fluid Power Association (BFPA)
Cheriton House, Cromwell Park
Chipping Norton, Oxon
OX7 5SR
United Kingdom
+44 (0)1608 647900
www.bfpa.co.uk

Canadian Fluid Power Association (CFPA)
2175 Sheppard Avenue East, Suite 310
Toronto, ON M2J 1W8
416-499-1416
www.cfpa.ca

Compressed Air & Gas Institute (CAGI)
1300 Sumner Avenue
Cleveland, OH 44115
216-241-7333
www.cagi.org

The FPDA Motion & Control Network
105 Eastern Avenue, Suite 104
Annapolis, MD 21403-3300
410-940-6347
www.fpda.org

Fluid Power Educational Foundation (FPEF)
P.O. Box 1420
Cherry Hill, NJ 08034-0054
856-424-8998
www.fpef.org

Fluid Power Institute (FPI)
Milwaukee School of Engineering
1025 N. Broadway
Milwaukee, WI 53202-3109
414-277-7191
www.msoe.edu/fpi

Institute of Electrical and Electronics Engineers (IEEE)
Operation Center
445 Hoes Lane
Piscataway, NJ 08854-4141
732-981-0060
www.ieee.org

International Society of Automation (ISA)
67 T.W. Alexander Dr.
P.O. Box 12277
Research Triangle Park, NC 27709
919-549-8411
www.isa.org

International Fluid Power Society (IFPS)
P.O. Box 1420
1930 E. Marlton Pike A2
Cherry Hill, NJ 08034-0054
856-489-8983
www.ifps.org

International Organization for Standardization (ISO)
1, ch. de la Voie-Creuse
CP 56
CH-1211 Geneva 20
Switzerland
+41 22 749 01 11
www.iso.org

National Electrical Manufacturers Association (NEMA)
1300 N. 17th Street, Suite 1752
Rosslyn, VA 22209
703-841-3200
www.nema.org

National Fluid Power Association (NFPA)
3333 N. Mayfair Road, Suite 211
Milwaukee, WI 53222-3219
414-778-3344
www.nfpa.com

Society of Automotive Engineers (SAE)
400 Commonwealth Drive
Warrendale, PA 15096-0001
724-776-4841
www.sae.org

Society of Manufacturing Engineers (SME)
One SME Drive
Dearborn, MI 48128
313-425-3000
www.sme.org

Index